1 MONTH OF FREE READING

at
www.ForgottenBooks.com

By purchasing this book you are eligible for one month membership to ForgottenBooks.com, giving you unlimited access to our entire collection of over 1,000,000 titles via our web site and mobile apps.

To claim your free month visit:
www.forgottenbooks.com/free613765

* Offer is valid for 45 days from date of purchase. Terms and conditions apply.

ISBN 978-0-267-59395-8
PIBN 10613765

This book is a reproduction of an important historical work. Forgotten Books uses state-of-the-art technology to digitally reconstruct the work, preserving the original format whilst repairing imperfections present in the aged copy. In rare cases, an imperfection in the original, such as a blemish or missing page, may be replicated in our edition. We do, however, repair the vast majority of imperfections successfully; any imperfections that remain are intentionally left to preserve the state of such historical works.

Forgotten Books is a registered trademark of FB &c Ltd.
Copyright © 2018 FB &c Ltd.
FB &c Ltd, Dalton House, 60 Windsor Avenue, London, SW19 2RR.
Company number 08720141. Registered in England and Wales.

For support please visit www.forgottenbooks.com

TRANSACTIONS

OF THE

MIDLAND INSTITUTE

OF

MINING, CIVIL, & MECHANICAL

ENGINEERS.

VOL. XI.

1887—9.

Barnsley:
R. E. GRIFFITHS, STEAM PRINTER AND LITHOGRAPHER, CHURCH STREET.

1889.

ADVERTISEMENT.

The Institute is not, as a body, responsible for the facts and opinions advanced by the authors of Papers, or of those who take part in the discussions.

CONTENTS OF VOL. XI.

	PAGE
Officers	1, 201
Life Members	2, 202
Honorary Members	2, 202
Ordinary Members	3, 203
Rules	7, 208
Report of Council	23, 212
Treasurer's Account	21, 216
General Statement	22, 217

GENERAL MEETINGS.

1887.

July 27.—Annual Meeting ... 11
    Discussion on Mr. ROWLAND GASCOYNE's paper on "The Easterly Extension of the Leeds and Nottingham Coal-field" ... 11
    Discussion on Mr. BERNARD E. CLARKE's paper on "An Arrangement for arresting the fall of Colliery Cages in cases of Breakage of the Rope" ... 12
    Discussion on the Committee's Observations on the Final Report of the Royal Commission on Accidents in Mines. 12
    Election of Officers ... 20
    Vote of thanks to the retiring President ... 28
    Annual Dinner ... 30

Aug. 30.—The President's Inaugural Address ... 31
    Supplementary paper on "The Easterly Extension of the Leeds and Nottingham Coal-field," by Mr. ROWLAND GASCOYNE ... 63
    Discussion ... 68
    Discussion on the Committee's Observations on the Final Report of the Royal Commission on Accidents in Mines. 73

Oct. 7.—"On Considerations arising out of Sections 51, 52, 53, and 54 of the Coal Mines Regulation Act, 1887," by Mr. GEORGE BLAKE WALKER ... 75
    Discussion ... 97
    Discussion on Mr. ROWLAND GASCOYNE's supplementary paper on "The Easterly Extension of the Leeds and Nottingham Coal-field" ... 97

CONTENTS.

	PAGE
Nov. 15.—Discussion on the Federation of the Mining Institutes of Great Britain ...	99
"On Hydro-carbon Explosives and their Value for Mining Purposes," by Mr. GEORGE BLAKE WALKER (Part I.) ...	101
Discussion on Mr. GEORGE BLAKE WALKER's paper on "Special Rules under the New Mines Act"...	117
Safety Lamp having a Shut-off Appliance exhibited by Mr. PATTISON ...	117
Dec. 14.—Discussion on the Federation of the Mining Institutes of Great Britain ...	119
Discussion on Mr. GEORGE BLAKE WALKER's paper on "Hydro-carbon Explosives" ...	122
Results of Experiments with Roburite, by Mr. RHODES	124
Discussion on Mr. GEORGE BLAKE WALKER's paper on "Special Rules under the New Mines Act"...	135
Model of a New Safety Cage exhibited by Mr. R. MILLER...	135

1888.

Feb. 21.—Discussion on the proposed Joint Fan Experiments .	137
Addendum to paper on "Hydro-carbon Explosives," by Mr. GEORGE BLAKE WALKER ...	138
Discussion...	144
Discussion on Mr. GEORGE BLAKE WALKER's paper on "Special Rules under the New Mines Act"...	146
Mr. Ashworth's Patent Hepplewhite-Gray Deputy's Safety Lamp exhibited ...	147
Mar. 27.—"On Foreign Mining Rents and Royalties," by Mr. H. B. NASH ...	149
Discussion on Mr. GEORGE BLAKE WALKER's papers on "Hydro-carbon Explosives" ...	157
April 27.—Discussion on Mr. NASH's paper on "Foreign Mining Rents and Royalties"...	163
Discussion on Mr. GEORGE BLAKE WALKER's papers on "Hydro-carbon Explosives" ...	170
Discussion on the proposed Joint Fan Experiments ...	171
May 29.—Meeting adjourned...	172
June 19.—Discussion on the Federation of the Mining Institutes of Great Britain ...	173
Discussion on the proposed Joint Fan Experiments ...	173
"Mining in the Middle Ages," by Mr. GEO. BLAKE WALKER	175

CONTENTS.

	PAGE
Discussion on Mr. GEORGE BLAKE WALKER's papers on "Hydro-carbon Explosives"	198
Vote of thanks to the Institution of Civil Engineers for Exchange of Transactions	199
Aug. 1.—Annual Meeting	212
Election of Officers	219
Discussion on Mr. GEORGE BLAKE WALKER's paper on "Mining in the Middle Ages"	221
Vote of thanks to retiring Officers	221
Discussion on Mr. GEORGE BLAKE WALKER's papers on "Hydro-carbon Explosives"	225
Discussion on Mr. H. B. NASH's paper on "Foreign Mining Rents and Royalties"	231
Annual Dinner	235
Sept. 25.—Election of President	237
"On Tonite as an Explosive when used with a Flame-destroying Compound," by Mr. W. HARGREAVES	239
Discussion	242
"Notes on Matters of Current Interest," by Mr. GEORGE BLAKE WALKER	247
Oct. 31.—The President's Inaugural Address	253
Discussion on Mr. W. Hargreaves' paper on "Tonite as an Explosive when used with a flame-destroying compound" in conjunction with Mr. GEORGE BLAKE WALKER's papers on "Hydro-carbon Explosives"	263
Discussion on Mr. GEORGE BLAKE WALKER's Notes on "Matters of Current Interest"	267
Discussion on Mr. H. B. NASH's paper on "Mining Rents and Royalties"	270
Dates of Meetings	272
Discussion on the Federation of the Mining Institutes of Great Britain (held in London)	273
Report of the joint Meeting of the North of England Institute of Mining and Mechanical Engineers, Midland Institute of Mining, Civil, and Mechanical Engineers, and the South Wales Institute of Engineers, on "Mechanical Ventilators, 1888"	301
Memoir of Mr. JOHN BROWN	315
Dec. 21.—"On Electricity as a Motive Power, with special reference to its application to Haulage in Mines," by Mr. GEORGE BLAKE WALKER	317

CONTENTS.

PAGE

"On an Electric Locomotive for Mines," by Mr. ALBION T. SNELL ... 333

"On Defective Detonators," by Mr. C. E. RHODES ... 336

1889.

Jan. 16.—Discussion on Mr. GEORGE BLAKE WALKER's paper on "Electricity as a Motive Power, with special reference to its application to Haulage in Mines," along with Mr. ALBION T. SNELL's paper on "An Electric Locomotive for Mines" ... 341

Feb. 19.—Visit to St. John's Colliery, Normanton ... 359

Discussion on Mr. GEORGE BLAKE WALKER's paper on "Electricity as a Motive Power, with special reference to its application to Haulage in Mines," along with Mr. ALBION T. SNELL's paper on "An Electric Locomotive for Mines" ... 359

March 20.—Discussion on Mr. GEORGE BLAKE WALKER's paper on "Electricity as a Motive Power, with special reference to its application to Hanlage in Mines," along with Mr. ALBION T. SNELL's paper on "An Electric Locomotive for Mines" ... 381

April 16.—Discussion on the Federation of the Mining Institutes of Great Britain ... 389

Discussion on Mr. GEORGE BLAKE WALKER's paper on "Electricity as a Motive Power, with special reference to its application to Haulage in Mines," along with Mr. ALBION T. SNELL's paper on "An Electric Locomotive for Mines" ... 393

"On the proposed Nicaraguan Canal and a new type of Lock for Ship Canals," by Mr. GEORGE BLAKE WALKER ... 394

May 22.—Discussion on the Federation of the Mining Institutes of Great Britain ... 401

"On Artificial Foundations and Methods of Sinking through Quicksand," by Mr. W. E. GARFORTH ... 407

June 18.—Visit to Allerton Main Colliery ... 419

Discussion on the Federation of the Mining Institutes of Great Britain ... 419

Discussion on Mr. GEORGE BLAKE WALKER's paper on "The proposed Nicaraguan Canal and a new type of Lock for Ship Canals" ... 421

Discussion on Mr. GEORGE BLAKE WALKER's paper on "Electricity as a Motive Power" ... 422

OFFICERS—1887-88.

President:

T. W. EMBLETON, The Cedars, Methley, near Leeds.

Vice-Presidents:

G. J. KELL, Beechwood House, Kilnhurst, Rotherham.
A. B. SOUTHALL, Monckton Main Colliery, Barnsley.
G. BLAKE WALKER, Wharncliffe Silkstone Colliery, Barnsley.

Council:

W. E. GARFORTH, West Riding Colliery, Normanton.
C. E. RHODES, Aldwarke Main and Car House Collieries, Rotherham.
J. LONGBOTHAM, Barrow Collieries, Barnsley.
C. HODGSON, Rockingham Colliery, Barnsley.
J. JARRATT, Houghton Main Colliery, Barnsley.
WM. HY. CHAMBERS, Denaby Main Colliery, Rotherham.
M. NICHOLSON, Middleton Colliery, Leeds
J. NEVIN, Dunbottle House, Mirfield, via Normanton.

EX-OFFICIO:

R. CARTER, C.E., F.G.S., Past President, Spring Bank, Harrogate.
T. CARRINGTON, C.E., F.G.S., Past President, Kiveton Park Colliery, Sheffield.
T. W. JEFFCOCK, C.E., F.G.S., J.P., Past President, Bank Street, Sheffield.
A. M. CHAMBERS, J.P., Past President, Thorncliffe Iron Works and Collieries, Sheffield.

Secretary and Treasurer:

JOSEPH MITCHELL, Assoc. Inst. C.E., F.G.S., Mining Offices, Eldon Street, Barnsley.

LIFE MEMBERS.

BRIGGS, ARCHIBALD, Whitwood Collieries, Normanton.
COOPER, S. J., Mount Vernon, Barnsley.
FITZWILLIAM, THE RIGHT HON. EARL, Wentworth Woodhouse, Rotherham.
INGHAM, E. T., Blake Hall, Mirfield

HONORARY MEMBERS.

ARMSTRONG, G. F., Yorkshire College, Leeds.
BELL, THOMAS, H.M. Inspector of Mines, Durham.
DICKINSON, J., H.M. Inspector of Mines, Manchester.
GREEN, Professor A. H., Yorkshire College, Leeds.
HALL, H., H.M. Inspector of Mines, Rainhill, Prescott.
MARTIN, J. S., H.M. Inspector of Mines, Lydney, Gloucester.
MIALL, Professor L. C., Yorkshire College, Leeds.
MOORE, R., H.M. Inspector of Mines, Glasgow.
ROBSON, J. M., H.M. Inspector of Mines, Swansea.
RONALDSON, J.M., H.M. Inspector of Mines, Pollockshields. Glasgow.
RUSSELL, R., Sea View, St. Bees, Carnforth.
RUCKER, Professor, Errington, Clapham Park, London.
SCOTT, W.B., H,M. Inspector of Mines, Wolverhampton.
STOKES, A.H., H.M. Inspector of Mines, Green Hill, Derby.
THORPE, Professor T. E., Science and Art Department, South Kensington, London.
WARDELL, F. N., H.M., Inspector of Mines, Wath-on-Dearne, Rotherham.
WYNNE, THOMAS, H.M. Inspector of Mines, Gnossal, Stafford.
WILLIS, JAMES, H.M. Inspector of Mines, Newcastle-on-Tyne.

LIST OF MEMBERS.

ACKROYD, A., Morley Main Colliery, Leeds.
ADDY, J. J., Carlton Main Colliery, Barnsley.
ASHTON, J. H., Waleswood Colliery, Sheffield.
BAILES, W., Cortonwood Colliery, Barnsley.
BAINBRIDGE, E., Nunnery Colliery, Sheffield.
BARNBY, REGINALD C., Monckton Main Colliery, Barnsley.
BATTY, W., Darley Grove, Worsbro' Dale, Barnsley.
BEACHER, G. F., Chapeltown, Sheffield.
BENNETT, J. T., Featherstone Main Colliery, Pontefract.
BICKERDYKE, C., Kippax Colliery, Leeds.
BLAKELEY. A. B., Soothill Wood Colliery, Batley.
BONSER, HAROLD, Newcastle-under-Lyme.
BROADHEAD, G., 7, Cedar Road, Carr Croft, Armley, Leeds.
BROWN, JOHN, Priory Place, Bristol Road, Birmingham.
BURNLEY, G. J., Birthwaite Hall, Darton, Barnsley.
CARRINGTON, T, Kiveton Park Colliery, Sheffield (PAST PRESIDENT).
CARTER, R., Spring Bank, Harrogate (PAST PRESIDENT).
CHAMBERS, A. M., Thorncliffe Collieries, Chapeltown, Sheffield (PAST PRESIDENT).
CHAMBERS, W. HOOLE, Rockingham Colliery, Barnsley.
CHAMBERS, WM. HY., Denaby Main Colliery, Rotherham (*Member of Council*).
CHILDE, H. S., Wakefield.
CLARKE, BERNARD E., Wharncliffe Silkstone Colliery, Barnsley.
CLARKE, R. W., Lidgett Colliery, near Barnsley.
CLAYTON, C. D., Broxholme House, Doncaster.
CLEGG, J. H., Monckton Main Colliery, Barnsley.
CLIFFORD, W., Club Garden Road, Sheffield.
COBBOLD, C. H., Wentworth Castle, Barnsley.
COCKING, A. T., Aldwarke Main Colliery, Rotherham.
COLVER, J. C., Eckington Colliery, Chesterfield.
COOPER, H., Wharncliffe Silkstone Colliery, Barnsley.
CRAIK, T., Church Street, Barnsley.
CRAVEN, JOHN, Westgate Common, Wakefield.
CRAWSHAW, C. B., The Collieries, Dewsbury.
CRESWICK, W., Sharlestone Colliery, Normanton.
CRIGHTON, JOHN, 2, Clarence Buildings, Booth St., Manchester.

EMBLETON, T. W., The Cedars, Methley, Leeds (PRESIDENT).
EMBLETON, T. W., Junr., The Cedars, Methley, Leeds.
FANGEN, S. AUGUSTUS, High green, near Sheffield.
FARRAR, J., Old Foundry, Barnsley.
FELL, J., Nunnery Colliery, Sheffield.
FINCKEN, C. W., Hoyland Silkstone Colliery, Barnsley.
FIRTH, J., Upper Fountain Street, Leeds.
FIRTH, W., Water Lane, Leeds.
FOSTER, G., Osmondthorpe Colliery, Leeds.
GARFORTH, W. E., West Riding Colliery, Normanton (*Member of Council.*)
GASCOYNE, ROWLAND, Mexbro', Rotherham.
GERRARD, JOHN, Inspector of Mines, Wakefield.
GREAVES, J. O., Westgate, Wakefield.
GREENWOOD, Professor, Technical School, Sheffield.
HABERSHON, M. H., Thorncliffe Collieries, Sheffield.
HAGGIE, F. W., Gateshead, Newcastle-on-Tyne.
HAGGIE, G., Sunderland.
HALL, M., Lofthouse Colliery, Wakefield.
HARGREAVES, J., Rothwell Haigh Colliery, Leeds.
HARGREAVES, W., Rothwell Haigh Colliery, Leeds.
HICKS, Professor, Firth College, Sheffield.
HIGSON, JOHN, Oak Bank, Lancaster Road, Eccles.
HINCHLIFFE, J., Bullhouse Colliery, Penistone.
HODGSON, C., Rockingham Colliery, Barnsley (*Member of Council*).
HOLLIDAY, T., St. John's Colliery, Normanton.
HOWDEN, THOS., Ironfounder, Wakefield.
JACKSON, A., Howley Park Colliery, Batley.
JACKSON, W. G., Whitwood, Normanton.
JARRATT, J., Houghton Main Colliery, Barnsley (*Member of Council*).
JEFFCOCK, T. W., Bank Street, Sheffield (PAST PRESIDENT).
JEFFCOCK, C. E., Birley Colliery, Sheffield.
JEFFERSON, J. C., 2, East Parade, Leeds.
JEFFREY, G. F., 44a, Park Square, Leeds.
JOHNSON, J., Carlton Main Colliery, Barnsley.
JONES, F. J., Rother Vale Colliery, Fence, near Rotherham.
KELL, G. J., Beechwood House, Kilnhurst, Rotherham (VICE-PRESIDENT).
LAMPEN, G. J., Kirkgate, Wakefield.

LONGBOTHAM, JONATHAN, Barrow Collieries, Barnsley (*Member of Council*).
LOWRANCE, J., Peel Square, Barnsley.
LUPTON, A., 6, De Gray Road, Leeds.
MADDISON, T. R., Wakefield.
MARSHALL, J. L., Monk Bretton Colliery, Barnsley.
MARSHALL, W., Liversedge Colliery, Liversedge, via Normanton.
MIDDLETON, E., Outwood House, Wakefield.
MILLER, R., Beech Grove, Locke Park, Barnsley.
MILLS, H. M., 15, Corporation Street, Chesterfield.
MIRFIELD COAL CO., Mirfield, via Normanton.
MITCALFE, F. D., Waratah Coal Co., Charlestown, N. S. Wales.
MITCHELL, JOSEPH, Mining Offices, Eldon Street, Barnsley (*Secretary and Treasurer*).
MITCHELL, T. W. H., Mining Offices, Eldon Street, Barnsley.
MOODY, GEORGE, 58, Park Road, Barnsley.
MOSLEY, MAJOR PAGET, 81, Warwick Road, Earl's Court, London, S. W.
NASH, H. B., Clarke's Old Silkstone Colliery, Barnsley.
NEVIN, JOHN, Dunbottle House, Mirfield, via Normanton (*Member of Council*).
NEWBOULD, T., Low Stubbin Colliery, Rawmarsh, Rotherham.
NICHOLSON, M., Middleton Colliery, Leeds (*Member of Council*).
PARRY, EVAN, Wharncliffe Woodmoor Colliery, Barnsley.
PATTISON, J,, Morley Main Colliery, Leeds.
PEAKE, C. E., Eskell Chambers, Nottingham.
PEARCE, F. H., Bowling Iron Works, Bradford.
PEEL, JOHN, Park House, Wortley, near Sheffield.
PITT, G. LANE, Harrogate.
PRINGLE, HENRY A., Barrow Colliery, Barnsley.
PURCELL, S., Prince of Wales Colliery, Pontefract.
RHODES, C. E., Aldwarke Main and Carr House Collieries, Rotherham (*Member of Council*).
RICHARDSON, A. M., 42, Swinegate, Leeds.
RITSON, JOSEPH, Strafford Colliery, Barnsley.
ROBINSON, J. G., Brick Manufacturer, Elland.
ROBERTS, SAMUEL, Junr., Park Grange, Sheffield.
ROUTLEDGE, R., Garforth Colliery, Leeds.
ROWLEY, W., 74, Albion Street, Leeds.
RYDER, W. J. H., Forth Street, Newcastle-on-Tyne.

MEMBERS.

SAINT, T. E. W., Mitchell Main Colliery, Barnsley.
SCOTT, F. W., Atlas Wire Rope Works, Reddish, Stockport.
SENIOR, A., Oak Well, Barnsley
SHAW, JOHN, Darrington Hall, Pontefract.
SHAW, G., Wath Main Colliery, Rotherham.
SHEPHERD, F. H., Naniamo, Vancouver's Island.
SHORT, W., Lambton Colliery, Newcastle, N. S. Wales.
SIMPKIN, J., Joan Royd Colliery, Heckmondwike.
SLACK, J., Fence Collieries, Rotherham.
SMITH, C. S., Shipley Collieries, Derby.
SMITH, H., Timber Merchant, Hull,
SMITH, SYDNEY A., Crown Buildings, Booth Street, Manchester.
SMITH, T. R., 52, Dingwall Road, Croydon.
SMITH, V., Strafford Colliery, Barnsley.
SOUTHALL, A. B., Monckton Main Colliery, Barnsley (VICE-PRESIDENT).
STEAR, J., Strafford Colliery, Barnsley.
STRINGER, G. E., Bath Buildings, Huddersfield.
STUBBS, THOS., Bradmarsh House, Ickles, Rotherham.
TEALE, W. E., Eccles, Manchester.
THIRKELL, E. W., Oaks Colliery, Barnsley,
THOMSON, J. F. Manvers Main Colliery, Rotherham.
TODD, W. G., Nunnery Colliery, Sheffield.
TYAS, A., Warren House, Sheffield Road, Barnsley.
TURNBULL, ROBERT, South Kirby Colliery, Pontefract.
WALKER, G. B., Wharncliffe Silkstone Colliery, Barnsley (VICE-PRESIDENT).
WALKER, HY. S., Lidgett Colliery, Barnsley.
WALKER, SYDNEY F., 195, Severn Road, Cardiff.
WALLACE, J., King Street, Wigan.
WALTERS, H., Birley Colliery, Sheffield.
WARD, W., Churwell Colliery, Leeds,
WARD JOSHUA, Wharncliffe Silkstone Colliery, Barnsley.
WASHINGTON W., Mitchell Main Colliery, Barnsley.
WEEKS, J. G., Bedlington Colliery, Morpeth,
WHITE, J. F., Westgate, Wakefield.
WILSON, W. B., Kippax Colliery, Leeds,
WOODHEAD, E., Low Moor Iron Works, Bradford.
WOODHEAD, L., Beeston Colliery, Leeds.

RULES.

1.—That the "SOUTH YORKSHIRE VIEWERS' ASSOCIATION" in future be called "THE MIDLAND INSTITUTE OF MINING, CIVIL, AND MECHANICAL ENGINEERS."

2.—That the objects of the Midland Institute of Mining, Civil, and Mechanical Engineers are to enable its Members to meet together at fixed periods, and to discuss the means for the ventilation of coal and other Mines, the winning and working of Collieries and Mines, the prevention of Accidents, and the advancement of the science of Mining Engineering generally.

3.—The Members of the Midland Institute of Mining, Civil, and Mechanical Engineers shall consist of Ordinary, Life, and Honorary Members.

4.—Ordinary and Life Members shall be persons educated or practising as Mining, Mechanical, or Civil Engineers, or persons having a direct interest in or the responsible management of operations connected with mining.

5.—Honorary Members shall be Mining Inspectors during the term of their office, and other persons who have distinguished themselves by their literary and scientific attainments, or who have made important communications to the society.

6.—The annual subscription for each Ordinary Member shall be £1 1s., payable in advance, and the same shall be considered due and payable on the first Tuesday in July, each year, or immediately after election. If the subscription be not paid within *three months* the defaulter's name shall be posted up in the Meeting Room:—And any Member whose subscriptions shall be in arrear for *six months*, shall be considered as withdrawn from the Institute, and his name erased from the list of Members after one month's notice from the Secretary.

7.—Members who shall at one time make a donation of £20, or upwards, shall be Life Members.

8.—Persons desirous of becoming Members shall be proposed at a General Meeting. The nomination shall be in writing, and signed by two Ordinary Members, and shall state the name, residence, and occupation of the person proposed. The proposal shall be hung up in the Room of the Institute, at Barnsley, for one month, and the election shall take place at the next Meeting.

9.—That the Officers of the Institute shall consist of a President (who shall be a Mining, Civil, or Mechanical Engineer), three Vice-Presidents (not more than one of whom shall be a Mechanical Engineer), eight Councillors (not more than three of whom shall be Mechanical Engineers), a Treasurer and Secretary, who shall constitute a Council, for the direction and management of the affairs of the Institute. The President, Vice-Presidents, and Councillors shall be elected at the Annual Meeting, and shall be eligible for re-election with the exception of any President or Vice-President who may have held office for the three immediately preceding years, and such four Councillors as may have attended the fewest Council Meetings during the past year; but such Members shall be eligible for re-election after being one year out of office. Voting papers, with a list of Officers, shall be posted by the Secretary to all Members of the Institute at least fourteen days previous to the Annual Meeting; such Voting Papers to be by them filled up, signed, and returned under cover either personally or through the post, addressed to the Secretary, so as to be in his hands before the hour fixed for the election of Officers. The Chairman shall, in all cases of voting, appoint Scrutineers of the lists, and the scrutiny shall commence on the conclusion of the other business of the Meeting. At Meetings of the Council, five shall form a quorum, and the minutes of the Council's proceedings shall be at all times open to the inspection of the Members of the Institute.

10.—That Presidents who have become ineligible shall be *ex-officio* Members of the Council so long as they continue Members of the Institute.

11.—A General Meeting of the Institute shall be held on the first Wednesday in every month, excepting the months of January and June, at two p.m., and the Annual Meeting in the month of July shall be held in the Room of the Institute at Barnsley, at which a report of the proceedings, and an abstract of the audited accounts of the previous year shall be presented by the Council. A special meeting of the Institute may be called whenever the Council shall think fit, excepting the months excepted, and also on a requisition to the Council, signed by ten or more members. The Council shall decide where the monthly meeting shall be held. The Members in any *District* wishing the Monthly Meeting to be held in it, shall make application in writing to the Council, for permission to hold such Meeting, and shall provide a suitable room and make all necessary arrangements for that Meeting free of expense to the Institute.

12.—Every question which shall come before any Meeting of the Institute, shall be decided by the votes of the majority of the Ordinary and Life Members then present.

13.—The funds of the Society shall be deposited in the hands of the Treasurer, and shall be disbursed by him according to the directions of the Council.

14.—All papers intended to be read to the Institute shall be sent for the approval of the Council, accompanied by a short abstract of their contents.

15.—The Council shall have power to decide on the propriety of communicating to the Institute any papers which they receive, and they shall be at liberty, when they think it desirable, to direct any paper read before the Institute to be printed and transmitted to the Members. Intimation, when practicable, shall be given at the close of each General Meeting, of the subject of the paper or papers to be read, and of the question for discussion at the next Meeting, and notice thereof shall be given by circular to each member. The reading of the papers shall not be delayed beyond such an hour as the President may think proper, and if the election of Members, or other business should not be despatched soon enough, the President may adjourn such business until after the discussion of the subject of the day.

16.—Members elected at any Meeting between the Annual Meetings shall be entitled to all the papers issued in that year, providing that subscriptions be not in arrear.

17.—The copyright of all papers communicated to, and accepted by, the Institute, shall become vested in the Institute, and such communication shall not be published, for sale or otherwise, without the permission of the Council.

18.—That each Member who may have taken part in the discussion upon any subject shall have a proof copy sent to him by the Secretary for correction ; such copy to be returned to the Secretary not later than three days from the date of its receipt.

19.—The Institute is not, as a body, responsible for the facts and opinions advanced in the papers, nor in the abstracts of the conversations at the Meetings of the Institute.

20.—The Transactions of the Institute shall not be forwarded to Members whose subscriptions are in arrear.

21.—No duplicate copy of any portion of the proceedings shall be issued to any of the Members unless by written order from the Council.

22.—Each Member of the Institute shall have power to introduce a stranger to any of the General Meetings of the Institute, and shall sign in a book kept for that purpose, his own name, as well as the name and address of the person introduced, but such stranger shall not take part in any discussion or other business, unless permitted by the Meeting to do so.

23.—No alterations shall be made in any of the Laws, Rules, or Regulations of the Institute except at the Annual Meeting, and the particulars of every such alteration shall be announced at the previous General Meeting, and inserted in its minutes, and shall be exhibited in the Room of such Institute at the Meeting previous to such Annual Meeting, and such Meeting shall have power to adopt any modification of suchproposed alteration or addition to the Rules.

24.—The author of each paper read before the Institute shall be allowed twelve copies of such paper (if ordered to be printed) for his own private use.

MIDLAND INSTITUTE OF MINING, CIVIL, AND MECHANICAL ENGINEERS.

ANNUAL MEETING.

HELD AT THE INSTITUTE ROOM, BARNSLEY, ON WEDNESDAY, JULY 27TH, 1887.

A. M. CHAMBERS, Esq., J.P., President, in the Chair.

The minutes of the last meeting were read and confirmed.

The following gentlemen were elected members of the Institute, having been previously nominated:—

Mr. A. T. COCKING, Aldwarke Main Colliery, Rotherham.

Mr. HENRY COOPER, Underviewer, Wharncliffe Silkstone Colliery, Barnsley.

Mr. REGINALD C. BARNBY, Mining Student, Monckton Main Colliery, Barnsley.

Mr. A. B. SOUTHALL and Mr. W. HOOLE CHAMBERS were appointed scrutineers to examine the voting papers and declare the result of the election of officers.

A paper was read on "Some experience of the Explosive Force of a Mixture of Air and Coal Gas," by Mr. W. CLIFFORD.

The PRESIDENT remarked that the paper had better be referred to the Council, and ask for further information from Mr. Clifford.

The meeting assented to this course.

ADJOURNED DISCUSSION ON MR. ROWLAND GASCOYNE'S PAPER ON "THE EASTERLY EXTENSION OF THE LEEDS AND NOTTINGHAM COAL-FIELD."

The PRESIDENT: The next thing is the adjourned discussion on Mr. Gascoyne's paper on "The Easterly Extension of the Leeds and Nottingham Coal-field." Have you been able to get further information, Mr. Gascoyne?

Mr. GASCOYNE: Only what I mentioned at the last meeting—a section of part of the boring at Owthorpe. I offered to get that, but

I did not understand it was accepted, because I think we expected a full detailed section from Mr. Bainbridge. I do not know whether that has come or not.

Mr. JOSEPH MITCHELL: Nothing has been received from Mr. Bainbridge.

The PRESIDENT: Has any gentleman any remarks to make on that subject? If not, I think we had better adjourn it again, because we should be able to get some further information. Don't you think so?

Mr. GASCOYNE: Yes, there is a lot of information to be got with respect to borings. I have a lot I should be glad to give if the Council think it worth while to put it in the Transactions. I kept the paper down to make it as clear and practical as possible and be as little geological as I could, so that it would be easier to deal with. There are sections at Beeston and Charnley Forest and different places that might add a little to the geological interest. If you think they would be of any use I would send them.

The PRESIDENT: I think they would be. I think the paper is one of great interest, and any further information would be very useful. With your approval we will adjourn the discussion and ask Mr. Gascoyne to give us his further information.

Carried unanimously.

DISCUSSION ON MR. BERNARD E. CLARKE'S PAPER ON "AN ARRANGEMENT FOR ARRESTING THE FALL OF COLLIERY CAGES IN CASES OF BREAKAGE OF THE ROPE."

The PRESIDENT: The next is a discussion on Mr. Bernard Clarke's paper on "An arrangement for arresting the fall of Colliery Cages in cases of Breakage of the Rope." You have that paper in the last Transactions, and a description. One of the amendments upon the Mines Regulation Bill which Mr. Pickard proposed was that the use of apparatus of this kind should be compulsory on all collieries. I think we have succeeded in disposing of that.

THE FINAL REPORT OF THE ROYAL COMMISSION ON ACCIDENTS IN MINES.

The PRESIDENT: The next thing is the discussion on the final Report of the Royal Commission on Accidents in Mines. I suppose that should be a discussion on the remarks which the Council of the Institute has made. The Council have been at very great trouble, have appointed a Committee, and have gone very carefully through the recommendations of the Royal Commissioners, and issued a Report which, I venture to think, is well worthy of the attention of the Institute,—very well worth reading and discussing. As one who drew up the Report of course I am not in a position to open a discussion upon it, but there is no doubt that a great many things that we recommended have become the basis of the present proposed legislation, though there are others which the Government seem to have dropped like a hot potatoe—I do not know for what reason. Still, I was in great hope that that Report might raise an interesting discussion in this meeting. I hope some of our friends have read it and are in a position to deal with it.

Mr. R. Carter: Has it been in the hands of members?

The President: Yes; it was issued in April. I should be glad if any member has read the Transactions and will give us his views on the recommendations of the Committee, at least upon the criticisms of the Committee.

Mr. R. Carter: Has this Report of the Council yet been sent to the Government department.

Mr. Joseph Mitchell: No.

Mr. R. Carter: The intention is that it should be sent?

The President: The Bill was too far advanced to hope it could receive any consideration in respect to the present Bill. I think that was a mistake probably. I thought it was sent.

Mr. R. Carter: The only value the observations possess is in becoming a historical record of what the Institute have had to say, amongst themselves at all events, upon the Report of the Commissioners.

The President: Yes.

Mr. R. Carter: The value of the observations of the Council may be that at present, but in the future they are likely to be of use. Is the Bill likely to pass?

The President: I think so.

Mr. R. Carter: I think they mean to.

The President: I think they mean to pass it if they can. Mr. Smith said the other night that as soon as the Irish Bill was got out of the way they would take the Mines Bill.

Mr. W. E. Teale: There are 40 or 50 clauses altogether yet to pass.

The PRESIDENT: They have been in committee, and it has some clauses which are entirely unworkable and which we have tried to get altered.

Mr. W. E. TEALE: Some of them will come on report

The PRESIDENT: One clause, as passed, means that every airway in any mine should not be less than five feet by four.

Mr. HINCHLIFFE: A very foolish thing as far as my experience leads.

The PRESIDENT: Utterly absurd; we have got the promise of the Home Secretary that that shall be altered.

Mr. HINCHLIFFE: There are many roofs that will not stand above 2 ft. 6 in. wide even in straight work.

The PRESIDENT: It came upon the question of the size of the communication between the upcast and the downcast pit, and it was moved by the miners' delegates that the size of that airway should be not less than 6 ft. by 5 ft. It was represented that that was a very large increase, and as a compromise a learned Professor from a University in Scotland proposed it should be 5 ft. by 4 ft. The old Act was 4 ft. by 3 ft., and he thought that would meet the case. Then the Home Secretary announced that that would apply to every communication between upcast and downcast, and that was passed —that all your airways are to be 20 feet. If you are only working a two feet seam you will have to make a 20 ft. airway. That is one of the things I think we should get altered.

Mr. HINCHLIFFE: It is worth while trying at any rate.

The PRESIDENT: We have got the promise of the Home Secretary that it shall only apply to the main way between the upcast and downcast, and shall not be retrospective, but only apply to future mines. The same way with regard to the distances of the upcast from the downcast shaft. That is fixed now by amendment under the new Act at 15 yards, and there are many pits in which they are nearer. That is not to be retrospective, but to apply to the future. There are some shafts down to an upper seam, and may go down to a lower; they may be allowed to stand on the basis of the old Act, which is, I think, 20 feet.

Mr. JARRATT: Ten feet.

Mr. W. E, TEALE: There seems to be great difference of opinion with regard to shot firing between the North and South Wales.

The PRESIDENT: It is a difference of opinion of all districts with South Wales.

Mr. W. E. TEALE: Yes; the other districts are against South Wales.

The PRESIDENT: The Northumberland and Durham people have moved a new amendment, and there is a meeting in London to-day about it, which I should have been present at but for this meeting coming in the way. I telegraphed my opinion as representing this district—there are three of us appointed a committee, Mr. Parker Rhodes, Mr. Charles Rhodes, and myself. I opposed any alteration of the Government Bill with regard to shot firing, because the Government Bill really comes within the practice of this district. The South Wales district have been doing what I consider a very dangerous practice, and I think they should not be allowed to go on with it, because if they do there is sure to be an agitation for another Bill within the next two or three years. The whole of the miners and coal owners of North Wales, including Staffordshire and those districts, are opposed to the amendments of South Wales, and miners and owners in South Wales are in favour of South Wales amendments, so there is a sort of double split, and it has done us very little good in trying to get our amendments passed.

Mr. JARRATT: What is the new proposition on behalf of Northumberland?

The PRESIDENT: Under the old Act if you had found inflammable gas within three months you had to have certain restrictions. The Bill of the Government proposes in the case of gas only that the restriction should only apply to a week; and in the case of gas in dry and dusty mines, to four weeks. The north of England amendment is that in both cases it shall only apply for twelve hours. We are in a very weak position in the House of Commons. We have only one practical miner as a representative of the owners, and that is Sir George Elliot. The other representatives of the owners hardly know anything of mining practically. One or two good men have died off.

Mr. R. CARTER: Mr. Knowles and other men have died off, and their places have not been taken by men of the same mining culture. I think the report is a very valuable document, if we may justly apprise our own productions. I feel regret it has not been made use of; but the time is past now. I went carefully through the draft of these observations, contrasting them with the Commissioners' Report, and I think the observations were not only important, but worthy the serious attention of the Home Secretary, which they would have got if he had got them in time. However, they now exist only in the history of our Institute as part of the Transactions. I hope at some future time that the observations which have originated in a very long investigation—the Commissioners themselves, I think, had been sitting for five or six years taking most important evidence, and they were indebted to no district more than to South Yorkshire and the Midland Institute especially for the help given them at the time—I do hope by some hook or crook we shall keep this subject before us; and it may be the opinions of this Institute, which have cost a great amount of labour and attention on the part of the oldest and most experienced members of the Institute, may not be without their value in taking up, as we shall have to take up before many months are over, the further consideration of the legislative interference and enactments with the great principles of coal mining. Your own observation with regard to the position of the coal owners in the House of Commons is one which we cannot regard with too much gravity and importance. The latter part of this great question is pretty well represented in the returns of last election, by which Mr. Pickard and others obtained a status there, to strengthen the hands of Mr. Burt, or weaken them, I don't know which, because as between Mr. Burt and Mr. Pickard there seems to be a good broad line of demarcation to be made, and whether the two will unite in promoting any practical effect upon the mind of the Government of the day I do not know; but at least it does seem an important question for the Institute, and for all parties interested in this most important question of the coal mining industry to keep their eye very wide open, especially as you remember we have lost so many good and able men from the House in the last year or two. One may hope when a general election comes men will be forthcoming who will take the place of such men as Mr. Knowles, and that the masters may be more adequately represented in the House of Commons, as every other interest is. The railway interest, the law, the army, the navy, have their members in the House, and as one of the most important interests of the country, it will be a great pity if the coal mining interest is not represented in the future. I can only say in passing these observations on this report into the history of the past, members will not lose sight of the fact that there is represented in these observations a very great amount of practical knowledge and practical acquaintance with all the details of mining by all the most approved systems we have been made acquainted with, and placed in juxtaposition with the Report of the Royal Commission the observations will be well worthy of taking up at any future time with all the respect and all the regard for their usefulness which members of the Institute for the time being may be disposed to give. I wish to pass these observations, as I should have been glad to have congratulated the Institute on having their footing as well established in the end of this enquiry as we had in the commencement. I am satisfied these observations must have their effect, and I trust the Council will see their way to put them before the Home Office before the thing is closed up. They should exist in the Home Office in reference to any future legislation taken up on this subject.

Mr. W. E. TEALE : I think it is a matter of great regret that the remarks of the Committee have not been brought prominently before the Home Secretary. The interest that he has taken in this Bill shows an intense desire to arrive at true results of what would be to the benefit of masters and men. He has tried to analyse as a non-practical man the arguments brought to bear in favour of one person or another. It seems to me with regard to the political representation of the mining interst that the workmen are far too strongly represented in the House, and the consequence is they have been able to bring a little coercion in the way of advancing their opinion, and there is a slight appearance of an intention on the part of the Home Secretary and Government as far as possible to meet their views. You can hardly blame them taking that course when we find there are no gentlemen above the position of the workmen's representatives who will go amongst the men and make themselves at home with them. I think it is a disgrace to the mining interest at large that we have no gentlemen of sufficient political interest on one side or the other who will make themselves at home with the men, and who will ventilate such subjects as are rammed down their throats, and let them know what the truth is. I was reading of two large meetings of miners—they had their representatives and Irish members of Parliament. It appeared to me the main objects of these meetings was not in the interests of the men, but to keep their votes in the hands of these representatives. I have made similar remarks before at kindred Institute's to this, and it is one I think should be taken up, and at some future time I shall bring it before the Institute. It is this—there seems to be a great want on the part of members of this Institute to do some little good in the way of getting up little meetings, trying to get the men together, showing an interest in their welfare, and explaining matters that appertain to them, not to take them to the high science of mining, but to such things as are useful to them. I had the pleasure of reading a paper at the invitation of a Church Committee some time since. Most of the men were colliers. I had great attention paid to my lecture, and it was very pleasant for me afterwards to receive the number of inquiries put to me. It struck me if the Institute would get a few gentlemen who have an hour or two to spare—say a couple of hours a month—if there is any question getting ripe you would be able to get one or two who would volunteer to give a little lecture in plain, simple language to these men, and the very fact of asking them to discuss the question would bring them in touch with you. It is a great pity that the Home Secretary has not had his attention called to the remarks made with regard to the Royal Commission. After so many years carrying on their business they at last were practically forced to conclude their report, and they hurriedly brought in some matters which were hardly fairly and duly considered. They did not take sufficient time, and they rather landed themselves in a mess in several respects. I am glad these have been pointed out by the Midland Institute. I feel satisfied if anyone will take the Royal Commission Report they will find a great deal to criticise, and some extraordinary contradictions. It is a pity these remarks of the Council have not been prominently brought before the Home Secretary, as he is not acquainted with mining matters, and, moreover, he has shown a disposition to try to grasp the question and to do the best he can for all parties concerned. I should like to hear whether anything has been done, or is proposed to be done, with regard to the representation of the Government in investigations in connection with accidents in mines. I had a conversation with the coroner in our district. The question of a Minister of Mines is out of the question; but he told me had had some conversation with the Home Secretary and others connected with the matter, and there was some idea of appointing a Commissioner who would act somewhat in the same respect as a Commissioner of Wrecks, and in case of an accident would come down and act as President of any investigation. The proposition, I understand, was, that he should be represented by two mining men, as assistants in the matter, and that the thing should be thoroughly gone into. It seems to me in the interest of the miners and everybody concerned, if an investigation were carried out on the same lines as in the case of wrecks, it would be a step in advance; it would point out where the difficulty occurred and that it could be provided for in the future.

Mr. EMBLETON: There is one thing to be said about what Mr. Teale has said as to having some one to come down and investigate cases of accident. Where can you get any barrister of so many years standing to understand what mining means, what ventilation means, what are the modes of working, or in fact, any one thing relating to the working of mines? There is the great difficulty. I remember this very distinctly, that in giving evidence on an explosion I was asked by a barrister as to the probability of being able to confine the gas to certain places, and he said, "Don't you think the best way to avoid these explosions is to wall up the goaf?" I said, "If you can tell me how to wall off the goaf I should be very glad, because I cannot understand how it can be done." Such a question as that shows perfect ignorance on the part of the person sent down to investigate the cause of explosions. The difficulty is to have a competent barrister. The competent person is a mining engineer, but there is this difficulty—the workpeople will say that he is prejudiced, and they will not have confidence in him.

The PRESIDENT: I shall try to find out the clause. There is nothing so formal as what Mr. Teale has indicated, but there is a clause under which the Home Secretary may appoint a Special Commissioner to come down, as he does now, in fact.

Mr. EMBLETON: When they come they know nothing about it.

Mr. TEALE: As it was conveyed to me it would not be so much a Commissioner as a person appointed by the Crown to take the chair at the investigation.

The PRESIDENT: There is nothing here of the kind.

Mr. TEALE: I wished to know if anything had been done in the matter. There have been some representations made with regard to Government Inspectors—that they, in their own district, should be the Commissioners. A second suggestion was made to appoint a man who would be an official of the Government on the principle of the Wrecks Commissioner, and simply to be president of the investigation instead of the coroner. He would come down and go into the question, having the assistance of some lawyer or barrister in the important questions. When each Government goes out or comes in some other barrister would be appointed, so there is not much chance of their knowledge of mining increasing. These Commissioners would be assisted by two mining engineers.

ELECTION OF OFFICERS.

The SCRUTINEERS reported the result of the election as follows:—

PRESIDENT:

Mr. T. W. Embleton.

VICE-PRESIDENTS:

Messrs. G. J. Kell, A. B. Southall, and G. B. Walker.

COUNCIL:

Messrs. W. E. Garforth, C. E. Rhodes, J. Longbotham, C. Hodgson, J. Jarratt, WM. HY. Chambers, M. Nicholson, and J. Nevin.

SECRETARY AND TREASURER:

Mr. Joseph Mitchell.

Mr. W. HOOLE CHAMBERS then read the annexed Statement of Accounts for the past year.

FINANCIAL STATEMENT.

THE TREASURER (JOSEPH MITCHELL) IN ACCOUNT WITH THE MIDLAND INSTITUTE OF MINING, CIVIL AND MECHANICAL ENGINEERS.

DR.

1887.		£	s.	d.
June 30.—To Subscriptions received		18	13	0
,, Sale of Transactions		80		
,, ,, Dinner Tickets		67	0	2
,, Amount received from Bank		91	0	0
,, Bank Interest		0	1	10
,, Balance due to Treasurer		26	10	5
		£251	18	11

CR.

		£	s.	d.
1886.				
July 1.—By Balance due to Treasurer		10	5	1
1887.				
June 30.—By R. E. Griffiths		55	0	0
,, C. Hammond (Annual Dinner)		140	0	4
,, Reporter		12	3	4
,, Room Cleaning		2	12	0
,, Insurance		0	15	9
,, Barnsley Gas Company		0	8	6
,, H. M. Edwards (Safety Lamps, &c.)		2	10	6
,, Hire of Rooms for Meetings in Leeds and Sheffield		2	7	6
,, Secretary's Salary		25	0	0
,, ,, Expenses		4	6	0
,, Rent of Room		30		
,, Stamps, Telegrams, Post Cards, and Sundries		11	3	7
,, Banking Company		95	16	4
		£251	18	11

Audited and found correct, ALFRED B. SOUTHALL, } *Auditors.*
W. HOOLE CHAMBERS,

July 25th, 1887.

THE MIDLAND INSTITUTE OF MINING, CIVIL, AND MECHANICAL ENGINEERS.
GENERAL STATEMENT.

DR.		£	s.	d.			£	s.	d.
1887. LIABILITIES.					1887. ASSETS.				
June 30.—To R. E. Griffiths..		76	3	9	June 30.—By Cash in Bank		20	5	0
,, Reporter		6	10	0	,, Arrears of Subscriptions		80	17	0
,, Barnsley Gas Company		0	1	7	,, Value of Transactions, 6089 at 1s. per copy		304	9	0
,, Balance being capital		322	15	8					
		£405	11	0			£405	11	0

Audited and found correct, ALFRED B. SOUTHALL, } *Auditors.*
W. HOOLE CHAMBERS, }

July 25th, 1887.

THE COUNCIL'S ANNUAL REPORT.

The Report of the Council was read as follows :—

The Council in presenting their Annual Report for the year ending June, 1887, to the members of the Institute, have to state that the following papers have been read :

" On the use of Rolled Steel Girders for supporting the Roof in Mines," by Mr. THOS. REEDER SMITH.

" On Elliott's Patent Multiple Wedge," by Mr. H. S. WALKER.

" On the Easterley Extension of the Leeds and Nottingham Coal-field," by Mr. ROWLAND GASCOYNE.

Observations of a Committee upon the Final Report of the Royal Commission on Accidents in Mines.

" Description of an Arrangement for arresting the fall of Colliery Cages in cases of Breakage of the Rope," by Mr. BERNARD E. CLARKE.

The following subjects have also been discussed :

Mr. JOHN NEVIN's paper " On the De Bay Ventilator."

Final Report of the Royal Commission on Accidents in Mines.

Mr. G. J. LAMPEN's paper " On Coal Cutting Machinery."

Mr. T. REEDER SMITH's paper "On the use of Rolled Steel Girders for supporting the Roof in Mines."

Mr. H. S. WALKER's paper " On Elliott's Patent Multiple Wedge."

Mr. ROWLAND GASCOYNE's paper " On the Easterley Extension of the Leeds and Nottingham Coal-field."

The Council still regret that the papers read are few in number, but hope that the members will contribute more papers during the present year.

Following up the remarks in last year's Report as to the ventilation of mines, a Committee has been appointed to report on the various fans, and it is proposed that experiments should be made on the following fans;—Guibal, Schiele, Walker's Improved Schiele, Waddle, Cockson, Leeds, De Bay, Capel, and Biram.

Another important subject has been brought before the Institute by Mr. Rowland Gascoyne, viz., the Easterly Extension of the Leeds and Nottingham Coal-field, and it would be of great advantage if other members would read papers on this subject, and give the Institute the results of the deep borings to the Silkstone and other seams, which have lately been carried out.

An invitation having been received from the North of England Institute of Mining and Mechanical Engineers, asking the members of this Institute to visit the Newcastle Exhibition during the first week in August, it is hoped that the members will endeavour to take advantage of the facilities so kindly afforded, as arrangements have been made to view the Exhibition, and works and collieries in the neighbourhood.

The number of members on the books of the Institute at the end of the year was 4 Life Members, 18 Honorary Members, and 142 Ordinary Members.

The yearly subscriptions are not as promptly paid as they should be, the item of arrears being £80 17 0, against £47 5 0 last year.

The New Mines Regulation Bill has had the attention of the Council, but it was thought that the members of the Institute connected with the Mining Association of Great Britain would be able to watch the interests which are represented by this Institute. A Committee of that Association, consisting of representatives of all the mining districts of the kingdom, has been carefully watching the progress of the Bill in the House of Commons, and has suggested many amendments in the interests of mining generally, which have received the best consideration at the hands of the Home Secretary, and will, we believe, prevent any undue and needless pressure upon the working of mines. It is hoped this Bill will be passed during the present session of Parliament, and will settle the mining laws for many years to come.

The PRESIDENT: Gentlemen, I have pleasure in moving this Report be received and adopted. It is indeed a matter of great difficulty to get papers on new subjects. I fancy many members think that some subjects which are of great interest to them are not of sufficient importance to bring before the Institute. That indeed is a great mistake, and I hope that some may be induced to give papers of a similar character to those which we have had during the past year, with reference to anything of use, any particular article, or the adoption of any particular improvement in mechanism, or any detail of the various processes used in connection with mining which tend to induce economy in the working of mines, for the condition of trade is such, it is quite certain that unless we can meet the depression by reducing the cost of production, we shall before long be left in the lurch. There is a process of exhaustion going on—a process of exhaustion from want of means, because the article we have to deal with is being reduced in price year by year. We are engaged in a struggle for existence, and it is only the fittest who will survive it. The Council, seeing the difficulty in getting papers, has been casting about for some other way of making the Institute of use to the mining engineers of the district, and a year or two ago adopted an investigation with regard to safety-lamps, which has been of great use and great interest. They have now turned their attention to the subject of fan ventilation. They have appointed a small committee to make experiments, and they have also appointed a committee of a more scientific character to elaborate these experiments—to make the necessary inferences from the observations which the first committee make, and put them in scientific form and present them to the Institute. I venture to think that the Report, when printed, will be of great interest and great use. If there is any other direction in which we can pursue investigations of this character we shall be glad to receive suggestions. We intend that committee shall go to work at once, and report at the earliest date. With reference to the question of further legislation with regard to mines, the conversation which we had a short time ago—I can only say that I have had very great pleasure at considerable inconvenience in giving a good deal of time and attention to the subject. I have been brought in contact with those promoting and conducting the Bill, and also with the miners' representatives. I can only regret one thing, that when we proposed there should be a joint meeting of colliery owners, mining inspectors, and miners' representatives, the miners' representatives very strongly objected to the mining inspectors being present, and, in fact, declined to meet the owners if they were present. I think that shows a disposition with respect to the subject which is not likely to promote the negotiations going on. At the same time we have met the miners' representatives, and the result of these meetings, which were continued over several days, has been to eliminate several subjects from discussion and opposition which otherwise would have been very strongly opposed in Parliament, and probably led to the consumption of a great deal of time. I think on the main provisions of the Bill, I may say, that, with one or two exceptions, the miners' representatives and coal owners have come to an agreement; but there are some questions which we shall be obliged to fight to the very utmost of our ability, and I can assure you the representatives of the owners are ready to give all the time and attention required to that subject. It is a most unfortunate thing that on one or two questions the owners of collieries and mining engineers are divided. That has made our whole representation in the House of Commons somewhat weak, and has not got us heard in the way in which we should be heard. The miners' representatives seem to be much more unanimous than the coal owners. We cannot help that. We must do the best we possibly can, and I hope this Mining Bill will be passed this year, and in such a form that there can be no agitation got up for further legislation for a good many years to come. I think if one or two points have to be given way upon, which may bear hardly upon a few mines in different districts, it will be better to give way to some extent on points of that kind, than be subject to agitation and further legislation with a Government that may be very much more difficult to deal with than the present Government. I beg leave to move that this Report and Statement of Accounts be adopted as the Annual Report and Statement of Accounts of the Institute.

Mr. R. CARTER: Mr. President, I have much pleasure in rising formally to second the adoption of the Report and Accounts, and at the same time to congratulate the Institute upon what, financially, seems to my mind, to be a very much improved state of things in the last year or two. Very little effort indeed will serve to put our finances in the position we all like to see, that is, not only with a balance to our credit at the bank, but also with a freedom of indebtedness to our Treasurer. That is very different to what I can remember a few years ago, and I gladly welcome it. I hope in the very early future that the Institute may be able to congratulate itself upon some return of activity to the coal trade, and that our members will be invigorated with greater zeal and spirit in their calling and in their devotion to this Institute. Following your remarks, I hope we shall see the effect of it in a good and sufficient contribution of papers by members, and that the Institute may flourish as it has not done in the past. I have much pleasure in seconding the motion that the Report be adopted.

The motion was carried.

The PRESIDENT: It is now my pleasure to retire from the office which you elected me to two years ago, and to introduce our veteran member, Mr. Embleton, to this chair. This is the year of jubilee—the jubilee of the Queen's reign. Mr. Embleton was, I believe, one of the first, if not the first, promoter of the Institute and of the Viewers' Association, indeed, which existed before this Institute was established. Therefore it was that the Council thought it would be a proper thing, and a graceful thing, to show our acknowledgment of the eminent services which he has rendered to the Institute for so many years by suggesting that he should be elected President for this year, especially as we have in view a visit to the town of Newcastle, of which town Mr. Embleton is a native. I am pleased to find that every member of the Institute who has sent in a voting paper, and I have no doubt those who have not sent in their voting papers, have been fully agreed, for their silence gives consent. I am pleased indeed to find that Mr. Embleton is elected President for the ensuing year. I have pleasure in congratulating him upon that appointment and in asking him to be good enough to take the chair, and to carry on, which I know that he will do, the affairs of the Institute, with that interest, and that consideration, and that enthusiasm, which he has always shown with respect to everything connected with the Institute. I beg to ask you, Mr. Embleton, to be good enough to take the chair.

The PRESIDENT ELECT took the chair amid applause and said: Gentlemen, I can only say I am very much obliged to you for electing me to the onerous and honourable office of President of this Institute. I know I have been President before; but I find that things are in a different state now to what they were then. We were then in our infancy; now we approach almost to manhood, and I again thank you for the honour you have done me. May I make one or two observations about the Report on the Mines' Regulation Bill which we have before us. It was a great pity, I must say, that that Report was not sent to the members of the House Commons or the Home Secretary, perhaps to both; but I think that no time should be lost in sending copies to those members of the House of Lords who are connected with mineral properties. Many of them are there and will reject, no doubt, clauses which go against their interests, and they will oppose them. The Bill then will receive a perfect examination and come out something what it should be. I say this, for this reason, that although the House of Commons may consider these things, and properly consider them, yet I think the House of Lords, seeing the ability of those who are members of that House, seeing the discussions that take place in that House as contrasted with those which take place in the House of Commons, I think I should rather trust the House of Lords to pass such a Bill as would regulate all these matters respecting mining, rather than the House of Commons. I remember long ago, and perhaps some of you may remember the old Radical Cobbett : he said on one occasion, " We ought to thank God we have a House of Lords." If these papers are sent to the House of Lords we shall have nothing to regret in having sent them there. I thank you very much for the honour you have done me.

Mr. R. CARTER: As the business for the day is drawing near its close, I think there is one duty we shall be anxious to discharge. It is a duty we owe to our retiring President, to thank him from the very bottom of our hearts for the very genial and successful manner in which he has fulfilled the office of President for the second year. I am sure that all those members of the Institute who have been able to attend its meetings more regularly than I have, the observation will occur with greater force, but enough has been presented to my own mind and judgment to satisfy me under what deep obligations our retiring President has placed us, for the very able, efficient, and painstaking manner in which he has devoted his very well cultivated mind to the duties of this position. I know the difficulties which he has had to contend with, difficulties which he has also to a considerable extent surmounted. But there are difficulties which arise out of the unavoidable condition of trade and other discouraging circumstances, which it is not in the power of the most sanguine President to find himself competing and overcoming. If it had been so, we had in our President, Mr. Chambers, a man who was in every way likely from his experience, judgment, and willing devotion to the interests of this Institute, to have accomplished all that was needed. You are as well acquainted as I am with the claims which he has upon us for the very zealous and very efficient manner in which he has fulfilled its duties, and I am sure you will join with me in tendering to him at this meeting an expression of our very heart-felt thanks for all that he has done in his capacity as President, and with sincere wishes that we may have the pleasure of seeing him on every future occasion, and be directed, guided, and assisted in all our deliberations with that willing judgment which he is capable of bringing to our service, and to our help and encouragement.

Mr. W. E. TEALE : After the very nice speech we have had from Mr. Carter one of our late Presidents, who has had some experience of the duties and difficulties of the post, and can speak with all the effect that comes to him so very easily, I need only say I have heard it expressed we never had a better President than the last. We have received every consideration at his hands, and I must congratulate him on the way in which he has caught the smooth end of any difficulty in discussion which has cropped up. I have great pleasure in seconding the motion.

The motion was carried unanimously.

Mr. A. M. CHAMBERS : Mr. President and gentlemen, I really feel I have not deserved all the very excellent commendations which have been given expression to by Mr. Carter and Mr. Teale, but at the same time I may say it has given me very great pleasure, though it has been some addition to one's work, to try and discharge the duties of President of this Institute. I felt it an honour to be elected to that post, an honour not to be enjoyed merely, but there was a certain amount of work one must give. I have endeavoured to maintain the dignity of the position, and to discharge the work devolving upon me to the very best of my ability. Whether I have succeeded or failed you know better than I do. At all events it has not been from any other desire than to promote the interests of the Institute in every respect. I may say sometimes in the occupation of this offiee, which has extended over two years, one has felt depressed with respect to the Institute, and I think this is the time to say it rather than when I was in the chair ; I do not think the Institute receives the amount of support from the members thereof that it ought to receive, and I really do not think the Council and President receive that support in the way of papers and of members coming forward with information for the benefit of the Institute that they have a right to expect. That has often given one cause of depression and fear that one has not been able to elicit the sympathies of members as they ought to have done. I do hope that our new President, our new old President will have more support in that respect, as it really is a beggarly account to give at the end of the year, that we have only had really three new papers during the year. We have an increasing number of members, and the character of members now proposed and elected to this Institute is, by the new rule which was adopted two years ago, confined to those who are really interested in mining, and who have everything to gain and nothing to lose by giving the results of their experience and getting the results of other people's experience, through the medium of the Institute. My only desire is, that the Institute may go on and prosper. It was an unfortunate thing our friends at Chesterfield were so blind as not to see the advantage of an amalgamation to make one really good Institute for the Midland mining district; but they broke the negotiations, we could not carry them further, and the result is we are as we are. At the same time it is encouraging that in many respects the Institute has improved. There is plenty of room for improvement further, and I do hope some of our friends who have not hitherto helped us with regard to information of the kind I have indicated will now come here and do something, and give us the result of their experience. I am much obliged to you for the hearty manner in which you have adopted the vote; I do not feel it is altogether deserved, but at the same time I do trust that every member of the Institute, whether President, member of the Council, or private member, will do his utmost to promote the interest and object we all have in view, which will, if promoted to their utmost extent, be of very large benefit to the colliery owners of the district, to the miners of the district, and more especially to the members of the Institute themselves.

The Annual Dinner was afterwards held at the Queen's Hotel.

MIDLAND INSTITUTE OF MINING, CIVIL, AND
MECHANICAL ENGINEERS.

GENERAL MEETING.

HELD AT THE QUEEN'S HOTEL, LEEDS, ON TUESDAY, AUGUST 30TH, 1887.

T. W. EMBLETON, Esq., President, in the Chair.

The minutes of the Annual Meeting were read and confirmed.

The following gentlemen were elected members of the Institute, having been previously nominated:—

Mr. GEORGE CRADOCK, Mechanical Engineer, Wakefield.

Mr. WALTER PEASEGOOD, Colliery Surveyor, Birley Collieries, Sheffield.

Mr. JOHN BENNETT, Park Hill Colliery, Wakefield.

Mr. WILLIAM BRIERLEY, Colliery Manager, Roche Colliery, Batley.

THE PRESIDENT'S INAUGURAL ADDRESS.

The President then delivered his Inaugural Address as follows:—

The beginning of coal mining, like most other things, is enveloped in the obscurity of the past. There is little doubt that coal, so far as can be gathered from scattered notices discovered here and there in various records, was first obtained by manual labour from the outcrops on hill sides or the margin of brooks, and rivers, or the cliffs of the sea coast. In such places the attention of the ancient people would be arrested by the contrast of the black mineral with the colour of the superincumbent and the adjacent strata. The use of coal and other mineral substances did not become general until the necessity for their use became apparent, nor were they at first extensively worked. In ancient Briton and Roman times coal could not have escaped the notice of the energetic people of the land, though the forests that then covered the greater part of England afforded abundant supplies of fuel. The Romans, as we shall see, had discovered coal and its uses, or had derived their knowledge from the Britons. In Anglo-Saxon and in Norman times, as manufactories of various kinds were established and increased in number and extent, the forests suffered in proportion.

The iron and glass works which were carried on extensively in the northern and eastern parts of England consumed increasingly enormous quantities year by year; the forests diminished, and the use of wood as fuel became more and more costly. This destructive process had the effect of directing public attention to the acquisition of mineral fuel for manufacturing and domestic purposes, and hence the excavation of coal rapidly increased; but it was only after the invention of the steam engine that the demand and supply enormously expanded, and has now attained dimensions which a century ago would have been deemed fabulous and altogether incredible.

Although the following extracts of the ancient working of coal were communicated to another Society, I think it desirable that they should be printed for the benefit of this Institute. I will endeavour to append to these extracts such information as I am able to collect respecting the development of the coal mines from the past to the present time, and as this is the Jubilee of Her Most Gracious Majesty it is important that the progress of the mode of working and raising coal should be noted, as a similar report on the progress of other industries has been before the public, not the least of which is the coal trade.

In the Saxon Chronicle of the Abbey of Peterborough there is the following passage: "About this time," A.D. 852, "the Abbot Coelred "let to hand the land of Sempringham to Wulfred, who was to send "to the monastry 60 loads of wood, 12 loads of coals, and peat and "other things."

Whether the meaning of the Saxon word is coal, as now understood, is subject to some doubt. The original words are "twælf "fothur Græfan" (i.e., 12 fothers of coal, about 15 cwts. each). In A.-S. this word (Græfa) may mean anything that is dug up. It is evident that Græfa was a different fuel from peat. There is no A.-S. for peat which is English only. The old British name is said to be Glo; the Welsh has Pwll Glo, Pit-coals. In Bosworth's A.-S. dictionary, Græfa is coal.

Coal is in Middle Eng., Col; Dutch, Kool; Icel. and Swed., Kol; Dan., Kul; Old High Germ., Chol, Cholo; Middle High

Germ., Kol; Mod. Germ., Kohl. The French houille, only wants a gutteral to assimilate it to the Northern word.

With reference to the word "coal," as so translated in the authorized version of the Holy Scriptures, the original word possibly means charcoal, which, like the Latin word may designate fuel derived from wood or coal itself.

Theophratus, who lived 288 years B.C., speaks of fossils called anthracite, which when broken up inflame and burn like charcoal, and are burned by the smiths.

From Pennant's tour in Wales, he states that a flint axe, the instrument of the aboriginals of our islands, was discovered stuck in certain pieces of coal exposed to day in Craig y Lare, in Monmouthshire, and in such a situation as to render it very acceptable to the unexperienced natives who in early times were incapable of pursuing the seam to any material depth.

Whittaker, in his History of Manchester, is of opinion that the primeval Britons used coal. This is evident, he says, from its appelation amongst us at present, which is not Saxon but British, and subsists amongst the Irish as their Guel, and amongst the Cornish in their Kolan to this day. In the great survey (the Domesday Book), carried out by William the Conqueror, there is no mention of coal or any other minerals. No instructions were given to the Commissioners to enquire into the extent and value of the mineral property of the central or northern counties. Strutt states that in opening some mounds in 1773, near Maldon, in Essex, bones, cinders, and charcoal were found; this he relates in his Book of Manners and Customs, Vol. 1, p. 60.

Marston, in his Natural History of Northamptonshire, speaking of the ruins of a castle at Castle Dykes, a high, entrenched hill, near Farthingstone, says he saw some huge lumps of cinder. This castle is supposed to have been built by Æthelfleda, about 913, it was destroyed by the Danes in 1013, and he concludes that there were stores of coal kept there. St. Augustine says, "is it not a wonderful thing "that though coals are so brittle that with the least blow they break "and with the least pressure they are crushed into pieces, yet no "time can destroy them, inasmuch as they who pitch land-marks "were wont to throw them underneath to convince any litigious per- "son who shall affirm, though ever so long after, that no land-mark "was there."

In Ireland coal was worked at an early period, as the following statement proves, although no precise date can be assigned for the work. At Ballycastle, in 1770, in a passage cut through one of the seams the old workings were entered; they were narrow, their sides were encrusted with sparry-like matter. This discovery led to a gallery which had been driven forward for several hundred yards. It branched into thirty-six chambers. The coal had been worked in a regular manner, pillars being left to support the roof. The remains of tools and baskets were found in the workings, but there is no statement of the kind of tools. Near Stanley, in Derbyshire, some years ago, some colliers driving in the Kilburn coal, broke into some old excavations, in which they found picks made out of solid oak. These implements were entirely destitute of metal, and were cut out from one solid piece of timber. Implements belonging to an equally early period are stated to have been found in old coal workings near Ashby de la Zouch, consisting of stone hammer heads, wedges of flint, with hazel withes round them, and also wheels of solid wood.

The Romans were acquainted with the use of coal. They had stations in places near the outcrop of coal seams, and coals and cinders have been found in the Roman towns and villas. Wigan in Lancashire, was a Roman station. Not far from that town, a bed of Coal, known as the Arley Main, crops out on the banks of the river Douglas. While driving a tunnel to divert the course of this river, the Arley seam, 6 feet in thickness, was found to have been mined in a manner hitherto altogether unknown. It was found to have been excavated into a number of polygonal chambers with vertical walls opening in to each other by short passages, and, on the whole, presenting a ground plan something of the appearance of a honeycomb. The chambers were regular both in size and form over an area of at least 100 yards in one direction, and were altogether different from any modern mining in the district. The arrangement and regularity of the work was peculiarly Roman, resembling in some measure their tesselations, or the ground plans of their baths and villas.

Whittaker, the historian, of Manchester, relates that in the West Riding of Yorkshire a quantity of Roman coins were found among many beds of coal cinders heaped up in the adjoining fields. These "coal cinders" were found to be slags from ancient ironworks.

I have not yet found any notice of early coal working in Yorkshire. Mr. T. Wright considers that the Shropshire coal-field was discovered by the Romans—not far from the borders of this coal-field stood the ancient Uriconum, now Wroxeter. During the excavations considerable quantities of coal, both in the raw state and partially consumed, were found; it had been used apparently in heating the ovens.

It is stated that coal was discovered in Belgium, near Liège by a pilgrim in the year 1189.

In the survey of Hugh Pudsey, Bishop of Durham, called the Bolden Duke, made in the year 1183, and which may be called the Domesday Book of that county, frequent mention is made of coal. The survey was made for the purpose of ascertaining the revenues of the whole bishoprick as they were then and the assized rents and customs as they then were and formerly have been. The following extracts will suffice :—

1. A certain collier holds one toft and one croft of four acres and finds coal for making the iron-work of the ploughs of Coundon.

2. The Carpenter at Wearmouth, who is an old man, has for his life 12 acres, for making ploughs and harrows. The Smith, 12 acres for the ironwork of the ploughs and harrows, and coal which he wins.

3. In Sedgfield.—The Smith, one oxgang for the ironwork of the ploughs which he makes, and he finds coals. The Carpenter, 12 acres for making and repairing the ploughs and harrows.

The customs mentioned in these extracts appear very curious, that the Carpenter and Smith should be rewarded for their materials and labour in making and repairing the ploughs and harrows, in consideration for a grant of land   Money no doubt was very scarce amongst the farmers and villagers of this period.

In the Magnus Rotulus recept., Dunelm Anno Antonii Episcopi xxv., there are the following entries :—

. " Minera Carbonum. Et de 12s. & 6d. de Minera Carbonum in " quarterio de Cestr."   (Chester-le-Street).

" Termnius Sancti Cuthberti Septembri."

1. Text.—Quidam Carbonari tenet j toftum et j croftum et iv. acras et invenit carbones ad ferramenta carucarum de Coundona.

2. Carpentarius qui senex est habit in vitasua xij. acras pro carucis et hercis faciendis. Faber xij. acras pro ferramacutis carucarum et carbonem quem invenit.

3. Faber j bovatum pro ferramentis carucarum quæ facit et carbonem invenit. Carpentarius xij. acras pro carucis et hercis faciendis et reparandis.

" Minera Carbonum. Et de Minera Carbonum 12s. & 6d. de "Minera Carbonum in quarterio de cestr." (Cestria).

It would appear from these notes that the coal must have been worked in the neighbourhood of Chester-le-Street before the date of the above payments.

At Lanchester, in the county of Durham, on exploration in the Roman station, the calcareous flooring of the baths was found to be mixed with coals and cinders. In the foundation of a circular Roman building, on the side of the Watling Street, about three miles south of Ebchester, coals and cinders were found buried together with a smith's hammer.

In Bishop Hatfield's Survey of the Manor of Collierley, in 1333-1335, the Bishop appoints a supervisor of his mines there. His successor makes a similar appointment in 1384. A coal mine is mentioned in the ordination of the Vicarage of Merryton, in the the county of Durham, in 1343, and in 1354 there is extant a notice of the sinking of pits at Ferryhill, in the same County.

If we refer to the coal mining in Northumberland, we find more copious notices of it than in any other county. On 1st Dec., 1239, King Henry III. granted a charter to Newcastle to dig coals and stones in the common soil of that town, in a place called the Castle Field (now the Town Moor) and the Forth. This is the earliest known date when permission was given to work coals at or near to Newcastle. It is evident that at that time the minerals belonged to the king and not to private persons. In fact the whole place then belonged to the king.

In 1330 the Prior of Tynemouth let a colliery called Heygrove, at Elstwick, for £5 a year; another colliery, also belonging to the Prior, in the East Field, was let for six marks a year; besides which he had one in the West Field, and another near Gallow Flat in the same estate in the years 1331-1334. Two men were drowned in the Gallow Flat Pit in May, 1658, the bodies were not recovered till 24th April, 1695, having lyen in the water for thirty-six years.

About 1333, Philippa, queen of Edward III., had estates in Tynedale, where she resided during Edward's stay in his campaigns in Scotland, and obtained a grant from her husband to work coal mines there.

License was granted to the Burgesses of Newcastle to dig coals and stones in the Castle Field, in 1351. On the 10th May, 1357, King Edward III. granted license to the men of Newcastle to work coal in the Castle Field and Castle Moor. He issued orders concerning the regulation of coal measures. At this time coal were also worked at Gateshead, on the opposite side of the river Tyne. This king gave permission for coal won in the fields of Gateshead to be taken across the Tyne in boats to Newcastle, in consideration of the owners complying with the usual customs of the port. After paying these dues, permission was given that the coal might be sent to any part of the kingdom, either by land or water, but to no foreign country except to Calais, which was then an English port. In this year coal was exported to London from Newcastle.

As in the county of Durham, so in Northumberland, coals had been worked by the Romans during the construction of their celebrated wall.

In 1762, when in digging up the foundations of the Roman station at Caervorran or Magna, some very large coal cinders were turned up, which glowed in the fire like other cinders, and were not known from them when taken out.

At Habiticum (Risingham) near to a spot where traces of a furnace were noticed, more than a cart-load of coals were found, which Mr. Shanks removed and used in his own grate.

At Walton House station, several rooms were found the floors of which, consisting of thick masses of strong cement, were supported on pedestals. There were many other curious floors found among the ruins, and some coal ashes.

When, in 1833, the eastern gateway (of the Station Cilurnum, The Chesters) was freed from the rubbish that then encumbered it, its eastern portal was found to have been walled up, and converted into a separate apartment; on the floor of the chamber thus formed there was found a cart-load of fossil coal, and in the furnace for heating the baths may now be seen the coke and the soot. At that time the Romans did not consume their smoke just as the people of to-day.

The following extract is from the " Roman Wall " by the Rev. J. Collingwood Bruce, LL.D., &c. " When the lower reservoir of " the Newcastle Water Company, in the neighbourhood of South " Beuwell (the Condurcum of the Romans) was formed in 1858 some "ancient coal workings were explored; the author examined them, "and though he and those whom he consulted saw no reason to sup- "pose that they were not Roman, no coin, lamp, or shred of samian, "was discovered to give authority to the conjecture. The seam of "coal was two feet thick. It was wrought by shafts sunk to the "depth of twelve or fifteen feet and at a distance of forty or fifty-five "yards from one another. Lines of excavation radiated in every "direction from the bottom of the shafts. The coal crops out on "the bank between the workings and the river Tyne so that the "mine could be drained by means of an adit."

The instances of ancient coal mining given above were in seams which outcropped on the banks of the river Tyne, and were drained by adits. Leland, in his Itinerary, has the following passage:—
"The vaynes of the se coles ly sometyme upon clives of the se, as "round about Cocket Island and other shors, and they as some will "be called se coals, but they are not so good as the coles that are "digged in the upper part of the lande."

Under the term "Sea Coal" a considerable trade was established with London, and it became an article of consumption then. However an impression arose that the smoke arising therefrom contaminated the atmosphere and was injurious to the public health; and it is said that the nice dames in London would not come into any house or room where sea coals were burned, or willingly eat of the meat that was either sod or roasted with sea coal fire.

The subsequent extracts from Harrison's Description of England prefixed to Hollingshead's Chronicle, edited in the year 1577, contain some very curious and interesting notices concerning the coal trade and his opinion of the use of coal.

"Of cole mines we have such plenty in the Northern and "Western parts of our island as will suffice for all the realme of "Englande. Sea cole will be good merchandise even in the Citie "of London, wherunto some of them alreadie have gotten readie "passage and taken up their mines in the greatest marchaint's "parlers."

This quaint writer goes on to contrast the manners of former times with his own, "Now we have many chimneys, and yet our "tenderlings complaine of rewmes, catarres, and poses; then we "had none but reredoses, and our heads did never ake. For as the "smoke of those days (wood) was supposed to be a sufficient harden-ing for the timber of the house, so it was reputed a far better medecine to keep the good man and his family from the quacke or pose, wherewith, as then very few were acquainted."

"There are old men dwelling in the village where I remain have noted the number of chimnyes lately erected, whereas in their young days there was not above two or three, but each one except great people made his fire against a reredosse in the halle, where he dined and dressed his meat." He further complains:—"When our houses were builded of willowe, then we had oaken men; but nowe that our houses are come to be made of oke, our men are not only become willowe, but a great many altogether of straw, which is a sore alteration."

We see that the complaint is not new in 1887, for it is still said that the last generation was better than the present.

In 1306 the complaint became so general that the Lords and Commons in Parliament assembled presented a petition to King Edward I., who issued a proclamation forbidding the use of this fuel and ordering the destruction of furnaces and kilns of all who should persist in using it. This prohibition was repeated at several subsequent periods. The use of coal in London was resumed a few years after its prohibition by the King in 1306, as we find in the " petitiones in Parliament " in 1321-2 a claim was made for ten shillings on account of coal which had been ordered by the clerk of the palace, and burnt at the King's coronation, but the payment for which had been neglected. However, owing to the scarcity and dearness of wood, coal was as it were forced into use in spite of proclamations, and prejudice gave way as the value of fossil coal became more known, and from that time its use became more extended. In the reign of Queen Elizabeth the coal trade flourished greatly and was regarded as an important source not only of local but national revenue by succeeding monarchs. In the reign of Charles I. coal was used all over the kingdom.

I will now endeavour as briefly as I can to give an account of the modes adopted in working the seams and for raising the coal.

The ancient pits of necessity were shallow and of small diameter, not exceeding 3 or 4 feet; others were oblong of various dimensions. The jack-roll was the only means of bringing the coal to the surface except the coal had been won by a free lever. Subsequently, when the depth of the coal increased, the horse-gin was substituted, at first by the employment of one horse. This power proved to be not sufficient, and in Northumberland and Durham the gins were enlarged, and relays of four full-bred horses were attached to each gin, working four hours in the day when employed to raise the quantity of coal required.

An increased depth demanded further inventions to overcome the difficulty. One of these was the application of water as a power. It consisted of a tub of water sufficiently heavy to more than counterbalance the ascending corf and coals, the water being discharged at the bottom of the shaft, and the empty tub ascended to the surface by the weight of the descending corf, to be refilled.

In August, 1749, a grant passed the Great Seal to Mr. W. Newton, of Burnop Field, and to Mr. T. Stokoe, of Bryan's Leap, in the county of Durham, both gentlemen of great experience in the coal works, for a new invented method of drawing coals and stones out of deep mines by water.

2nd June, 1753. A machine at this time was going at Chater's Haugh, in the county of Durham, belonging to W. Peareth, Esq., invented by Mich. Menzies, Esq. (for this he had obtained an Act of Parliament to secure the property to himself), by which coals were drawn up, not by the power of horses, but by the descent of a bucket of water of a weight superior to that of the coals drawn up, lifting a corf of above 600 lbs. weight out of a pit 50 fathoms deep in two minutes.

6th Nov., 1778. A new constructed machine for drawing coal was set agoing at Willington Colliery on the river Tyne. Its performances exceeded the most sanguine expectations, uniformly drawing 30 corves of 20 pecks each (equal to $8\frac{1}{2}$ tons) in one hour from a depth of 101 fathoms. At Waterloo Colliery, near Leeds, the above method of drawing the coal was used almost recently. The name given to this colliery was in consequence of the pits being sunk in the year of the great victory. Several years ago I saw the disused apparatus.

I think it is needless for me to record the history of the steam engine from its infancy to its present state of perfection as used in different forms, for this history is well known.

Mode of Working.

The mode of working, even from the earliest times, as seen in the records of ancient workings, seems to have been board and pillar. In these records mentioned below, no dimensions are given either of the width of the board or the size of the pillar; but I will endeavour to supply this want in some measure, though the workings are not very ancient. At Middleton Colliery, near Leeds, I explored the old workings in the Little Coal there. I found the boards vary from 3 to 4 yards wide, and the pillars from 3 to 4 feet in breadth and 6 yards in length.

At Butternowle, in the county of Durham, I had, in order to decide a particular question, to descend an old shaft by means of ladders tied together. On reaching the seam it was seen that it had been worked on the board and pillar system; the boards were 9 feet wide and the pillars 6 to 7 feet broad and 8 to 10 feet long. There was this peculiarity. It appeared that no timber had been used to support the roof in the boards, but a few inches of the top coal had been left, and this and part of the lower coal was dressed to the form of an arch. The slits were similarly arched, the junction of the two arches meeting as accurately as if both passages had been arched with brickwork.

As time passed on the breadth of the board was increased, and the length of the pillar sometimes exceeded 20 yards. This mode of laying out the workings continued until recently both in Northumberland and Durham.

Continuing the description of the mode of coal getting, I may mention what is called the Barnsley method with its various modifications.

For the longwall system we are, I think, indebted to the Midland counties, of which method there are several classes.

It is only from actual experiment that we can decide how the coal shall be worked; there is no royal road.

As to Shafts and Tubbing.

At first the sides of the shaft were not protected. Afterwards, where necessary, the deeper shafts were sheeted, that is, the sides were secured by boards nailed to cribs attached to the sides of the shafts. Then followed ashlar and brickwork. Tubbing was made with 3-inch planks, the edges of which were planed radiating with the diameter of the shaft, and secured to cribs. Cast iron segments were substituted, bolted together by flanges projecting into the shaft and laid on wooden wedging cribs, and interspersed with wood cribs at certain intervals. This mode was sure to fail, as witness the accidents at some collieries, notably the shaft at the Garforth Colliery a few years ago. Cast iron wedging cribs came into use, with segments of iron tubbing having the space between the segments inlaid with pieces of deal about $\frac{1}{4}$ inch thick. This did not effectually render the tubbing water-tight, and to make the tubbing perfect it was necessary to stop any leakage by driving in pitch pine wedges. Latterly I have found that filling in the space between the sides of the shaft and the tubbing with dry soil passed through a riddle of $\frac{1}{4}$ inch mesh is equally as effective as wedging, and to be a more economical process. In some instances ashlar walling laid in cement has been adopted with great success, and is less costly than iron.

Guides in Shafts.

The first attempt to introduce guides into shafts was by Mr. John Curr, "Colliery Agent to his Grace the Duke of Norfolk at "Sheffield Park and Attercliffe Common Collieries, in the county of "York." This would be about the year 1787. He obtained for this invention His Majesty's Royal Letters Patent. Mr. Curr says that by his invention " a greater quantity of coal can be drawn in a given " time and from equal depths than has been hitherto done, and the " dashing and wear of the corves, and waste at their meetings, are " entirely prevented." His plan was to place in the upper and lower parts of the shaft one set of conductors as far as necessary, and in the vicinity of the meetings the passing of the corves was effected by a long switch which opened when the ascending corf passed and closed when it descended, in a similar way in which wagons pass each other on an inclined plane at the surface when laid with three rails above and one below meetings. To diminish the friction of the guides, rollers were fixed to what he calls triangle balls, that is, the the bar which supports the corf by two chains. A platform was drawn on an inclined plane over the shaft, on which the corf was landed, while the empty corf was placed on another platform and lowered on to the platform from which the loaded corf had been withdrawn. The original engraved details of Curr's patent are signed "J. Buddle, jun." (see Plates I. and II.) This plan was adopted in many pits in Yorkshire up to 1860, with various modifications. It would therefore appear that previous to 1787 coals had been drawn to the surface with the ropes and corves free in the shaft. I may here mention a method in use only a few years ago at Tynings on the Hill, in Somersetshire. The shaft was about 4 feet in diameter. The coals were drawn in a large barrel, the middle diameter of which was rather less than the shaft; it was about 5 feet deep. On arriving at the surface an inclined platform was drawn over the shaft; a chain was attached to a ring at the bottom of the barrel, and by a winch it was lifted over and emptied into the screen. Indeed, in the North of England and perhaps elsewhere, shafts not fitted with guides, were so used for many years after the above date; indeed, till about 1833, in the North guides were not used, the ropes and basket corves hanging free in the shaft. Oftentimes the ropes twisted together and work was stopped until the ropes swung free. The men descended and ascended by placing a leg in a loop made on the chain by attaching the hook to a link of the chain above them. But now cages are attached to the winding rope and the men walk into them. The cages on reaching the surface are rested on "keps," and these are withdrawn when the cage is lowered. Subsequently to Curr's plan, iron rods were used instead of wood, screwed together like boring-rods; and on the invention of the iron-wire rope this was substituted for the iron rods. In deep mines the common railway bar constitutes the guide.

PUMPING.

The first power to raise water were windmills and waterwheels, the latter connected with some fast flowing brook. Many designs for these contrivances may be seen in Agricola's book as used in Germany. The first pumps were made of trunks of trees bored out to the requisite diameter and fitted together by spigot and faucet joints. Indeed, all the pipes for supplying water to towns were made in this way. Then came in the use of cast iron pumps, still called in some districts "pump trees," a name derived from the original make; and then pumps were made of large size, sometimes exceeding 2 feet in diameter, with an accompanying size of engine.

Railways.

Previous to the introduction of railways, the coals, more especially where the mines had no connection with the seaboard or with a river, were distributed by carts, and in remote parts droves of ponies, mules, and asses were driven on the old jagger roads from the mines for the supply of distant districts. This mode I have seen in the Auckland district so late as the year 1820, and no doubt the same conveyances were in use in other mining localities. Even in 1758 the coal was carted from Leeds to the neighbouring towns, even as far as Ripon.

Railways took their rise in the North, caused, no doubt, by the distance from the collieries from the place of shipment of the coal. One wagon and one horse was the rule.

The first essay was not successful, as witness the following extract from an old work:—

"1649. Master Beaumont, a gentleman of great ingenuity and "rare parts, adventured into our mines (Northumberland) with his "£30,000, who brought with him many rare engines not known then "in these parts, as the art to boore with iron rodds to try the deep- "nesse and thicknesse of the coal, rare engines to draw water out of "the pit, wagons with one horse to carry down coals from the pits "to the staithes to the river. Within a few years he consumed all "his money, and rode home on his light horse."

These railroads were made of wood about 6 inches square laid on cross sleepers. These were not found very durable, although the wagon wheels were made of wood. To increase their durability, thin plates of malleable iron were nailed down on the working surface, these not proving much better, for in dry weather the iron was apt to be detached from the wood, causing numerous accidents. These were followed by cast iron, in pieces of 3 to 4 feet long, resting on stone blocks with cast iron chairs. Next came the malleable iron fish-bellied rails, 15 feet long, made at the Bedlington Ironworks from Longridge's patent, weighing 32 lbs. to the yard. These became too light for the increased size and weight of the locomotives and heavy trucks, and the strength was increased to 80 lbs. and upwards to the yard. A supposed improvement took place by facing the rails with steel, and ultimately steel took the place of all other inventions.

Ventilation.

In the beginning there was no means of ventilating the mines except by natural ventilation, for the mines were shallow and there was no inflammable gas and only occasionally choke-damp. As the mines became deeper, it was the custom to place a fire-pan, as it was called, at the bottom of the shaft to produce the ventilation. Necessity called for further means, and furnaces of various descriptions were adopted, the air from the mine passing directly over the fire. Accidents followed this plan, and the return air was made to pass up the shaft by a special drift, and the furnace was fed with air that had no communication with the workings of the mine. The workings becoming more and more extended, demanded a larger current of air to remove the gases, and in order to produce this extra amount of air the intensity of the furnace was increased to such an extent that often the furnace drift became seriously injured, and resort was had to the fan.

The first fan erected in Yorkshire was by the late Mr. Birom, the inventor of the anemometer, at one of Earl Fitzwilliam's collieries, about the year 1833.

The kinds of fans are now many; the inventor of each is, as usual, anxious to push his own. I hope, however, that the committee appointed by the Institute to obtain information respecting the performances of the various fans will during their investigations be able to lay before the members their report, from which may be gathered some valuable information.

Lighting the Mines.

The primitive light was the candle, sometimes so small that 60 of them made a pound. After an explosion, the opening of the mine before the invention of safety-lamps was partly facilitated by the use of the steel mill, as it was called. It consisted of a steel wheel and flint. The wheel was fastened by a belt to the person using it. The velocity of the wheel was increased by gearing, and on the application of the flint to the circumference of the wheel a shower of sparks was produced, sufficient to give light for the men to proceed in their examination. The use of this mill was the cause of some explosions. The candle was succeeded by the Davy lamp and about the same time the Stephenson lamp, and afterwards by the

Clanny. Various improvements have taken place in these lamps, but they are all condemned by Act of Parliament to the limbo of other once useful things.* It is needless to name their successors, for they are legion. So much progress has been made in electric lighting that we may hope that this will in the future be a more safe light than the safety-lamp.

I think I should not omit to mention respecting the former state of the coal trade in Leeds as compared with its present condition, and it is useful to compare what happened in past years with what exists now.

The agents of the coal owners now sedulously solicit orders in every quarter; at the time I am about to mention the consumers solicited the coal owner. An Act was obtained in 1758 by the owners of the Middleton Estate, near Leeds for the purpose of laying down a wagon road for the conveying coals to Leeds. It was estimated that in the above year the annual consumption of coal in Leeds was 22,500 tons, and the price to be paid was fixed at 4¾d. a corf weighing 210 lbs., or at the rate of 4s. 2⅔d. a ton.

A second Act was passed in 1779. In that Act it is stated that the quantity of 22,500 tons annually was not sufficient to supply Leeds. This Act goes on to say, "May it therefore please your "Majesty, at the humble petition and request of your Majesty's "loyal subjects, Charles Brandling and the inhabitants of the "town and neighbourhood of Leeds, in the county of York, that it "may be enacted," &c., that the quantity to be supplied quarterly shall be 11,250 tons, or 45,000 tons annually, at 5½d. per corf as before, equal to 4s. 10¾d. per ton (this advance is stated to be on account of the increased expenses). This Act contains a schedule of prices for conveying the coal into different parts of Leeds.

A third Act was obtained in 1793. New winnings had been made, and in consequence of "the advanced price of labour and the "price (allowed by the Act of 1779) being much lower than the price "demanded and paid at all coal works in the neighbourhood, and if "the wagon way and repository should be given up it would be a "material hurt and detriment to the manufacturers of the said town

---

* NOTE.—Since the above was written the House of Lords has by an amendment authorized the use of these lamps, and the House of Commons has agreed to that amendment.

"and parish of Leeds, and be a cause of great distress to the in-"habitants in general." It was therefore enacted that the price to be paid should be 13s. 1d. a wagon containing 24 corves, equal to 45 cwt., being at the rate of 5s. $9\frac{7}{9}$d. a ton. All persons except the inhabitants of the parish of Leeds were prohibited, under a penalty, from purchasing coal so brought down. The quantity of coal to be brought down from Middleton was 20,000 wagons, equal to 45,000 tons. On each day in the week 64 wagons were to be delivered.

The last Act was passed on 24th March, 1803. The same arguments were used—the extra cost of working, and to the inhabitants the cause of great distress if the supply of coal should be discontinued. The price therefore was advanced to 16s. the wagon of 45 cwts., being at the rate of 7s. $1\frac{1}{3}$d. per ton. Eighty wagons were to be brought down daily (Good Friday, Christmas Day, and Fast Days by proclamation excepted), or about 56,000 tons a year.

	Quantity. Tons.	Price per Ton. s. d.
We have therefore in 1758	22,500	4 $2\frac{2}{3}$
in 1779	45,000	4 $10\frac{3}{4}$
in 1793	45,000	5 $9\frac{7}{9}$
in 1803	56,000	7 $1\frac{1}{3}$

In 1758 colliers earned 1s. 4d. a day, and usually 6 men were employed in one pit; and in this year of Jubilee hundreds in one pit. In 1830, 4s. to 4s. 6d. was the daily wages.

In this particular wagon road wood rails were used; at first the wagon wheels were of wood, and during a dry summer the wagons passed through a pond in order to keep the wheels in proper order.

The inhabitants of Leeds and its vicinity do not now draw their supply of coal from Middleton alone, but under the changed conditions since the above Acts were passed a large supply is obtained from far distant coal-fields. The manufacturers have no cause now to fear "a material hurt and detriment to the said town and parish "of Leeds," or the inhabitants "a cause of great distress."

I hope I have endeavoured to show in some measure, from the earliest available information, the advance of the coal trade and its appliances up to this Jubilee year of Her Most Gracious Majesty.

AN ACCOUNT

OF

An improved Method of Drawing COALS,

AND

Extracting ORES, &c. from MINES;

WHEREBY

A *greater* Quantity can be drawn, in a given Time, and from equal Depths, than has been hitherto done; and the *Dashing* and *Wear* of the CORVES, and *Waste* at their *Meetings*, are entirely prevented.

ALSO OF A

New, easy, and expeditious Method of *Landing* and *Loading* the Coal, &c. so drawn,

Into *Waggons* or other Carriages, without the Assistance of an *Horse*, and with very little Labour to the *Bankmen*.

By *JOHN CURR*,

Colliery Agent to his Grace the Duke of Norfolk at *Sheffield-Park* and *Attercliff-Common* Collieries, in the County of York;

Where the above Inventions have been carried into full Practice, have been *Eighteen Months* in Use, and where Gentlemen may have ocular Proof of their Facility of Operation.

Every other Kind of Information respecting the Propriety and Advantages of their Application, will be given by the Inventor, who, in order to secure to himself a moderate Share of the Emoluments to arise from such his Inventions and Improvements, has for that purpose obtained

HIS MAJESTY's ROYAL LETTERS PATENT.

NEWCASTLE:

PRINTED BY S. HODGSON.   1789.

COPY of the LETTERS PATENT.

GEORGE the Third, by the Grace of God, of Great Britain, France, and Ireland, King, Defender of the Faith, and so forth. To all to whom these Presents shall come greeting. Whereas JOHN CURR, of the Parish of Sheffield, in our County of York, Gentleman, hath, by his Petition, humbly represented unto us, that he hath with much Study, Labour and Industry, invented a new Method of drawing or raising Coals, Lead, Tin, or any other Minerals out of Mines; by which the danger and damage frequently occasioned by the ascending and descending Corves running foul of each other in their Meetings, and beating the Sides of the Pit or Shaft, are prevented, and by which one, two or more Corves are drawn or raised at the same Time; and also of landing or delivering of the Corves upon the Banks or Top of the Pit, by means of a Platform upon Wheels or Carriage running over the Pit Top, done chiefly without the Assistance of the Banksmen; and also of making the Corves overturn themselves in order to discharge their Burthen without Assistance. That the Petitioner is desirous of securing to himself the Benefit of the said Invention; the Petitioner therefore most humbly prayed wee would be graciously pleased to grant unto him, his Executors, Administrators and Assigns, our Royal Letters Patent, under the Great Seal of Great Britain, for the sole Benefit and Advantage of his said Invention, within that Part of our Kingdom of Great Britain called England, our Dominion of Wales, and Town of Berwick upon Tweed, for the Term of Fourteen Years, pursuant to the Statute in that Case made and provided. And wee being willing to give Encouragement to all Arts and Inventions which may be for the Public Good, are graciously pleased to condescend to the Petitioner's Request. Know ye therefore, that wee of our especial Grace, certain Knowledge, and meer Motion, Have given and granted, and by these Presents for us, our Heirs and Successors, do give and grant unto the said John Curr, his Executors, Administrators and Assigns, our especial Licence, full Power, sole Privilege and Authority, that he the said John Curr, his Executors, Administrators and Assigns, and every of them, by himself and themselves, or by his and their Deputy or Deputies, Servants or Agents, or such others as he the said John Curr, his Executors, Administrators or Assigns, shall at any Time agree with, and no others, from Time to Time, and at all Times hereafter, during the Term of Years herein expressed, shall, and lawfully may, make, use, exercise, and vend his said Invention, within that Part of our Kingdom of Great Britain called England, our Dominion of Wales, and Town of Berwick upon Tweed, in such manner as to him the said John Curr, his Executors, Administrators and Assigns, or any of them, shall in his or their Discretion seem meet. And that he the said John Curr, his Executors, Administrators and Assigns, shall and lawfully may have and enjoy the whole Profit, Benefit, Commodity and Advantage, from Time to Time, coming, growing, accruing, and arising by reason of the said Invention, for and during the Term of Years herein mentioned. To have, hold, exercise and enjoy the said Licence, Powers, Privileges and Advantages herein before granted or mentioned to be granted unto the said John Curr, his Executors, Administrators and Assigns, for and during, and unto the full End and Term of Fourteen Years from the Date of these Presents next and immediately ensuing, and fully to be compleat and ended according to the Statute in such Case made and provided. And to the End that he the said John Curr, his Executors, Administrators and Assigns, and every of them, may have and enjoy the full Benefit, and the sole Use and Exercise of the said Invention according to our gracious Intention herein before declared, Wee do by these Presents for us, our Heirs and Successors, require and strictly command all and every Person and Persons, Bodies Politic and Corporate, and all other our Subjects whatsoever, of what Estate, Quality, Degree, Name, or Condition soever they be, within that said Part of our Kingdom of Great Britain called England, our Dominion of Wales, and Town of Berwick upon Tweed aforesaid, that neither they nor any of them, at any Time during the Continuance of the said Term of Fourteen Years hereby granted, either directly or indirectly do make, use, or put in practice the said Invention, or any part of the same so attained unto by the said John Curr as aforesaid, nor in any wise counterfeit, imitate or resemble the same; nor shall make or cause to be made any addition thereunto, or substraction from the same, whereby to pretend himself or themselves the Inventor or Inventors, Devisor or Devisors thereof, without the Licence, Consent, or Agreement of the said John Curr, his Executors, Administrators or Assigns in writing, under his or their Hands and Seals first had and obtained in that behalf, upon such Pains and Penalties as can or may be justly inflicted on such

Offenders for their Contempt of this our Royal Command; and further to be answerable to the said John Curr, his Executors, Administrators and Assigns, according to Law, for his and their Damages thereby occasioned. And moreover wee do-by these Presents, for us, our Heirs and Successors, will and command all and singular the Justices of the Peace, Mayors, Sheriffs, Bailiffs, Constables, Head-boroughs, and all other Officers and Ministers whatsoever, of us, our Heirs and Successors for the Time being, that they, or any of them do not nor shall at any Time hereafter during the said Term hereby granted, in any wise molest, trouble or hinder the said John Curr, his Executors, Administrators or Assigns, or any of them, or his or their Deputies, Servants or Agents, in or about the due and lawful Use or Exercise of the aforesaid Invention, or any thing relating thereto. Provided always, and these our Letters Patent are and shall be upon this Condition, that if at any Time during the said Term hereby granted it shall be made appear to us, our Heirs or Successors, or any six or more of our or their Privy Council, that this our Grant is contrary to Law, or prejudicial or inconvenient to our Subjects in general, or that the said Invention is not a new Invention, as to the public Use and Exercise thereof, in that said Part of our Kingdom of Great Britain called England, our Dominion of Wales, and Town of Berwick upon Tweed aforesaid, or not invented and found out by the said John Curr as aforesaid, then upon Signification or Declaration thereof to be made by us, our Heirs or Successors under our or their Signet or Privy Seal, or by the Lords and others of our or their Privy Council, or any six or more of them under their Hands, these our Letters Patent shall forthwith cease, determine and be utterly void to all Intents and Purposes, any thing herein before contained to the contrary thereof, in any wise notwithstanding. Provided also, that these our Letters Patent, or any thing herein contained, shall not extend or be construed to extend to give Privilege unto the said John Curr, his Executors, Administrators or Assigns, or any of them, to use or imitate any Invention or Work whatsoever which hath heretofore been found out or invented by any other our Subjects whatsoever, and publicly used or exercised in that said Part of our Kingdom of Great Britain called England, our Dominion of Wales, or Town of Berwick upon Tweed aforesaid, unto whom Letters Patent or Privileges have been already granted, for the sole Use, Exercise and

Benefit thereof. It being our Will and Pleasure that the said John Curr, his Executors, Administrators and Assigns, and all and every other Person and Persons to whom like Letters Patent or Privileges have been already granted as aforesaid, shall distinctly use and practise their several Inventions by them invented and found out, according to the true Intent and Meaning of the same respective Letters Patent, and by these Presents. 𝔓𝔯𝔬𝔳𝔦𝔡𝔢𝔡 likewise, nevertheless, and these our Letters Patent are upon this express Condition, that if the said John Curr, his Executors or Administrators, or any Person or Persons which shall or may at any Time or Times hereafter during the Continuance of this Grant, have or claim any Right, Title or Interest, in Law or Equity, of, in or to the Power, Privilege and Authority of the sole Use and Benefit of the said Invention hereby granted, shall make any Transfer or Assignment, or any pretended Transfer or Assignment of the said Liberty and Privilege, or any Share or Shares of the Benefit or Profit thereof; or shall declare any Trust thereof to or for any Number of Persons exceeding the Number of Five; or shall open or cause to be opened, any Book or Books for Public Subscriptions to be made by any Number of Persons exceeding the Number of Five, in order to the raising any Sum or Sums of Money under Pretence of carrying on the said Liberty or Privilege hereby granted; or shall by him or themselves, or his or their Agents or Servants, receive any Sum or Sums of Money whatsoever, of any Number of Persons exceeding in the Whole the Number of Five, for such or the like Intents or Purposes; or shall presume to act as a Corporate Body; or shall divide the Benefit of these our Letters Patent, or the Liberty and Privileges hereby by us granted into any Number of Shares exceeding the Number of Five; or shall commit or do, or procure to be committed or done, any Act, Matter or Thing whatsoever, during such Time as such Person or Persons shall have any Right or Title, either in Law or Equity, in or to the said Premises, which will be contrary to the true Intent and Meaning of a certain Act of Parliament made in the sixth Year of the Reign of our late Royal Great-Grandfather, King George the First, intituled (An Act for the better securing certain Powers and Privileges intended to be granted by his Majesty, by two Charters of Assurance of Ships and Merchandizes at Sea, and for lending Money upon Bottomry, and for restraining several extravagant and unwarrantable Practices therein mentioned). Or in case the said Power,

Privilege or Authority shall at any Time hereafter become vested in, or in Trust for more than the Number of Five Persons or their Representatives at any Time (reckoning Executors or Administrators as and for the single Person they represent, as to such Interest as they are or shall be intitled to in Right of such their Testator or Intestate), that then, and in any of the said Cases, these our Letters Patent, and all Liberties and Advantages whatsoever hereby granted shall utterly cease, determine and become void; any thing herein before contained to the contrary thereof in any wise notwithstanding. Provided *also, that if the said John Curr shall not particularly describe and ascertain the Nature of the said Invention, and in what manner the same is to be performed, by an Instrument in Writing under his Hand and Seal, and cause the same to be inrolled in our High Court of Chancery within One Calendar Month next and immediately after the Date of these our Letters Patent, that then these our Letters Patent, and all Liberties and Advantages whatsoever hereby granted, shall utterly cease, determine and become void; any thing herein contained to the contrary thereof in any wise notwithstanding.* And lastly, wee do by these Presents for us, our Heirs and Successors, grant unto the said John Curr, his Executors, Administrators and Assigns, that these our Letters Patent or Exemplification thereof, shall be in and by all things good, firm, valid, sufficient and effectual in the Law, according to the true Intent and Meaning thereof, and shall be taken and construed and adjudged in the most favourable and beneficial Sense, for the best Advantage of the said John Curr, his Executors, Adminstrators and Assigns, as well in all our Courts of Record as elsewhere, and by all and singular the Officers and Ministers whatsoever of us, our Heirs and Successors, in that Part of our said Kingdom of Great Britain called England, our Dominion of Wales, and Town of Berwick upon Tweed aforesaid, and amongst all and every the Subjects of us, our Heirs and Successors whatsoever and wheresoever, notwithstanding the not full and certain describing the Nature or Quality of the said Invention, or of the Materials thereto conducing and belonging. In witness whereof wee have caused these our Letters to be made Patent. Witness ourself at Westminster the Twelfth Day of August, in the Twenty-eighth Year of our Reign.

By Writ of Privy Seal,
WILMOT.

COPY of the SPECIFICATION.

TO all whom these Presents shall come greeting. 𝔚𝔥𝔢𝔯𝔢𝔞𝔰 his present Majesty King George the Third, hath by his Royal Letters Patent under the Great Seal of Great Britain, dated at Westminster the Twelfth day of August last past, given and granted unto John Curr, of the Parish of Sheffield, in the County of York, Gentleman, his Royal Licence, Power and Authority, for the sole making, using, exercising, and vending, for the term of Fourteen Years, a certain invention of the said John Curr, of a new Method of drawing or raising Coals, Lead, Tin, or any other Minerals, out of Mines, and of landing such Minerals, and for other the Purposes in the said Letters Patent more particularly mentioned; in which said Letters Patent is contained a Proviso that if the said John Curr shall not particularly describe and ascertain the nature of his said Invention, and in what manner the same is to be Performed, by an Instrument in writing under his Hand and Seal, and cause the same to be inrolled in his Majesty's High Court of Chancery, within One Calendar Month next and immediately after the Date of the said Letters Patent, that then the said Letters Patent, and all Liberties and Advantages whatsoever thereby granted, should utterly cease, determine and become void, any thing therein contained to the contrary notwithstanding, as by Reference to the said Letters Patent may more at large appear. 𝔑𝔬𝔴 𝔴𝔦𝔱𝔫𝔢𝔰𝔰 these Presents, that the said John Curr, by this present Instrument in writing under his Hand and Seal, 𝔇𝔬𝔱𝔥 describe and ascertain, by the Aid of the Plans and Figures herewith given, the Nature of his said Invention, and the Manner in which the same is to be performed, as follows, that is to say,

Figure 1, represents a Side-view of the Pit or Shaft, with the Corves near the Meetings. The upright Timbers therein, *A A*, extend from the Bed of the Mine, out at the Top of the Pit or Shaft, and which, with the parallel or counterpart Timbers, (better understood by the *Profile*, Figure 2) constitute what may be called Conductors, as they serve to conduct the Triangles *B B*, to which the Corves are suspended, up and down the Shaft by the Groove or Channel which they form. The Ends of the Triangles are fitted up with three Rollers or Wheels, to prevent as much as possible any Friction in their Movement; two of them work against the inside Face, and one in the Groove or Channel of the Conductors, as is more clearly shown by the Plan of the Pit on an enlarged Scale, Figure 4: And the Chains of these Triangles are so constructed, as to prevent any Inclination to turn round, which keeps the Corves in their due Position.

Figure 2, represents a contrary or End-view of the Pit or Shaft, and the Conductors, with the Corves near the Meetings, $A\ A\ A\ A$, shew the Conductors, which expand or widen at the Meetings, to let the Corves pass clear of each other. The sundry cross pieces, $F\ F\ F$, &c. represent the Buntons or Bearers to which the Conductors are bolted. D is a Director to the Triangles in their Descent down their respective sides; it works upon a Pin at n, and the ascending Triangle throws it over to the opposite Side, where it rests and gives the Direction to the descending Triangle. E is another Director near the bottom of the Conductors, working upon a Pin at p, which makes the empty Corves conveniently drop always upon the same Space at the Bottom of the Pit.

Figure 3. A Plan of the upper Part of the Pit or Shaft upon an enlarged Scale, shewing the Position of the Corves $a\ a$, $A\ A\ A\ A$ the Conductors, and $F\ F$ the Buntons or Bearers to which they are bolted or screwed.

Figure 4. A Plan of the central Part or Meetings of the Pit or Shaft, shewing the Position of the Corves $a\ a\ a\ a$, at that Point. $B\ B$ is a Plan of the Triangles, with their Rollers for Wheels at each End, and by which the Manner of working in the Conductors is plainly shewn.

Figure 5. A Plan of the lower part of the Pit or Shaft, and Position of two of the Corves, $a\ a$, there.

Figure 6. An Elevation or Side-view of the Head-geers and Apparatus above Ground for landing the Corves; C the Carriage which runs over the Shaft, and on which the Corves are landed; D the Road upon which the above Carriage runs. The Mode of rendering the Carriage and Road useful and convenient may many Ways be varied; but that here adopted will be found very convenient and expeditious.—In the Interval whilst the Corves are drawing, the Banksman, by Means of the Jack-roll i, raises the back Part of the Carriage-road D at c, and sets the Stops q (shewn only by dotted Lines) upon the Rest h, and at the same Time gives the Lever k hold of the back Part of the Carriage-road, so that the Carriage standing on an inclined Plane when within the Snecks *e*, was it not for such Catches or Snecks, would run forward over the Pit Top. *f* is the Axis or Centre on which the Carriage-road moves, and is fixed to the Sills *G* of the Main Standards by a Bolt and Thimble at the End, which receives the Iron Axis that the Carriage rests on. The back End *c* of the Road *D* is heavier than the fore Part, even when the Carriage is upon it. *t* shews four Boards for the empty Corves to run down on to the Carriage, and, while the Corves are drawing, are hanging down, as the dotted Lines *t* represent it. *H* is a Scaffold or Stage, on to which (during the drawing) the two empty Corves, intended next to be sent down, are taken up along an inclined Road made for that Purpose, and are left there ready to take the Place of the loaded Corves when drawn upon, being first hooked to the small Chain *o*. These Corves are placed in a declining Position, by Means of four cast Iron Rails sunk down in the Scaffold, as the Plan describes; and as the Scaffold is weakened by the Insertion of the Iron Rails, the small Rod *x* is fixed to strengthen it. The Rope *T*, by turning the Jack-roll *i*, draws up the Carriage-road *D* at *c*; and the Rope *V*, by the said Roll, is worked immediately back again, that the Frame *l l* may fall down. The Carriage *C*, during the drawing of the Corves, is detained by the Snecks *e*; and when the Corves are drawn up to the Top of the Pit, the End-rollers *B* of the ascending Triangles force back the Springs *z z*, which strike off the Stops *q* that stood upon the Rest *h*. The Frame *l l* then falling down, and its lower Cross-bar striking the Tail of the Snecks *e*, the Carriage runs over the Pit down the descending Road, as appears by the Plan. When the full Corves are landed upon the Carriage *C* and unhooked, the Banksman, with his Hand, moves the horizontal Lever *m*, which draws off the upright Frame *n*, against which the empty Corves rested; (a Front View of which is shewn by Figure 13) and thereupon the empty Corves, already placed, as before mentioned, on the Stage or Scaffolding *H*, rush forward on to the Carriage *C* covering the Pit, and push away the full Corves, and are detained themselves on the Carriage by the small Chain *o*. The Banksman then disengages the empty Corves from the Chain *o*; and having hung them on to the Chains of the Triangle, pulls the Cord *p*, which strikes off the single Lever *k*, that held up the back Part of the Carriage-road; and having drawn the empty Corves towards him from off the Carriage, the back

Part of the Carriage-road being the heavier as before mentioned, it falls down upon the Stop w, and the Carriage runs again into the Snecks e. The Banksman then pulls the Cord r, which draws off the Stops or Supporters, s, of the Triangles, which Stops upheld them at a proper Height for unhooking the Corves; which being done, the Corves are ready to go again.

Figure 7. DD, a Platform of the Carriage-road; ff shew the Ends of the Iron Axis on which the Road is supported.

Figure 8. A Platform of the Carriage C before mentioned; ee shew the Catch for the Snecks; and vv the Part that rests on the fixed Landing-board, when it is over the Pit Top, to give it more solid Bearing.

Figure 9. A Plan of the Ground Sills or foundation of the Head-geers and Apparatus above Ground; and into which the four upright Legs II, and the two back Supporters K, are mortised. L shews the Place of the Upright in which the Jack Roll i works, and to which Upright the Centre of the Lever k is fixed.

Figure 10. A Front View of that Part of Figure 6, marked d, which draws back the Supporters of the Triangles marked s; which Supporters are described more particularly by Figure 16.

Figure 11. A Front View of the Jack Roll i in Figure 6, by which the back Part of the Carriage-road, D, is drawn up at c, as before mentioned.

Figure 12. A Plan of the Frame by which the back part of the Carriage-road is drawn up with the Assistance of the Jack Roll, and marked ll, in Figure 6, and in which Frame the Stops or Rests (marked q, in Figure 6, working upon a Pin, and upholding the Carriage-road till struck off as before mentioned) are fixed. W is a Side View of one of these Stops; and that the Frame may not fall down too low, it is hung by two small Chains, continued from the Top of the Head-geers or landing Frame, to the Top of the said Frame ll, Part of which Chains are here shewn by UU.

Figure 13. A Front View of the Stops or Rests for the empty Corves, when placed on the Scaffold or Stage H in Figure 6, which are drawn aside by the Lever m in Figure 6.

Figure 14. The Spring Hook at the End of the small Chain o that detains the Corves on the Carriage.

Figure 15. A plainer View of the Springs zz in Figure 6, which strike off the Stops *q*, and by which the Carriage is set at Liberty to run over the Pit.

Figure 16. A more plain View of *s* in Figure 6, which supports the Triangles at a proper Height.

Figure 17. A Front View of one of the four Boards, on which the empty Corves run gently down from the Scaffold or Stage on to the Carriage, and marked *t* in Figure 6; when the Carriage is on its Sneck, the Boards hang perpendicularly down.

Figure 18. A Front View of the Wheel *g*, in Figure 6, which communicates with the Jack Roll, and by lapping round its Axis, the Chain of the Frame, Figure 12, raises the back part of the Carriage Road.

Figure 19. A Side View of the Turning-over Machine; *O* the Axletree on which it works in the Frame *M*: It is cranked, as more plainly appears in Figure 21. The full Corf *a* being pushed forward on to the Frame *N* (the fore Part whereof is constructed so as to form an inclined Plane) by its own Weight and Motion, assisted by the said inclined Plane, is overturned; the small Rope or Chain *X* stops it, when so far overturned as to empty it; and when emptied the Weight or Balance *R* on the under Side of the Frame brings it back again. The two Hooks *b*, at the Front End of this Machine, catch the Spending of the Corf; and the Iron Frame *Q*, in Figure 21, more clearly shewn crossing over the Top of it, prevents the Corf turning out of the Frame. It is to be understood, both in this and the Carriage Road for landing, that there is a Margin standing up to keep the Wheels of the Corves in their Channel and Direction.

Figure 20. Is a Plan of the Turning-over Machine, and the Frame in which it works.

Figure 21. An end view of the Turning-over Machine.

To draw any even Number of Corves at one Time, the Shaft and every other Part of the Machinery may be made upon the Principles of the Plans herein set forth, which represent two Corves drawn at a Time; observing that the Size of the Pit and other Parts of the Machinery must be always adapted to the Number and Size of the Corves to be drawn: But if any odd Number, as three, is proposed to be drawn at any one Time, the Corves must be suspended upon the Triangle, so that the down Rope shall run in the vacant Space between the Corves; if one only, the Rope must fall on one Side of it, otherwise the Corves and Rope will interfere with each other when drawing, and the down Rope will obstruct the landing of the Corves above. In 𝔚itness whereof the said John Curr hath hereunto set his Hand and Seal, the Sixth day of September, in the Twenty-eighth Year of the Reign of our said Sovereign Lord, George the Third, and in the Year of our Lord Christ, One Thousand Seven Hundred and Eighty-eight.

ADDRESS

To the OWNERS of COLLIERIES and PROPRIETORS of MINES.

WHENEVER new Inventions or Improvements are offered to the Public, the Motives that suggested, as well as the ultimate Advantages to be derived from their Application, are Duties incumbent on their Authors to state, in the most satisfactory manner they are capable of.

Several Years Application to the Business of Collieries in the Neighbourhood of Sheffield, &c. enable me to offer the following *Facts* relative to that Branch, of such Magnitude in the System of Trade, and the Manufactures of Great Britain.

1. Of the rapid Progress of Manufactures, and their concomitant Population, the increased Consumption of *Coal* is an obvious Consequence:—And accordingly we now find, that the Bulk of that Article which lay within the Limits of Day-levels, or could be procured by the Application of Hydraulic Machines, at reasonable Expences, is nearly exhausted; and that the regular and adequate Supply in Future, must be obtained from far greater Depths, and at such increased Expenses, as nothing but the effective Operation of very powerful and well-adapted Engines and Machines can so far lessen, as to bring within the Reach of the public-spirited and industrious Manufacturer; and afford the adventurous Proprietor of Mines a Share of Emolument, in *some* Degree proportionate to the Hazard of Embarkation in these uncertain and expensive Undertakings.

2. CUSTOM having given its Sanction to such a limited Mode of Working Collieries in this Part of the Kingdom, that at many Coal

Works, no less than 6 or 7 Pits are actually *in Work* for a Supply of Coals, which, by the new Method, here pointed out may be afforded by ONE: Where the Beds of Coal lie at great Depths, where the incumbent *Strata* are *hard*, where large Feeders of Water are met with, and where the *black* and *inflammable Damps* are both troublesome and dangerous, the weighty Expence and Time required for the opening and keeping in Work such a Number of Pits, are too obvious to be further insisted upon; besides, the original Cost, and subsequent Support of the extra Stock of Gins, Ropes, Labour, &c. &c. are Objects of such Magnitude as to be deservedly attended to.

3. The Roads necessary to be made and upheld for the Leading of the Coals from such a Number of Pits, and the Waste and Spoil of Ground *unavoidably* occasioned by these Roads and Pit-hills, especially where the Land is *valuable*, becomes a further Object of Attention; but which, my *Inventions* are peculiarly adapted to obviate.

To prevent needless Repetitions, I shall proceed to a Statement, without Exaggeration, of the several Advantages to be derived from the Use of my Machinery, as follow:—

1st. By adopting this Mode of *Drawing Coals* a *new Colliery* is *much sooner* established in *full* Work, than by the common Method of sinking and opening a greater Number of *Pits* in Succession; which is not only a great Saving in the *original Winning Charge* of Labour, Materials, and Accommodations, but in the Articles of *Interest, Wear and Tear,* &c. &c. besides the Advantage to be derived from an early Sale of a large Quantity of *Coal*, where certain annual Rents, or absolute Purchase-money for the Mine, takes Place on the Commencement of the Undertaking.

2d. By being able to deliver the Coals to the Carriage in the same State as they are produced in the Workings, without suffering the least Breakage, through the Processes of Hurrying or Putting, Drawing, &c. &c. and consequently of offering to Sale the greatest possible Proportion of *large* or *round* to *small* or *slack*; an Advantage that may be justly reckoned the grand *Desideratum* in every *Colliery*.

3d. By having all the *Work* of a Colliery, or, at least, for a great Number of Years, drawn at one *Shaft*, the necessary Accommodations for carrying on the Sale, (and the *receiving* of the Money in extensive *Land-sales*) are easily adapted and upheld; and the whole Business so contracted as to be managed in the most compendious and satisfactory Manner:—The *Covering-in* of the Machine, so as to save the Workmen and Materials from the Inclemency of the Weather, and keep on the Works during a severe Storm are Objects of no small Moment, where a certain Quantity of Coals is daily required.

4th. Where the Land is valuable under which a Colliery is opened, the Damages arising from the Waste and Spoil of Ground occasioned by a great Number of Pits, and the Roads, &c. necessary to be occupied among them, will amount to a considerable *annual* Sum; which, with the additional Expence of filling-up the Old Shafts, levelling and leading-off the surplus Rubbish, so as to leave the Surface, at the Expiration of the Term, in a State fit for the Plough, (and which is often attended with a Charge exceeding the Fee-simple of the Land) will be so far obviated by my Machine, and fewer Pits, as to be an additional Recommendation in its Favour.

5th. The Time I have had this Machinery in Use has shewn a very great Saving in the *Wear* of ROPES; as the Conductors, by keeping the Corves *always* in the same position, entirely prevent their *twisting*, so destructive (especially in *Wet Shafts*) to that *expensive* Article in the List of Colliery Materials.

I forbear a further Detail of other *demonstrative* Advantages, as the invention is so far from being *ideal only*, and requiring a Multitude of Words to set it off, that it is reduced to *real* practice in his Grace the Duke of Norfolk's Collieries at *Sheffield* and *Attercliff Common*, in the County of YORK; where an uniform Success in its Application for *two years* past, during which Time, the *Minutiæ* of mechanical Invention have been so far matured, as to enable me, with Confidence, to request the attention of *Gentlemen* to ocular Proof; where every Information will be given, and Licence for its Erection on the most fair and equitable Terms,

By their most obedient humble Servant,

Sheffield Park, JOHN CURR.
November 1, 1789.

Mr. A. M. CHAMBERS: I have great pleasure in rising to move a vote of thanks to our President for the paper he has read to us. From the information he has given us with regard to the past, I think some of us would say that they were really the "good old times," and we should be very thankful now if the price of coal could be fixed by Act of Parliament so that there might be good wages for the miner, and a good profit for the coal owner. We cannot fail to get great advantage from comparing the past with the present. We have had from the President a paper of a very exhaustive kind on the conditions of the coal trade in the past. We know what they are now. We feel the pressure of the times, and though the quantities are so large that there can be no complaint from the public generally, yet, at the same time, from the point of view of those who have to produce the coal, and those who have to carry out the arrangements for producing the coal, there is considerable cause of dissatisfaction, and such a pressure is brought upon the responsible parties, as makes being engaged in the coal trade a burden rather than a pleasure. But the Parliament now-a-days seems to be rather for the restriction of the coal trade than its development. The Mines Regulation Bill, as it came from the Government, seemed to be fair and reasonable on the whole, yet passing through Committee of the House of Commons you have seen what extreme amendments have been proposed, some carried with very large majorities of the House, simply on philanthropic grounds, without any regard to the cost they may entail or the trouble they may bring; and that they do not in the least degree increase safety in working. We really have every reason now to complain of interference from the Legislature, rather than to be thankful for it. We have, however, to face the conditions by which we are surrounded, to do our best to conform to them, and to make our mines a source of some profit to the owners. I am personally obliged to Mr. Embleton for the information he has afforded us, and every member of the Institute will be glad to join in giving him a hearty vote of thanks.

Mr. G. J. KELL: I have great pleasure in seconding the proposition.

Mr. JOSEPH MITCHELL: Mr. Embleton, I am sure we are all deeply grateful to you for the trouble you have taken in collecting and giving us a history of the coal trade from the earliest period, in

INEERS.

'S ACCOUNT OF AN IMPROVED METHOD OF DRAWING COALS

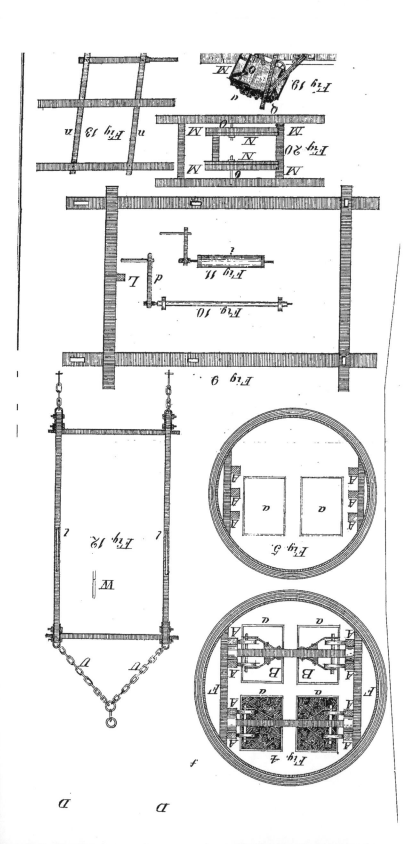

JRR'S ACCOUNT OF AN IMPROVED METHOD OF DRAWING COALS
EXTRACTING ORES, &c., FROM MINES. 1789.

Plate II

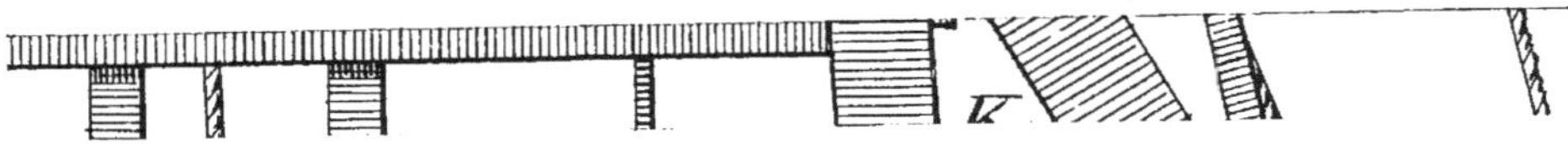

I V I

other parts of the kingdom as well as Yorkshire. It will be duly appreciated by the members of the Institute, and be a lasting benefit to the members. We shall have great satisfaction in looking back to the time when you were re-elected for the third time President of this Institute. Personally thanking you for the paper you have read and congratulating you upon your being re-appointed President in the jubilee year of Her Most Gracious Majesty.

The resolution was carried unanimously.

The PRESIDENT: I thank you very much for the way in which you have received this address. It is rather a disjointed affair altogether to my mind, but unfortunately I could not make it better. Where you find extracts in these old books, many of them take a long time to find out, and they do not seem to be connected at all; but if you take it as a whole, and read those parts which, in order to shorten the time I have omitted, you will see there is a connection. The Romans themselves were considerable miners of coal, iron, and lead, and used these minerals in their works, as may be seen in the remains of their bridges. The way in which they constructed their bridges shows how perfectly they constructed their work. The stones were lifted just as we lift heavy stones now. The stones themselves in their bridges were connected with iron cramps; perhaps one stone had four or five cramps each run in with lead, which showed them to be good workmen both in iron, lead, and coal. I can only say, gentlemen, if it has given you as much pleasure to hear this read, as it has given me to compose it, I am thoroughly satisfied

The following paper was then read:—

ON THE EASTERLY EXTENSION OF THE LEEDS AND NOTTINGHAM COAL-FIELD.

(SUPPLEMENTARY PAPER BY Mr. ROWLAND GASCOYNE).

This subject having at the last few meetings of this Institute been continuously put down for discussion with little or no success, the writer feels that some explanation is due for the step he has taken in again troubling the members with the subject. The probable extension of the coal measures in an easterly direction under the newer measures, is, however, of such importance, that while the subject is being dealt with, it is perhaps better to deal with it as fully as the information at our disposal will admit. The writer also hoped that the introduction of the subject would bring forward sections of borings and other explanations, which perhaps might tend to throw fresh light, and might perhaps result in a more definite boundary being fixed of the easterly extension of the coal measures. In this the writer has to some extent been disappointed, and as the President, on behalf of the Council, has expressed a desire for all available sections bearing on the subject to be recorded in the Transactions of the Institute, the writer has pleasure in supplementing the sections given in the paper by others, which either throw further light on the subject or give some information with respect to the thickness of the geological formations known to overlie the coal measures. There is one section given in the first paper, that of South Scarle, which has a most important bearing on the subject, but unfortunately the thickness of coal measures proved is so small, that it is impossible to say to what division of the coal measures the beds at the bottom of the bore-hole belong. Some are of the opinion that they belong to the upper coal measures, principally on account of their red colour, but the writer has already dealt with this view in his first paper, whilst others consider the red beds older than coal measures. The writer has noticed that the last core from the bore-hole showed near its base a change in colour from red to blue, which may perhaps denote the limit downwards to which the staining process has penetrated. If this is so, it tends to show that these beds cannot be said to be upper coal measures; but as already stated, are probably stained, although with such a vast thickness of marl slates overlying these beds, there may perhaps be something said against the staining theory. It is, however, encouraging to be able to point out the true colour of these beds, even should the red beds be considered to be stained, whilst it will likewise go a long way towards proving to what particular formation they belong—a problem the writer is attempting to solve. With respect to the view that these Scarle beds are older than the coal measures, the writer does not see much in its favour. Before such a view can be considered tenable, it will be necessary to show that there are grounds for assuming that the eastern coal measures overlaid by the newer formations, rise quicker to the east than the exposed coal measures do to the west, but we have no evidence of this; what facts have been observed go to show that instead of an earlier rising, the probabilities are that the measures rise more slowly to the east than they do to the west, so that in all liklihood there is a fair stretch of the coal measures to the east of the Scarle boring. The writer, whilst admitting that there are difficulties surrounding the identification of the 10 ft. of beds at the bottom of the Scarle boring, considers that everything points to their being of coal measure age, and until something tangible to combat that view has been discovered, it would be imprudent to consider them as belonging to any other formation. Some appear to be of opinion that the coal measures do not rise to the east, but those who adopt that theory do not state upon what grounds they arrive at that conclusion. Mr. G. B. Walker, at the discussion which took place in Sheffield, pointed out that it was possible that the coal measures terminated in an easterly direction against a bar of older rocks, and in that case could not be considered to form a basin—a view Mr. Walker tried to substantiate by the deterioration of the Silkstone seam. But the Silkstone seam deteriorates in all directions, is in itself a most unreliable seam, whilst the thickness and quality of the Barnsley seam at the more easterly collieries does not in the least favour an easterly determination of the coal measures, although locally, as at Shireoaks, the seam may be somewhat less in thickness. This view is also shown to be erroneous by the fact as stated by Mr. Greaves, that at Frystone Colliery the basin shape has already set in the seam of coal there worked, rising gently in the east at an angle of 1 in 77, with a tendency to rise even faster in an easterly direction. There is therefore considerably more to be said in favour of the basin shape than against it, in fact, everything that is reliable favours that view, but the writer would be glad, notwithstanding, to hear anything that can be urged against it. The writer is not aware that there was anything else raised against the views set forth in the paper which calls for special attention, but there was one statement contained in the paper to which the writer desires to call attention, viz., that the concealed coal-field occupied an area of something like 2,000 acres, instead of 900, as stated by the Royal Commission appointed to enquire into the probable duration of our coal supply. Seeing that several members of this Institute took part in various ways in the preparation of that report, it would seem by their silence that they admit the area of the concealed coal-field therein stated to be in error, whilst the area given by the writer as obtained from the actual result of borings, is far more likely to be correct, in fact, the writer considers it to be under rather than over the mark.

Appended are several sections made in different parts of the coal-field :—

SECTION OF THE STRATA BORED THROUGH AT BEESTON, NEAR NOTTINGHAM, COMMENCED 1ST APRIL, 1872.

No.	Strata.	Thickness.		
		yds.	ft.	in.
1	Earth	0	1	6
2	Gravel and sand	4	0	2
3	White sand	0	1	0
4	Red clay	1	0	0
5	White bands	0	1	6
6	Red clay	5	1	10
7	White bands	0	1	2
8	Sand rock	2	0	0
9	White bands with gypsum	7	1	2
10	Brown rock bands	0	1	4
11	Clay and hard bands	1	1	0
12	Red clay and white bands	4	1	4
13	Clay and thin bands	2	2	0
14	Hard bands of gypsum and rock	0	1	6
15	Clay	2	0	0
16	Strong white bands	0	1	0
17	Clay and hard white bands	3	1	6
18	Stone clunch	1	0	0
19	Red and white bands	1	2	0
20	Hard cank stone	1	0	0
21	Hard rock bands	3	2	0
22	Thin bands of clay	4	1	0
23	Sand rock and thin bands	11	1	0
24	Hard red and white bands	4	0	0
25	White rock	0	2	0
26	Band	0	0	3
27	White rock	0	2	0
28	Band	0	0	4
29	White rock	0	2	6
30	Band	0	0	4

Section of Strata at Beeston (continued).

No.	Strata.	Thickness.		
		yds.	ft.	in.
31	Rock and band	11	1	7
32	Hard clunch rock	0	0	9
33	Sand rock	4	2	3
34	Red clay	0	2	0
35	Rock bands and clunch	6	1	0
36	Rock bands	15	0	0
37	Hard band	0	2	0
38	Red and white rock	8	0	0
39	Red sand	0	2	0
40	Gypsum and clay	0	0	4
41	Conglmerate and bind with pebbles	5	0	8
42	Conglmerate rock	5	0	0
43	Red sand rock	12	2	0
44	White rock	2	1	8
45	White clay	4	2	4
46	Red sand rock	6	0	10
47	Hard red rock	2	1	2
48	Shales, sandstones, and ironstone	21	0	3
49	Coal	0	0	6
50	Bind and ironstone	14	2	$11\frac{1}{2}$
51	Coal	0	1	$5\frac{1}{2}$
52	Rock and bind	18	0	10
53	Coal	0	1	1
54	Rock, bind, and ironstone	37	1	9
55	Bind and shale	36	0	3
56	Coal	0	0	3
57	Rock and bind	54	0	9
	Depth of bore-hole	336	0	0

Section of Strata at Owthorpe, 3 miles W. by S. of Langar Church.
From 1876 to 1880.

No.	Strata.	Thickness.	
		ft.	in.
1	Lower Lias, clays and limestones	12	6
2	Rhœtie shales	34	6
3	Keuper Marls	653	0
4	Lower Keuper sandstone	59	0
5	Bunter sandstone	339	0
6	Red clays, &c., belonging coal measures	18	3

SECTION OF STRATA AT OWTHORPE (CONTINUED).

No.	Strata.	Thickness. ft. in.
7	Coal	0 3
8	Blue shale	26 8
9	Coal	2 4
10	Blue shale and sandstone	154 5
11	Coal	2 3
12	Clunch	3 7
13	Coal	1 0
14	Blue and black shale	36 6
15	Coal	0 3
		1343 6

BORING FOR COAL AT REEDNESS, NEAR GOOLE, ABOUT 1835.

	ft. in.
Drift, &c.	69 8
Keuper Marls	272 2
Bunter (?)	787 2
Total depth from surface	1029 0

Mr. NEVIN: The bore-hole at Beeston is on the west of Nottingham; I think you said the east?

Mr. GASCOYNE: It is on the west.

Mr. BONSER: As to the three beds or seams not being identified as remarked by Mr. Gascoyne, I had something to do with that bore-hole myself, and identified the three beds we passed as the three seams that exist between the black shale and the Kilburn seam, and I have no doubt that the Kilburn seam would be found. The coal found in a southerly direction bears out the correctness of Mr. Gascoyne's theory. All the boring operations in that direction which have taken place, have proved more or less the accuracy of that theory. But I cannot agree with Mr. Gascoyne in saying that in the borings at Beeston the coals passed through were not identified. If we take the borings at Wilford, on each side of the Trent, which prove the Clifton collieries, and compare them with the measures at Beeston, we find the measures at Clifton with a slight covering on them of 15 or 16 yards under the red marls; at Beeston if you take the average dip of the district, you will find that the Kilburn seam will outcrop a little more to the west of Beeston than the borehole put down by Mr. Charlton of Chilwell Hall. It is ten or twelve years ago. I put myself at a great deal of trouble to go through the matter, and I identified the three small seams as those existing just above the Kilburn seam but below the Black shale or Silkstone coal. I took my sections chiefly from those of collieries in that neighbourhood, that had been sunk to the actual seams of the district. That bears out the uniformity, on the south side at any rate, of the theory of the basin shape and character of the South Yorkshire and Nottingham coal-field. My observations are simply addressed to the identification of the particular seams passed through at Beeston.

Mr. A. M. CHAMBERS: What are the coal seams worked at Clifton?

Mr. GASCOYNE : The deep soft and the deep hard.

Mr. A. M. CHAMBERS : The deep hard is the Parkgate; but the deep soft I do not recognise.

Mr. GASCOYNE : It is a seam that is not in Yorkshire.

Mr. A. M. CHAMBERS: But I always thought that the Kilburn seam was below these; I took it to be below the Silkstone?

Mr. GASCOYNE : Yes.

Mr. A. M. CHAMBERS : How are you going to find it at Beeston?

Mr. GASCOYNE : What Mr. Bonser means is that owing to the fault these measures are thrown up, bringing these lower coal measures nearer to the surface, and the upper coal measures have been denuded.

Mr. A. M. CHAMBERS : When you come to Blackwell?

Mr. BONSER : That is to the north.

Mr. A. M. CHAMBERS : It is also to the west.

Mr. BONSER: This boring is in the extreme west.

The PRESIDENT: Can Mr. Bonser supply us with a section?

Mr. BONSER: Mr. Gascoyne has furnished sections. The exception I took was to Mr. Gascoyne's statement that it was to the east of Nottingham, which was a clerical error; it is to the west of Nottingham. The other was that it was not identified; but I identified them clearly with the three seams above the Kilburn coal, which is the bottom workable coal of the Nottinghamshire series.

Mr. GASCOYNE : I should like to suggest that Mr. Bonser furnish sections of the sinkings at Bramcote and Trowell, and also at Cossall, which he says rather helps me, or rather absolutely proves the identification of these particular seams of coal. Mr. Bonser said that he identified the coal by the Cossall sinkings which have not been made more than three or four years, and the Beeston boring was put down in 1872. Another thing I wish to say, at Woolaton Colliery, the nearest to these borings, they have sunk their shaft to prove the Kilburn seam of coal, but could not find a seam at all where the Kilburn seam should be. I have a number of those sections from Woolaton, which are more likely to compare with Beeston, than Cossall or Trowell, and I cannot find the least resemblance between the two. I should be glad to have a detailed copy of the sections of Ilkeston, Trowell, and Cossall, which seem to prove conclusively the position of the beds in these coal measures at Beeston. You will find they are very different—the thickest is only 1 ft. $5\frac{1}{2}$ in., another 6 in., and another 1 ft. 1 in. At the time I thought probably the true section of the seam was not shown by the diamond drill, as it sometimes happens it does not bring up a full core of the thickness of the coal. At Woolaton Colliery they passed through a four foot seam of coal with the diamond drill without ever seeing it, and it is now the leading seam they are working. It may be these seams of coal are thicker than they are shown on the Beeston section.

Mr. A. M. CHAMBERS: I should like to ask if in any of the workings of these southern collieries there are signs of rising to the south?

Mr. GASCOYNE: They rise to the south at Stanley; in fact they sweep round.

Mr. A. M. CHAMBERS: Where is Stanley?

Mr. GASCOYNE: Five miles south of Ilkeston. It is the most southerly colliery working, and there the measures sweep round to the south, and at Woolaton the coal measures dip due north.

Mr. A. M. CHAMBERS: In the same way the coal measures in the basin at Sheffield sweep round and rise to the south, but they sweep back again and rise to the west.

Mr. GASCOYNE: Throughout the coal-field there are local changes, but if you take the average you will find that the strata shows the basin shape.

Mr. BONSER: If this discussion is to be continued, and any information I can throw upon it will be acceptable, I shall have great pleasure in submitting a short paper on my own observations in these matters. I think in all these questions of geology some gentlemen take a microscopical view of the question, and some take a telescopic view. If you take the whole district it may be in one direction, but if you take small pecularities you may be mixed anywhere. It is the safest to take a broad view and allow for little differences, and if the Institute likes I will give them the why and the wherefore of my opinion. Mr. Gascoyne points out that the sinkings at Cossall were taken after the borings at Beeston, but the sinkings at Woolaton only bore out my theory of the borings at Beeston, and strengthened my belief.

The PRESIDENT: The Institute will be obliged if you will afford further information. As far as I see Mr. Gascoyne's paper does not at all give us any more information than we had before as to the extension of the coal-field eastward. That is my impression, but I may be mistaken. If Mr. Gascoyne could by any means get information from those bore-holes mentioned in one of his papers, that would give some idea whether there is any increase of rise to the east in the underlying measures. We should feel obliged to Mr. Bonser for further information.

Mr. A. M. CHAMBERS: The observation about microscopical and telescopical views of the question rather makes me think we take too many telescopical views, and do not take the microscopical views we can do. It is only by accurate observation, and noting these observations, and comparing them with one another, and so correcting the data we have, that we get at the truth of the matter. We are in danger of too great generalization; we want these particular observations, and when we have got them together we can get information which may be relied upon.

The PRESIDENT: Don't you think using the telescope first—using the form of speech as Mr. Bonser used it—don't you think that should be used first, and having ascertained there are certain things, the microscope would come into operation.

Mr. A. M. CHAMBERS: You cannot pierce with the telescope.

The PRESIDENT: I know what Mr. Bonser means. If you take a survey of any part of the country you must lay down your long lines first, and fill up the details afterwards. That is what Mr. Bonser meant by telescope and microscope.

Mr. GASCOYNE: With respect to my paper, I did not expect I should be able in the second paper to bring forward any new light with respect to the easterly extension—that had been very fully gone into in the first paper. But there have been several objections to that view stated to me, not when the paper has been discussed, but by members of the Institute, and I thought it would give me an opportunity of defending those views, and of giving them an opportunity of showing, if they desire to do so, where they object to my views. Then again, I was asked to give additional sections. These sections I have included in the paper, and if there are any others that can be given, they will add considerably to the value of the information included in the Transactions of the Institute. With respect to the sinkings at Woolaton, seeing Mr. Bonser has broached the Trowell and Cossall sinkings, I shall be glad to give these to the Institute to be included in the Transactions; but it will only refer to these particular sinkings. It is not of much consequence, but will complete the information. I understand Mr. Bonser to say that the Woolaton sinkings confirm his view of the Beeston sections. I should be glad to know how he makes them agree with the Woolaton sinkings.

Mr. BONSER: Trowell.

Mr. NEVIN: The ten feet of coal measure shales at the bottom of the South Scarle boring have never been identified; for anything known they might be lower age shales, or any part of the Yoredale shales?

Mr. GASCOYNE: If they are Yoredale shales there will have to be a more sudden uprising to the east than to the west, and what we know show they do not rise so fast to the east as to the west. A line from Nottingham to Leeds will form the axis of the coal-field, if you allow the same thickness on the easterly side of the line, that we know and have proved on the westerly side, and then the coal measures will go beyond Lincoln. Therefore, it is improbable these can be anything else but coal measures. These shales contain ironstone, and I have not found any ironstone in the Yoredale shales; and the ironstone looks much like coal measure ironstone. Everything seems to point to them being of coal measure age, and not older shales such as the Yoredale.

The PRESIDENT: Is it the opinion of the meeting, now that the discussion seems to be ended, that the second paper should be printed in the Transactions?

Mr. A. M. CHAMBERS: I beg to propose that, and that Mr. Gascoyne be thanked for the further information that he has given to us.

Mr. J. JARRATT: I beg to second that.

ADJOURNED DISCUSSION ON THE COMMITTEE'S OBSERVATIONS ON THE FINAL REPORT OF THE ROYAL COMMISSION ON ACCIDENTS IN MINES.

The PRESIDENT: The next business is the adjourned discussion on the Committee's observations on the Final Report of the Royal Commission on Accidents in Mines. I really thought for my part that this discussion having been carried on so long was completed, until I found this notice on the paper. I think at the last meeting at Barnsley it was suggested that a copy of the paper should be forwarded to the Home Secretary, and that it should be communicated to the House of Lords. That was also mentioned. Has that been done?

Mr. JOSEPH MITCHELL: To the Home Secretary, but not to the House of Lords. Shall we send to the Lord Chancellor?

The PRESIDENT: Or to Lord Fitzwilliam or some peer connected with coal mines.

Mr. A. M. CHAMBERS: If they will only be in the House, but I fear they have dispersed themselves and will not come back.

Mr. JOSEPH MITCHELL: We might send to Lord Cross.

Mr. A. M. CHAMBERS: I do not know which peer takes the home business in the House of Lords. I tried but could not ascertain.

Mr. JOSEPH MITCHELL: We could only send to Lord Cross.

The suggestion was agreed to.

The PRESIDENT: Might I ask about safety-lamps—whether the three lamps mentioned in one discussion in the House of Commons are absolutely excluded or not. My impression at the time was they were absolutely excluded.

Mr. A. M. CHAMBERS: You mean the proposal that the lamps which should not be used should be the Stephenson, the Clanny and the unprotected Davy. That is absolute as far as it stands now, but on report the amendment proposed by the Mining Institute is that the word "unprotected" shall be put before "Stephenson," so that only the unprotected Stephenson, the unprotected Clanny, and the unprotected Davy are affected.

The PRESIDENT: I asked that the members should know how these matters stand; taking the newspapers as correct is not at all times satisfactory.

Mr. A. M. CHAMBERS: Even the reports in the *Times* of the discussions have been very inaccurate.

Mr. GARFORTH moved a vote of thanks to the President.

Mr. M. NICHOLSON seconded the vote, which was carried unanimously.

The PRESIDENT: I am much obliged.

MIDLAND INSTITUTE OF MINING, CIVIL, AND MECHANICAL ENGINEERS.

GENERAL MEETING.

HELD AT THE BULL HOTEL, WAKEFIELD, ON FRIDAY, OCTOBER 7TH, 1887.

T. W. EMBLETON, Esq., President, in the Chair.

The minutes of the last meeting were read and confirmed.

Mr. JOHN EDWIN CHAMBERS, Mining Engineer, Huddersfield Road, Barnsley, was elected a member of the Institute, having been previously nominated.

The following paper was then read:—

CONSIDERATIONS ARISING OUT OF SECTIONS 51, 52, 53, AND 54 OF THE COAL MINES REGULATION ACT, 1887,

BEING THE PROVISIONS REQUIRING THE ESTABLISHMENT OF SPECIAL RULES AT ALL COLLIERIES WITHIN THE OPERATION OF THE ACT.

BY GEORGE BLAKE WALKER.

THE passing of the new Mines Regulation Act involves some alterations of importance, but on the whole it may be regarded as legalising and requiring the adoption of a system of management which has gradually been introduced voluntarily into the best collieries, rather than requiring radical and novel changes. On the whole most colliery managers will, I believe, regard the Act of 1887 as an improvement on the Act of 1872, and as an honest attempt on the part of the Home Secretary to produce a measure which, while ensuring that every precaution for the safety of the workpeople shall be taken, should not injuriously hamper the coal trade.

The changes which have been made, however, are by no means unimportant, but it may be said that their leading characteristic arises from the recognition of the fact that collieries are now-a-days on a far larger scale than they were even twenty years ago, and the new provisions are in many cases contrived to provide for the existing state of things. The complete failure of the Act of 1872 to secure the "daily supervision" of the mine by the manager was caused by the inherent impossibility of the requirement that the person fitted to be the manager of a large colliery could, among his multifarious duties, exercise that constant daily supervision of the underground operations which no doubt the Act of 1872 was drafted to secure. The new Act admits the impossibility of his doing this at any except quite small collieries, and recognises under the title of "under-manager" the under-viewers who at large collieries have hitherto performed this duty. This is an eminently practical and common-sense step, and will at once relieve the chief manager of some of the weight of responsibility for minute details with which he is now burdened, and give the under-viewers a more responsible and dignified status.

It is not, however, the object of this paper to discuss the new Act in general, but to consider in how far it is desirable to reconsider at the present time the Special Rules which have to be framed in accordance with Section 51 of the new Act. Nominally, it is required of each owner or manager to draw up Special Rules adapted to the requirements of his own collieries; but as a matter of fact this privilege has hitherto been waived in favour of a uniform set of rules for each district, drawn up by a committee in consultation with the Government Inspector. No doubt from many points of view it is highly desirable to have one uniform set of rules for all the collieries within a certain district; but to secure this something is undoubtedly sacrificed. The Special Rules must be drawn sufficiently widely to include in the same net fish of all sorts and sizes. Large and small collieries, gaseous and non-gaseous collieries, thick seams and thin seams, are all to be worked under one and the same set of rules. Hence a considerable number of useful rules suitable to the circumstances of some collieries must be excluded because they would be unsuitable to others. This is a point on which there is certain to be diversity of opinion between the Inspectors and the managers, and how the matter will be arranged it would be useless at present to predict.

Now the question is, on what principle are the Special Rules to be drawn up? On the last occasion the new Special Rules were based on those previously existing, and there is of course much to be said for altering the existing rules as little as possible. But it must not be forgotten, that in commencing to work under the new Act, many things will require alteration, and if improvements are possible and would be of advantage, now is the time to make them.

What I desire in bringing this matter before the Institute, is to arouse interest among colliery managers on this most important question, so as to evoke the expression of opinion before the limited time expires in which the revision of the Special Rules will have to be taken in hand. To give point to my own ideas, and to afford a basis for discussion, I have prepared a draft set of rules, such as seem to me to be an improvement on those now in use. But I wish particularly, to guard myself against conveying the impression that I regard them as other than a *basis for discussion*. As such they may be useful, and they are put forward for this end alone.

Three principles may I think be laid down which will be generally accepted as desirable:—

1st. That in drafting the new Special Rules there should be as little repetition as possible of matter already provided for in the General Rules.

2nd. That the duties of the different members of the colliery staff be defined with as much precision as possible.

3rd. That the arrangement of the rules should be carefully made, so that the greatest obtainable clearness may be secured.

With regard to the first—the General Rules form part of the law of the land, and upon the exact language in which they are expressed any legal proceedings would actually be based. To re-express the same rules in other language can do no good, and may very easily do harm. If the re-habilitated General Rule is less precise than before, the looseness of the Special Rule will be no protection; if more stringent, we are hampering ourselves with bonds which Parliament has not decreed. The first principle then I wish to urge in the framing of the new Special Rules, is that they should be purely *supplementary* to the General Rules, and that no Special Rule shall cover the same ground as a General Rule.

Secondly. That "what is everybody's business is nobody's business" is an old and trite proverb, and one which may be usefully borne in mind in connection with the matter in hand. The present

Special Rules are most unsatisfactory in this respect, especially those which commence with "The underground-viewer or his deputy shall * * * *" This is a phrase which may have been well enough at one time, but it is altogether out of date to-day. Every large colliery has a highly organised staff, the duties of each member of which are clearly defined by practice, and which have become so not by choice, but by necessity. It is surely highly desirable that a precise definition of the duties of the principal officials should be clearly defined in the Special Rules. By so doing the manager is not relieved of his proper responsibility, but good managenent and discipline of the pit is greatly assisted, and an increased sense of responsibility rests upon his subordinates.

Thirdly. A clear and consecutive arrangement of the rules is most important. In especial it is desirable that the rules affecting most particularly the ordinary workmen, should be placed in one group under a suitable heading. In spite of the Education Act, a pamphlet of twenty or thirty closely printed pages is a formidable thing to the average collier—a task he probably shrinks from or defers to a more convenient season. Some of us may instruct our underviewers to turn down a certain page when giving out copies of the rules, and draw the men's attention particularly to the rules which concern them most, but how much better would it be to have these rules distinctly set forth in a particular place, where they can be seen and read in a few minutes—and *remembered*. We cannot make things too simple for many of the men we employ, and it is even more important to make the rules clear to the dull and stupid than to the experienced and intelligent workman. "They that are whole have no need of the physician," &c.

It now remains only for me to state succiutly the lines on which in my opinion the new Special Rules should be drawn up.

The principles which I have already laid down should of course be clearly kept in view, namely, the non-repetition of any matter provided for by the General Rules; the definition of the duties of the various members of the supervising staff, and the consecutive arrangement of the rules affecting different persons or classes of workmen.

It is a difficulty in carrying out the latter principle that there has never been the same subdivision of staff work in this district which has always been adopted in the counties of Northumberland and Durham; and the terms employed in Yorkshire are variously understood at different collieries.

Notwithstanding the variety of designation, certain duties are performed at all collieries. These may be roughly stated to be as follows:—

(1.) *General Direction of the Mining Operations.*—Here there is no difference of designation, because the term "manager" is laid down in the Act of Parliament. But the Act further provides that a manager shall exercise personal daily supervision (which means, I suppose, being at the pit every day it is at work); and, since this would involve an impossibility in the case of the manager of a large colliery or collieries, it is further provided that these duties of daily supervision may be exercised by a person who is styled "under-manager," if the manager is himself unable to perform them.

So far, then, the terms of the Act decide the designation of those who have charge of the general direction of the mining operations.

(2.) Under the manager, or under-manager, there is always an official who immediately supervises the work people employed in the mine. In a small colliery there may be one such person, but in large collieries there will be one for each district or each seam. Such a person is known in the North of England, Scotland, and in other districts as the "overman." In Yorkshire the most common designation is that of "deputy." But, unfortunately, the word "deputy" is not only confined to the official in charge of the seam, or district, but to any person who occupies an official position below ground; and, amongst others, to those who make the examination of the mine, whose proper designation is that of "fire-triers" or "firemen."

In large collieries, also, the management of the ventilation is an important matter, and one requiring careful and constant supervision. A person styled the "master wasteman" is, in the North of England, usually appointed to specially look after the ventilation; but here, so far as I know, it is not generally customary to appoint a special official. In some cases, the duties are exercised by the manager or underviewer in person, or by each overman or deputy for his own district; but it is not necessary, because the duties of such an official are defined in the Special Rules, that a separate person need necessarily be appointed. This is a matter of arrangement, which may be modified according to circumstances; but the separate definition of the duties of superintending the ventilation (whether intrusted to a single individual or not) is a convenient arrangement.

The duty of examining the mine is performed by fire-triers or firemen, and may be set forth under a separate head.

I propose, then, to classify the rules under the following heads:—

1. General Direction of the Mine.
2. Daily Supervision of the Mine.
    (a) By Manager or Under-manager.
    (b) By Overmen or Deputies.
3. Ventilation.
4. Examination of the Mine.
5. Rules for Underground Workmen.
6. Safety Lamps.
7. Examination of Machinery, Ropes, and Tackle.
8. Rules to be observed at Shafts and on Engine Planes.
9. Rules to be observed by Surface Workmen.
10. General.

The Special Rules being intended to embody (in the language of the Act) such regulations "for the conduct and guidance of the persons acting in the management of a mine, or employed in or about the mine, under the particular state and circumstances of such mine, as may appear best calculated to prevent dangerous accidents, and to provide for the safety, convenience, and proper discipline of the persons employed in or about the mine," it must be admitted to be almost impossible to so draft a set of special rules which shall be so particular as to secure the perfection of organization at one mine without introducing provisions which would constitute positive hardships at other mines where different circumstances exist. If, therefore, a uniform set of rules is to be adopted throughout the district, only those can be adopted which are of pretty general application. I wish to draw attention to the wording of Clause 51 of the new Mines Act, which I have just quoted, because it lays special stress upon the Special Rules being framed "for the conduct and guidance of the *persons* acting in the management of a mine," etc. The Act contemplates, evidently, that by means of these Special Rules the special duties of each person employed in the management of a mine shall be clearly and distinctly defined; and, in my opinion, no Special Rules will be satisfactory in which such precise definition of individual duty is wanting. On the other hand, we should be careful to avoid superfluous regulations. We should confine ourselves to such rules as "are calculated to prevent dangerous accidents, and to provide for the safety, convenience, and proper discipline of the persons employed in and about the mine." In venturing to add, as an appendix to this paper, a draft set of Special Rules, I should like to say that these points have been made the touchstone in every case: they are the outcome of the thought and experience of some years, and have not been hastily put together on the spur of the moment. The subject is one which has frequently, during the past three or four years, been discussed in detail between myself and other colliery managers with whom I am intimate. Previous drafts have been made, and those rules which I append to this paper, however imperfect they may be, are the fruits, not of my own experience and judgment only, but in great part of my friends'. We have felt the subject to be one of the gravest importance, and one which, when the proper time should come, ought to be thorougly thrashed out by all the managers in the district. Otherwise, nothing would have induced me to appear to wish to force myself and my own views into prominence by reading this paper, and whatever else they may condemn, I beg the members of the Institution not to lay the sin of vanity to my charge.

In conclusion, I would only mention that, though I have named 3 hours as the time within which the mine must be examined before the commencement of a shift, I consider that, for non-gaseous mines, this time is unnecessarily restricted, and it will, I presume, be a matter to be settled, according to circumstances at each colliery, by the manager and the Chief Inspector. As to the use of explosives, I think it would be unwise to add anything in the Special Rules to the elaborate provisions of General Rule 12, upon which there has been so much controversy. Whatever rules may eventually be adopted for the district as a whole, I trust that modifications adapted to the circumstances of individual collieries will be permitted and incorporated, so that at every colliery the Special Rules may indeed be those "best calculated to prevent dangerous accidents, and to provide for the safety, convenience, and proper discipline of the persons employed in or about the mine."

APPENDIX.

GENERAL DIRECTION OF THE MINE.

1. The Manager having the supreme direction of the colliery both above ground and below, is responsible for the general safety of the works, machinery, ropes, and tackle, and for the adequate ventilation of the mine. It is his duty to provide for the carrying out of all the provisions of the Mines Regulation Act, for the publication of the rules and regulations set forth therein, and for the general safety of all persons employed in or about the colliery.

2. The Manager shall appoint fit and proper persons to be Under-Managers, Deputies, fire-triers, enginemen, banksmen, furnacemen, lamp-keepers, and others, and shall see that the persons so appointed to positions of responsibility faithfully discharge the duties required of them; and if any person so appointed prove to be unfit to satisfactorily perform the duties of his position, the Manager shall remove him from such position of responsibility and appoint a fit and proper person in his stead. No person of intemperate habits shall be deemed a fit and proper person to fill a post of responsibility in any mine.

3. The Manager shall cause the ventilation of the mine to be constantly maintained, and shall cause measurements of the principal air currents to be made not less frequently than once a week by a competent person or competent persons, and shall cause a record of such measurements to be recorded in a book to be kept at the office of the mine for the purpose, and signed by the person or persons making the same.

4. The Manager shall see that an adequate number of proper and efficient safety-lamps are provided at the mine, and that the same are maintained in good condition and repair; and that competent persons are appointed to clean, examine, and repair them. Every lamp must be of such construction as to be safe if carried against an explosive atmosphere travelling at the rate of the highest velocity ordinarily existing in the mine.

5. The Manager shall see that the cages in which workmen descend and ascend are provided with a sufficient cover overhead to protect such workmen from falling bodies.

6. The Manager shall see that an ample stock of timber, and other requisite material is provided, not only on the surface at the colliery, but underground, in situations convenient to the men who have to use them.

7. The Manager shall cause all abandoned and dangerous places to be fenced off; sumps, where they exist at the bottom of working pits, to be covered with scaffolds; and shall provide for the proper ventilation of the sumps below such scaffolds; and at all intermediate points, in any winding shaft, at which men or coals are sent up, a sufficient gate or fence shall be provided and kept closed when the cage is not being loaded.

8. The Manager shall appoint the stations required under General Rule 4, at the entrance of the mine, and shall cause the said stations to be made conspicuous either by whitewashing or otherwise, so that no person can inadvertently pass them. He shall also appoint stations where safety-lamps may be re-lighted, but such stations may not be in the return air.

9. The Manager shall cause the ventilating fan or furnace to be visited at least every alternate day by a competent person, who, if he finds anything wrong with the ventilating apparatus, shall immediately apprise the Manager of the fact.

10. The Manager shall cause safety-lamps to be used in any pit or district in which gas exists, or may be reasonably expected to make its appearance from time to time, even if such pit or district is unusally free from accumulations of gas, and shall cause notice boards to be conspicuously placed at the entrance of such pit or such district, and no person shall be allowed to go beyond such notice board with any light except a properly examined and securely locked safety-lamp.

11. The Manager shall appoint a qualified deputy to be in charge of a pit or portion of a pit during the time that any workmen are employed in the workings of such pit or portion of a pit. Such deputy must not leave the pit till he is relieved by another deputy, or until the workmen are out of the pit.

12. The Manager shall take such steps as shall tend to thorough discipline being observed throughout the mine.

13. The Manager shall himself exercise constant daily supervision in the mine underground, unless he shall appoint in writing some person holding a first or second-class certificate to be Under-Manager, in which case the constant daily supervision shall be exercised by the Under-Manager. The duties devolving upon the person who exercises the daily supervision are set forth under the next section, but where no Under-Manager is appointed these duties must be performed by the Manager in person.

14. The Manager is primarily responsible for the carrying out of all the General Rules contained in the Mines Regulation Act, 1887.

15. The Manager shall cause a printed copy of the General and Special Rules and of the Abstract of the Mines Regulation Act, 1887, to be supplied *gratis* to any workmen who may apply for it, and shall also see that a sheet copy be conspicuously posted on the pit bank.

DAILY SUPERVISION OF THE MINE.
(A) BY THE MANAGER AND UNDER-MANAGER.

16. (1.) The duty of the Under-Manager (where one is appointed) is to see that the rules and the orders of the Manager are carried out, and to act for the Manager in his absence. As the representative of the Manager, he has supreme control and direction of all work and workmen within the limits defined by the Manager when at the colliery, and the same powers as the Manager when the latter is absent. Where there are two or more Under-Managers at the same colliery, one of them must be nominated by the Manager to take the chief direction in his absence.

17. (2.) Where personal daily supervision is not exercised by the Manager in person it must be exercised by the Under-Manager, who must attend the pit every day unless absent through sickness or other unavoidable cause; see that the rules are closely and rigidly carried out by all persons employed; maintain effective discipline throughout the mine; see that the ventilation is adequate, and that proper provision is made for the safety of all persons employed.

18. (3.) The Manager or Under-Manager shall daily examine the reports of the fire-triers, deputies, master-wasteman, and others, and shall at once attend to the removal of any source of danger; he shall, as far as practicable, inspect personally such parts of the mine, shafts, machinery, or tackle as may be reported to him to be unsafe or in any way need his attention. He shall advise with daily and instruct the deputies and other officials on all points necessary to enable them to carry out the duties of their departments, and to keep himself informed of the condition of all parts of the colliery.

19. (4.) The Manager or Under-Manager shall see that all the reports required under the Mines Regulation Act, 1887, are properly and regularly kept by the persons whose duty it is to make the respective reports.

20. (5.) The Manager or Under-Manager shall see that printed or painted boards are conspicuously erected at the various stations appointed in accordance with General Rule 4, at the points beyond which safety-lamps may alone be taken. He shall cause any place or part of the mine which may for the time being be dangerous, either by reason of fire-damp or other cause, to be fenced off either with timber or rails crossed, and shall have a *Danger* or *Fire* board hung on the fencing so that no person may inadvertently enter the same.

21. (6.) The Manager or Under-Manager shall cause printed or painted notices to be placed at all places from which signals are given, both at shafts and on engine-planes, shewing the signals in use at such shaft or on such engine-plane, and the number of persons who may ascend or descend at one time.

22. (7.) The Manager or Under-Manager shall see that a sufficient quantity of suitable timber is obtained for use and is sent into the various parts of the pit and delivered at the respective sidings or pass-byes, or at other points convenient to the workmen.

23. (8.) The Manager or Under-Manager shall at all times see that the ventilation in all parts of the mine where men are employed is adequate for their health and safety, and if at any time the ventilating current in any part of the mine is so laden with noxious gases as to be unhealthy or unsafe, he shall cause the men working in such places to be withdrawn, and the same fenced off until the ventilation is restored to a proper state.

24. (9.) The Manager or Under-Manager, in making arrangements for the examination of the mine under the provisions of General Rule 4, shall assign to the fire-trier or fire-triers such portion of the mine as he can carefully and thoroughly examine within a space not exceeding   hours.

DAILY SUPERVISION OF THE MINE.
(B) BY THE DEPUTIES OR OVERMEN.

25. (1.) The Deputy is a person appointed by the Manager to have the responsible daily charge of the underground workings of a pit or portion of a pit; subject, however, to the Under-Manager, where one is appointed. The Deputy shall have authority over all underground workmen in his pit or district. He shall be in the pit every day during the whole of the shift during which he is in charge, and shall not leave the pit so long as any workmen remain in the workings, or until he is relieved by another Deputy.

26. (2.) The Deputy must see that the rules are closely and rigidly carried out by all persons employed in his pit or district; maintain discipline; carefully watch the ventilation; and constantly examine the state of the working places, gates, levels, horse-roads, engine-planes, etc., and see that these are in a fit and safe state for working and passing therein.

27. (3.) The Deputy shall, immediately on descending, examine the reports of the fire-triers who have been engaged in examining the pit or district prior to the commencement of the shift, and countersign such reports.

28. (4.) The Deputy shall immediately inspect personally such parts of the pit or district as may be reported to him to be unsafe, or in any way need his attention, and cause any defect or source of danger to be remedied. If he deem it necessary, he shall send immediate information of the defect or danger to the Manager or Under-Manager.

29. (5.) The Deputy shall, during each shift on which he is on duty, examine carefully any place which has been fenced off on account of the existence of any danger, such as an accumulation of fire-damp or the like.

30. (6.) The Deputy shall visit every working place in his district at such intervals as may be ordered by the Manager, who shall give him written instructions as to this.

31. (7.) The Deputy shall, in visiting the working places, carefully examine the state of the roof and sides, and shall cause the miners to set sufficient timber to protect them from falls of roof or side. He shall also see that every miner has a sufficient supply of props, sprags, and other timber, and if he finds any place insufficiently timbered the Deputy shall cause the workmen in such places to cease working and set timber until the place is, in his opinion, safe. The Deputy shall see that there is a reserve supply of timber near all sidings or pass-byes, or in other places convenient for the workmen.

32. (8.) The Deputy shall see that all air-courses and wind-gates leading to the working places and from one working place to another are maintained of such size as the Manager may direct.

33. (9.) The Deputy shall not allow explosives to be used without the consent of the Manager or Under-Manager, and where the use of explosives is permitted the Deputy shall make such arrangements as will meet the requirements of General Rule No. 12.

34. (10.) The Deputy must see that all trap doors and sheets are kept in good working order, and fall to of themselves, that brattices are not more than 10 feet from the face of rise headings, nor less than 18 inches from the side.

35. (11.) The Deputy shall cause all refuge or man-holes on engine-planes or horse-roads to be kept clear of rubbish or material, and since places where rubbish and material can be placed will be necessary, he shall with the consent of the Manager or Under-Manager provide receptacles at such intervals as may appear necessary.

36. (12.) The Deputy shall cause any place which is disused or which it is dangerous to enter to be fenced off across its whole width, and shall hang a *No Road* board on the fencing, so that no person may inadvertently enter the same.

37. (13.) The Deputy shall constantly inspect the safety-lamps in use by workmen, and shall cause any defective lamp to be instantly extinguished, and report to the Manager or Under-Manager any case of injury to a safety-lamp.

38. (14.) The Deputy shall take special care of young boys in his pit or district. He shall explain to them how to perform their duties, advise them how to avoid unnecessary risks, and as far as possible keep them out of harm's way. The Deputy shall not allow an inexperienced boy to drive a vicious or dangerous horse.

39. (15.) The Deputy shall see that no horse is allowed to work in a place of insufficient height.

40. (16.) The Deputy shall see that back-stays or drags are provided for attaching to all trains ascending inclined planes.

41. (17.) The Deputy shall not allow one person to work alone unless other workmen are within call; and he shall not allow an inexperienced person to be in charge of a place (see General Rule 39).

42. (18.) The Deputy, where no special person is appointed, shall constantly notice the condition of ropes used on inclined planes where the breakage of a rope would allow the load to run back, and endanger persons in the inclined plane.

43. (19.) The Deputy, before leaving the pit or district of which he is in charge, shall meet the Deputy coming on the following shift, and personally inform the on-coming Deputy as to the state of the pit or district.

44. (20.) The Deputy shall personally communicate with the Manager or Under-Manager daily.

45. (21.) The Deputy shall at the close of his shift make a report of the state of the pit or district, and of his proceedings, in a book provided for the purpose, and shall sign such report.

VENTILATION.

Where a special person is appointed by the Manager to have charge of the ventilation of the workings, and the maintenance of the return air-ways, the duties of such person shall be as follows:—If no person be specially appointed these duties to be carried out by the Deputy—

46. (1.) The person so appointed shall have (under the direction of the Manager and Under-Manager) the entire charge of the ventilation. He will have to see that an adequate quantity of pure air is passed through all the working places throughout the mine; and will be responsible for the maintenance of all return air-ways. He must see that the ventilation is properly conducted throughout the mine, and shall be responsible for all return air-ways after the air leaves the last working place, and for abandoned workings.

47. (2.) The return air-ways shall be proportioned to the quantity of air passing through them, but in any case they shall not be less than 12 square feet as a minimum.

48. (3.) He shall travel all the principal return air-ways frequently —not less than once a week.

49. (4.) He shall measure all the main currents of air and splits throughout the pit with an anemometer or other instrument once a week, and record such measurements in a book to be kept at the office of the mine for the purpose.

50. (5.) The separation doors between the returns and intakes shall be kept locked.

51. (6.) Whenever it is necessary to stop a fan, or to put out a furnace, or to slacken either for more than three hours, a competent person, appointed by the Manager, must remain in the pit himself during such stoppage. The person so appointed must carefully observe the state of the ventilation, and visit any places where gas may be expected to make its appearance or to accumulate. During the stoppage of the fan or furnace no persons must be allowed to remain in the workings, and only safety-lamps must be used underground.

52. (7.) A competent person, appointed by the Manager, shall examine every upcast shaft once in twelve months, and make a detailed report on its condition to the Manager.

53. (8.) No lights other than locked safety-lamps must be used in return air-ways.

EXAMINATION OF THE MINE.

54. (1.) The fire-triers shall be persons appointed by the Manager to make the examinations required under General Rule 2 before the commencement of a working shift. It is their duty to carefully and thoroughly examine, within   hours of the time for commencing work, all working places, gates, road-ways, etc., in the district assigned to them, and the goaf edges so far as they can do so with safety; to fence off any places which they may find in a dangerous state, and to make a true report of the result of their examination on their return to the station.

55. (2.) No person shall be allowed to proceed beyond the appointed stations till the districts beyond such stations have been examined as required under General Rule 2.

56. (3.) The fire-triers having made their examination as above and returned to the stations, shall pass in each individual person into the district, examining the safety-lamps to see that they are locked and in perfect order, and shall inform the workmen concerned of any alteration which may have been made in any place to which such workmen may be proceeding, or of any danger to be guarded against. Any workman whose place is fenced off shall be kept back till his place is made safe, or shall be sent to work in some other place at the discretion of the Deputy.

57. (4.) Before commencing an examination, fire-triers must carefully examine their safety-lamps and see that they are in perfect order, and in the course of their examination they must notice them frequently to see that they have not sustained any injury.

58. (5.) In trying for gas, the lamp must be slowly and cautiously raised, and the elongation of the flame or the appearance of a blue cap carefully observed. Gas must not be allowed to explode in the lamp if it can be avoided.

59. (6.) If, when the fire-trier arrives at a working place, he finds that the roof is weighting, he shall not proceed to examine it, but shall set additional timber if he can do so with safety, and fence off the place, mentioning it in his report. He shall afterwards go in with the men, and if the place is quiet, he shall make his examination, and if he finds all right allow the men to work. If the place be still weighting he shall withdraw the men.

60. (7.) In examining goaf edges, fire-triers must beware of hanging stones, and not run unnecessary risk.

61. (8.) A fire-trier after examining a working place, shall write the date with a piece of chalk on the flag at the gate-top or other conspicuous place.

62. (9.) When fire-triers are appointed to fire shots, they must do so in accordance with the provisions of General Rule 12.

WORKPEOPLE.

63. (1.) No persons shall be allowed to descend or work at any colliery until they have been admitted as workmen by the Manager or Under-Manager, and have been entered in the book kept at the mine as a register of the persons employed therein. Each workman will receive *gratis* on application a printed copy of the Abstract of the Mines Regulation Act, 1887, and of the General and Special Rules.

64. (2.) Every workman when he has been set on will have some definite work assigned to him, and he will be placed under the orders of the Deputy or Foreman of the department in which he is to work, and he will be required to strictly obey the lawful orders of such Deputy or Foreman, and the rules in force at the mine.

65. (3.) All persons descending or ascending a drawing shaft shall do so in a covered cage, and without corves, tools, or materials of any kind.

66. (4.) No person shall give a signal at any drawing shaft or on any engine-plane unless he is specially authorised so to do by proper authority.

67. (5.) All persons descending or ascending shafts or travelling along engine-planes shall do so under the directions of the banksmen, hanger-on, or the persons in charge of the engine-plane.

68. (6.) No person must go beyond a station until authorised to do so by a Deputy or fire-trier.

69. (7.) Every miner must constantly examine the roof and sides of the place in which he works, and of the gate leading thereto, and take every precaution necessary for safety by the proper timbering of the place.

70. (8.) If the timbering put up by the miner is in the opinion of the Deputy or other official inadequately or badly applied, the miner must immediately set or re-set such timber as the Deputy may direct.

71. (9.) Miners shall undercut or hole the coal in a workmanlike manner, and shall set sprags to support the coal so undercut at intervals of not more than 6 feet.

72. (10.) Safety-lamps must be used with great care and frequently examined during the shift, and if accidentally damaged instantly extinguished. Safety-lamps must be suspended by the ring and not placed upon the floor, or upon any ledge, lid, or clog. Safety-lamps must not be hung within 2 feet of the swing of the pick. Lamp hooks with sharp points must not be attached to safety-lamps.

73. (11.) Any person perceiving gas in his lamp must carefully draw down the flame with the pricker, and warning those who may be working near, withdraw from the place and inform the Deputy.

74. (12.) Any person discovering any stoppage or derangement of the ventilation, injury to air-crossings, regulators, doors, sheets, stoppings, brattices, or air-pipes; finding any obstruction in any air-course, weakness in the roof, deficiency of timber, a dangerous feeder or accumulation of gas or water; or a naked light where safety-lamps are ordered to be used, must immediately take steps to inform the Deputy in charge of the shift, and if danger is apprehended warn the men and boys employed in that part of the pit.

75. (13.) A person doing heading or strait work must not leave coal, slack, or rubbish in the face, so as to obstruct the ventilating current, and must not leave a corf standing so as to impede the flow of the air.

76. (14.) A person passing through a door or sheet must see it closed again without delay, and any person accidentally injuring a door or sheet shall rectify the damage with the least possible delay, and inform the Deputy at once.

77. (15.) No person shall take any lucifer matches or tobacco pipes into any part of a mine where safety-lamps are required to be used.

78. (16.) No person shall take gunpowder or other explosive into the mine unknown to the Deputy, and all explosives must be made up into cartridges and conveyed in a secure case or canister containing not more than 5 lbs. All shot-holes shall be drilled in a careful and skilful manner; and no person shall drill a shot-hole except he be an experienced workman, and has received authority to do such work from the Manager, Under-Manager, or Deputy.

79. (17.) All workmen shall present themselves for work when required, and shall obey all the lawful orders of the Manager and of persons to whom superintendence may be entrusted by him. Every person shall perform the work assigned to him in a diligent and workmanlike manner; he shall not go into any part of the mine than that in which he has to work, nor leave his work before the proper time for ceasing work without the permission of the Deputy, except in cases of accident, danger, or sudden sickness.

80. (18.) No person shall ride in or upon any underground train unless he have permission to do so from the Manager or Under-Manager, except at such times as may be appointed for men to ride to and from work, at which times the trains shall be run at a slow speed, and the engineman made aware that men are riding. No person shall interfere with the working of any underground engine plane, or with the loading of any cage, or give any signal, except for the purpose of averting danger, or with the permission of the Manager, Under-Manager, or Deputy.

SAFETY-LAMPS.

81. (1.) The lamps provided for use in places where the use of naked lights are prohibited shall be of such construction and design as shall satisfy the requirements of General Rule No. 9.

82. (2.) All lamps when taken into the mine for use shall have been recently and thoroughly cleaned. Every part shall be perfect and be accurately fitted, and there shall be a proper pricker or other means of raising or lowering the flame.

83. (3.) Every safety-lamp shall be securely locked before being taken into the mine for use.

84. (4.) A competent person or competent persons shall be appointed at each mine to have charge of the safety-lamps used therein. He shall be responsible for the issue of all safety-lamps to the workmen in a perfect and clean condition, and shall see that every lamp is carefully trimmed and securely locked.

85. (5.) He shall satisfy himself by constant examination that all lamps in use conform to the standard decided upon by the Manager. He shall withdraw from use any lamp which is in any way defective, and shall bring under the notice of the Manager any lamp returned in a damaged condition.

86. (6.) All lamps must be taken to pieces daily and the parts separately cleaned. When lighted and ready for use, each lamp must be carefully examined by the lamp-keeper in charge; and where apparatus for testing with gas is provided, each lamp shall be tested in such apparatus.

87. (7.) Every safety-lamp shall be carefully examined a second time at the station by the Deputy or fire-trier, as provided in Special Rule.

88. (8.) A safety-lamp which, when returned to the lamp cabin, bears evidence of having been allowed to burn in gas, must be reported to the Manager.

89. (9.) Every collier who employs a trammer where safety-lamps are used is to be held responsible for the safety, proper usage, and return of any lamp entrusted to the trammer's care; but this shall not relieve the trammer from such responsibility as is necessarily imposed upon the actual user of the lamp.

EXAMINATION OF MACHINERY, ROPES, AND TACKLE.

90. (1.) The Manager shall appoint a competent enginewright to have charge of the machinery above and below ground, and a competent person or competent persons to make the examinations specified in General Rule No. 5. Should such person or persons find any defect likely to cause danger to life or property, he shall immediately report the same to the head enginewright, who shall cause the said defect to be immediately remedied.

94 THE COAL MINES REGULATION ACT.

91. (2.) The ropes used for raising and lowering persons shall have a breaking strain of not less than 10 times the maximum working load; they shall be securely fastened to the drum, and when the rope is paid out there must not be less than two rounds of rope upon the drum.

92. (3.) The capping of ropes used for raising and lowering persons must be equal to the resistance of a strain equal to 10 times the maximum working load, and every such capping must be passed by the head enginewright before being used for raising and lowering persons.

93. (4.) All chains, rings, disconnecting hooks, etc., used on cages where men are raised and lowered, must be tested from time to time unless periodically annealed.

94. (5.) The head enginewright shall cause all boilers to be thoroughly cleaned and examined at least once in every   weeks.

SHAFTS AND ENGINE-PLANES.

95. (1.) The Manager shall cause proper signalling apparatus to be erected between the engines used for winding and hauling and the openings into the shafts or the stations on the engine-planes for which they are used, and shall appoint sufficient and complete codes of signals for regulating the working of the cages or trains in such shafts or on such engine-planes. Every such code of signals shall be painted or printed in large characters upon a board conspicuously exhibited in every engine-house, and at every point at which signals are given.

96. (2.) When men are ascending or descending a shaft or riding on an engine-plane, the engine must be carefully run at a reduced speed.

97. (3.) The Manager shall appoint competent persons to give signals at shafts and on engine-planes, and no persons not so authorised must give signals except in cases of great urgency, and in the absence of the persons so appointed.

98. (4.) When men are ascending or descending a shaft they shall not take tools or bulky materials with them, and no corf shall be in the same compartment of a cage as that in which they are travelling.

99. (5.) An engineman shall not move his engine without first receiving a signal, and having received it shall start his engine in the direction indicated; but if he has reason to doubt that the correct signal has been given he shall wait for it to be repeated.

100. (6.) No engineman shall leave the hand-gear of his engine while it is in motion for any purpose whatever; and no engineman shall allow any unauthorised person to come into his engine-house or interfere with his engine.

101. (7.) Any engineman observing any defect in his engine, which he cannot himself rectify, must send at once for the head enginewright.

102. (8.) A notice shall be conspicuously placed at the pit head and the pit bottom stating how many persons are allowed to ascend or descend in one cage. Persons about to ascend or descend must enter the cage under the direction of the person in charge of the signalling, and shall not do so without his permission.

SURFACE WORKMEN.

103. (1.) The bank inspector shall have control of the drawing pit top, and of the banksmen, screeners, and waggoners.

104. (2.) All workmen shall attend their duties punctually at the hours appointed, and shall not leave their posts without leave.

105. (3.) Smoking about the screens and pit banks is strictly prohibited, and lamps and gas lights must be carefully watched lest they should set fire to the erections.

106. (4.) The bank inspector must see that all entrances to shafts are kept securely fenced off at all times except when the cage is at the top, and in particular shall see that everything is left safe when the pit ceases to draw; and that all unnecessary lights are extinguished.

107. (5.) The bank inspector shall not allow any person to descend the pit out of the ordinary course of work without proper authority.

108. (6.) No stranger must be allowed to descend the pit or be upon the bank without the sanction of the Manager.

109. (7.) Boys under eighteen must not be allowed to move railway trucks.

110. (8.) If a person meets with an accident in or about a mine the bank inspector must send instantly for a surgeon, or send the injured person to the infirmary, and at once inform the Manager thereof.

GENERAL.

111. (1.) No person in a place of trust shall absent himself from duty or appoint another to do his work without the sanction of the Manager, Under-Manager, Enginewright, or Bank Inspector.

112. (2.) Wages are to be paid in money at a "pay office" of the mine.

113. (3.) No person shall wilfully damage, or without proper authority remove or render useless any fence, fencing, casing, lining, guide, means of signalling, signal, cover, chain, flange, horn, break, indicator, steam-gauge, water-gauge, safety-valve, or other appliance or thing provided in any mine in compliance with this Act. Any person who, by a wilful or criminally negligent act, causes damage to the property of his employers, or endangers the lives of his fellow-workmen, shall be liable to an action at law for the recovery of such damages as the court may direct."

114. (4.) No person shall be allowed to work either above or below ground if he is to any extent under the influence of liquor.

115. (5.) Any workman refusing to carry out the orders of those in authority in or about the mine, shall be sent out of the pit or off the premises until he has seen the Manager, and his offence has been dealt with.

116. (6.) Any person fighting or quarrelling in the pit, using obscene or foul language and interfering with the comfort of his fellow workmen, illusing any horse, or otherwise misconducting himself, shall be sent out of the pit and suspended until the offence has been investigated by the Manager and dealt with.

117. (7.) Any person wilfully injuring the property of his employers, defacing or removing any notice, or interfering with the work of the mine or any persons employed therein, shall be suspended from his work until his offence has been inquired into and dealt with by the Manager.

118. (8.) The Manager, Under-Manager, Deputies, Enginewrights and other officers are particularly enjoined to comply with and enforce the rules, and any person knowing of a breach of any of the rules shall immediately inform the Manager or Under-Manager.

119. (9.) Any person committing a breach of any of the foregoing Special Rules is guilty of an offence against the Act.

The PRESIDENT: There is only one observation I wish to make —I perfectly agree that the same set of rules will not do for all collieries. That I think is quite certain. I found that to be so in 1872. At the collieries I had the management of the the custom was, long before the Act was obtained, for one man, chosen by the men who worked there, to go and examine all the workings with a safety-lamp. But the men worked with candles, and so this alteration of the rule was made—that after the examination was made with a lamp an examination should be made with a candle. I did so on this principle, as I told Mr. Morton at the time, the argument I used was that it would be better to have one man killed than twenty. That was inserted in the rules of that colliery and is not one of the rules of any other colliery I have seen. This shows, that even at the time when Mr. Morton was inspector, there were occasions when special rules, different to those in force at other collieries, were the rules of collieries as you suggest.

Mr. A. M. CHAMBERS: I propose that the best thanks of the Institute be given to Mr. Walker for his paper, and that it be printed and circulated in the Transactions of the Institute. I do not wish to make any detailed remarks upon it now.

Mr. W. G. JACKSON: I beg to second it.

The resolution was carried unanimously.

Mr. LONGBOTHAM: I suppose it is intended to take the discussion on Mr. Walker's paper at next meeting, after the distribution of the Transactions?

The PRESIDENT: Yes.

ADJOURNED DISCUSSION ON MR. ROWLAND GASCOYNE'S PAPER ON THE "EASTERLY EXTENSION OF THE LEEDS AND NOTTINGHAM COAL-FIELD."

The PRESIDENT: The next business is the discussion on Mr. Gascoyne's supplementary paper on the "Easterly Extension of the Leeds and Nottingham Coal-field."

Mr. JOSEPH MITCHELL: Mr. Bonser promised at the last meeting to give some further information, and write a short paper, but he has not yet done so.

The PRESIDENT: Nothing can be done about it?

Mr. JOSEPH MITCHELL: It will have to stand adjourned.

The PRESIDENT: Then the discussion on Mr. Gascoyne's paper will be adjourned because Mr. Bonser's paper is not ready.

Mr. LONGBOTHAM: I beg to propose a vote of thanks to the President.

Mr. MARSHALL: I second that.

The PRESIDENT: Gentlemen, I am much obliged. I have not had much to do to-day, but we have a very important discussion to come.

MIDLAND INSTITUTE OF MINING, CIVIL, AND
MECHANICAL ENGINEERS.

GENERAL MEETING.

HELD AT THE VICTORIA HOTEL, SHEFFIELD, ON TUESDAY, NOVEMBER 15TH, 1887.

T. W. EMBLETON, Esq., President, in the Chair.

The minutes of the last meeting were read and confirmed.

The following gentlemen were elected members of the Institute, having been previously nominated :—

Mr. JOHN POLLARD, Mining Engineer, Wakefield.
Mr. SEPTIMUS H. HEDLEY, Mining Engineer, East Gawber Colliery, Barnsley.

THE FEDERATION OF THE MINING INSTITUTES OF GREAT BRITAIN.

Mr. JOSEPH MITCHELL (Secretary) read the following letter from the Secretary of the North of England Institute :—

"Newcastle-on-Tyne,
"October 7th, 1887.
"*The Secretary of the Midland Institute.*
"Dear Sir,
"I am instructed to inform you that the following gentlemen have been appointed a Committee 'to further the scheme of Amalgamation proposed in Mr. Bunning's paper':—
"J. B. Simpson, J. Daglish, A. L. Steavenson, T. Forster Brown, A. Sopwith, Jas. Willis, T. J. Bewick, Lindsay Wood, G. B. Forster, Jno. Marley, W. F. Howard ;
and I am desired to ask you if you will be so kind as to assist in the matter by appointing a Committee of your members to discuss the scheme.
"Your kind attention will exceedingly oblige.
"Yours faithfully,
"THEO. WOOD BUNNING."

The PRESIDENT: Most of the members have seen the paper written by Mr. Bunning, which was read at the meeting of the North of England Institute of Mining and Mechanical Engineers at the Royal Exhibition at Newcastle-upon-Tyne. It was proposed that there should be an amalgamation of the various Societies such as ours, the head-quarters of which should be in London, in order, as I may shortly say, to endeavour to prevent similar papers being published by different Societies, and if it is your pleasure I will move that a Committee should be appointed for that purpose. Perhaps some one will second that.

Mr. T. W. JEFFCOCK: I shall be very glad to second it.

The motion was carried, and it was agreed that the President, Messrs. G. B. Walker, T. Carrington, A. M. Chambers, and T. W. Jeffcock should be the Committee.

The PRESIDENT: There is a letter from Mr. Bonser, which is shortly this—that he cannot throw any additional light on the subject of the "Easterly Extension of the Leeds and Nottingham Coal-field," and expressing his opinion that the data now laid before the meeting are of a remote date, so that he declines to go any further with the question. With respect to Mr. Bonser's paper on "Winding and Hauling Ropes," the Council have had it under consideration, and they think it is too diffuse and greatly needs condensation, and the Council have given authority to the Secretary to write Mr. Bonser to that effect. I shall ask Mr. Walker to read his paper.

The following paper was then read:—

HYDRO-CARBON EXPLOSIVES AND THEIR VALUE FOR MINING PURPOSES.

WITH A REPORT ON A SERIES OF EXPERIMENTS WITH ROBURITE.

By GEORGE BLAKE WALKER, F.G.S.

PART I.

MINING has always been and must ever remain a perilous occupation, and although the various dangers which attend the working of collieries have been met by scientific discovery and invention, new ones seem to come into prominence as the old ones are overcome. It was hoped that by the invention of the safety-lamp, and by means of more thorough ventilation, explosions would cease to occur, but so far from this they are still far too frequent, and they are more disastrous in their consequences as time goes on. It is now generally believed that gunpowder is a fruitful source of accidents, and its use was very severely restricted by the Mines Regulation Act of 1872. An explosion in a coal mine is a most difficult thing to trace back to its source. The havoc it creates is so great that it is exceedingly difficult to tell where the original explosion took place, but it became an almost accepted explanation at one time to attribute almost every explosion to the use of gunpowder if it were known that gunpowder was used in the mine where the accident happened. As the evidence often seemed to be entirely at variance with this theory when it could not be shewn that any shot had been fired in a situation where there was any probability of gas having been present, this presumption was most unsatisfactory. A series of accidents which happened between the years 1880 and 1885, however, possessed some facts in common which simultaneously directed the thoughts of Mining Engineers both here and in Germany to the question of whether it were possible that coal dust, and not so much fire-damp, had in these cases been the cause of the mischief.

The elaborate experiments instituted by the German Government on dusts brought from a large number of collieries in the Westphalian and Rhenish coal-fields, established beyond the possibility of doubt the fact, that many dusts do possess extraordinary explosive qualities, and are thus a fruitful means of spreading, if not of originating, explosions. The same question has been thoroughly worked out by two of our own Inspectors, Messrs. W. A. and J. B. Atkinson, and in the work they have published on the explosions at Seaham, Tudhoe, Trimdon Grange and other places, they have convinced most mining men that the serious effects of these accidents are traceable to the ignition and explosion of the coal dust which had accumulated in the haulage roads of these collieries.

The discovery of this source of danger, hitherto unsuspected, has greatly impressed the public mind; and in the Mines Regulation Act, which has just been passed, the use of explosives in dry and dusty mines is prohibited, except under elaborate precautions; and shot firing in the haulage roads is absolutely forbidden when more than ten persons are underground.

With respect to the kind of explosives which alone may be used in fiery and dusty mines, the same Act provides, under Section G (2) of the 12th General Rule, that the explosives employed must be used with water or other such contrivance, as to prevent it from imflaming gas, or it must be of such nature that it cannot inflame gas.

Explosives are usually roughly grouped under two broad divisions —*Gunpowder* and *Nitro-compounds*, the former being the less, and the latter the more violent. The composition of powders is too well known to detain us, but of the Nitro-compounds it may be said that their leading characteristic is that the so-called nitro-group NO_2 plays the part of the oxydising agent. They are more or less true chemical compounds resulting from the introduction of the nitro group into certain organic bodies. The principal explosives in this class are Tri-nitro-glycerine, Tri-nitro-cellulose or gun-cotton, and Tri-nitro-phenol. Nitro-glycerine being a liquid was naturally both dangerous and inconvenient, and it was soon discovered that it could only be used in connection with absorbent substances, such as infusorial earths, in which form we know it as dynamite. Infusorial earth is, of course, an inactive absorbent and will contain about 75 per cent. of Nitro-glycerine. The considerable proportion of the inactive base is, of course, a great drawback in theoretical effectivenecs. A great advance was made when Nobel discovered Blasting-gelatine, the gelatine affording an active instead of an inactive base. The following tree-table shews the relation of the various Nitro-glycerine compounds :—

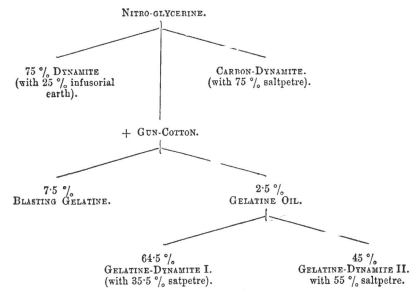

There are other dynamites into which Nitro-glycerine enters in different proportions, but they are unimportant for our present purpose. We must now turn to another group of explosives, which are composed of separate components which are first mixed just previous to use. To this group belong Sprengel's explosives, as well as *Hellhoffite*, and lastly *Roburite*.

Hellhoffite is in the main a mixture of concentrated nitric-acid with hydro-carbons or their nitro relatives so far as these can be disengaged, as for instance Benzol and Nitro-benzol. By adding infusorial earth, an explosive similar to dynamite is produced, which is known as *Carbonite*.

Roburite is a development of the idea which first took shape in Sprengel's compounds, but also containing chlorine with an oxidising agent. The contained products yield, when mixed with 2 to 3·5 parts of the oxydising agent (generally ammonium-nitrate), powerful explosives varying in power and violence according to the proportions of the components. The effect of the chlorinising appears to be that on the one hand the Nitro-products are kept drier and, in the presence of the oxidising agent, give prompter reactions, and on the other hand in consequence of the high temperature produced by the union of the Chlorine and Hydrogen a still higher effectiveness. Finally the products of explosion are incapable of supporting combustion, hence the initial flame or rather the intense heat produced by detonation is immediately quenched, when the explosion takes place under confinement. The following is the precise composition of Roburite:—

$$\text{Roburite} = \text{Chlorodinitro-Benzol.}$$
$$= C_6 H_3 Cl.(NO_2)_2$$
$$\text{or} \left\{ C_6 H_3 \begin{array}{c} NO_2 \\ NO_2 \\ Cl. \end{array} \right\}$$

in which 3 atoms of Hydrogen have been replaced by 1 atom of Chlorine, and 2 molecules of nitric peroxide (NO_2).

Three main considerations enter into an estimate of the value of an explosion, from the miner's point of view. These are:—

1. Its rending power.
2. Its safety and simplicity for handling and keeping.
3. The effects of its explosion on the conditions obtaining in coal mines.

As to efficiency. An explosive is suited for different purposes by the comparative rapidity or slowness of its action. Thus if the object be to shatter or tear out a piece of hard crystalline rock, the action should be as rapid as possible. If the object be to disengage stratified rock with as little breakage as possible, a slower action is best. Rocks possess in varying degrees the properties of elasticity and rigidity. Thus, shales are elastic and sandstones are rigid. It is evident that different qualities in the explosive must be employed to obtain the best results in blasting elastic and rigid rocks. For the latter, it is important that the explosive should be as much concentrated as possible, and should occupy every cranny of the bore-hole; while for the former, a more extended disposition of the explosive will often be most effective. Gun-cotton being practically incompressible does not adjust itself to the form of the hole; gelatine dynamite, on the other hand, does so readily, hence its eminent suitability for siliceous and hard rocks; roburite, being granular, though it adapts itself to the hole, is less concentrated than gelatine, and for a given size of shot-hole will occupy more space. Gunpowder has always been admittedly a most satisfactory agent where slow action is required. The power of an explosive is roughly measured by the relative volume of the gases resulting from explosion. According to Herr Georgi (*Mittheilungen über die theoretische Bewerthung und practische Untersuchung der Sprengstoffe*—Freiburg, 1887), the relative volumes of the resultant gases from dynamite, blasting gelatine, and roburite are as follows :—

    Dynamite ... ... 881.
    Gelatine ditto ... 1200.
    Roburite ... ... 1400.

These figures only differ slightly from those of Herren Trauzl and Klose, quoted in the same work. According to Trauzl, roburite has *theoretically* 9 times the force of gunpowder, but practically its effect may be taken at 4·1 as compared with powder.

As is well known, most kinds of nitro-compounds require detonators containing at least 1 gramme of fulminate. This is the case with blasting-gelatine, gun-cotton, hellhoffite, and roburite. We must bear in mind that to obtain a perfect explosion, the detonation must produce a degree of heat adequate to the complete detonation of the whole charge. If the heat attained is insufficient, a portion of the charge will *burn* instead of exploding, and the effect will be proportionately less.

We have next to consider convenience and safety in handling and storing.

It is of the first importance that explosives should not be liable to explode in consequence of mechanical action, such as shaking, as this makes them unsafe to transport by railway, and so must heighten their cost. Thus in America, nitro-glycerine can only be conveyed in a frozen state. They must also not be liable to chemical change,—deterioration as we may call it ; and finally they must not be hygroscopic, that is greedy of moisture. If they are the latter, they will absorb moisture from the surrounding air, and rapidly become unfit for use.

In these respects dynamite and blasting-gelatine are fairly satisfactory, but roburite answers to all these requirements even more perfectly. It has been attached to quick-running machinery placed under a steam-hammer, and spread on a rail over which a heavy locomotive has passed. It can be burned on an open fire, and if wetted it can be dried with safety, and loses none of its power.

Our third division was " The effects of explosion on the conditions obtaining in coal-mines." These effects are in the main two— the ventilation of the air, and the production of flame.

It is well known that most explosives produce, in theory, carbonic acid and nitrogen, together with steam. It is also known in practice, however, that the theory is not always borne out, but that other gases, such as sulphuretted hydrogen with powder, and hydrogen with nitro-glycerines, as well carbonic oxide (CO) are largely produced. Carbonic acid and nitrogen, though incapable of supporting life, can be diluted and rendered harmless, but carbonic oxide and sulphurred hydrogen are positively poisonous. They are, of course, produced by the imperfect combustion of the charge. In this respect the resulting products of roburite are singularly innocuous. Neither carbonic-oxide nor sulphuretted hydrogen are produced at all. Definite chemical compounds, like nitro-glycerine and gun-cotton, cannot be altered in their action when exploded, or flame prevented or quenched save by some mechanical means, such as Settle's water cartridge; but in roburite, Dr. Roth has, as it were, *built up* the *organic* constituent of roburite so as to possess within itself this inestimable safe-guarding quality. Although it is difficult to determine with accuracy the precise nature of the changes which occur when roburite is exploded, we may say that the proportion of the oxydising agent (ammonium nitrate) is so large that it is almost certain that all the carbon is burned or oxydised into carbonic-acid gas, which, of course, will not support combustion, and therefore will not give flame; while the hydro-chloric acid gas is a still further quenching agent. In the combustion of gunpowder, and the explosion of gun-cotton or dynamite—even when detonated with mercuric fulminate—there appears to be always a very considerable proportion of carbonic-oxide produced, and it is the burning of some of this into carbonic-acid gas by combination with the oxygen of the air, which produces the vivid flame noticed in practice in firing these substances.

The question of the proportion of large coal got where explosives are used for bringing down coal, is most important. We shall see presently what our own experience has been in this matter. Dynamite is admittedly too violent in its action and shatters the coal. Herr Georgi gives the following figures as the result of his experiments at Zaucherode —

Explosive.	Length holed.	Yield.						Explosive. Per 100 hectos.		Cost. Per 100 hectos.*	
		Small Coal.	Large Coal.	Total.	Average.	Per centage of Small.	Quantity.	Cost.	Average.	On Large Coal.
	Metres.	hl	hl	hl	hl	%	kg.	Marks.†	Marks.	Marks.
Grain Powder..	46·5	355	935	1290	27·7	27·5	0 47	0 32	11·12	9 47
Roburite	42·5	355	1070	1425	33·5	25·0	0 38 / 0·21	} 0·95	9 91	8 41

* A Hectolitre=6102 cubic inches.   † A Mark=1 shilling.

PART II.—EXPERIMENTS.

In 1882 the Prussian Government appointed a Commission to enquire into the causes of explosions in coal mines, and to report on the means best calculated to prevent them. The Commission early realised the fact that, so far as Germany was concerned, to prohibit the use of explosives in underground operations would be to close a large number of collieries, and they did not despair of the ability of modern chemical science to supply an explosive which should be at once powerful, convenient, and safe. One of their first proceedings was to erect at Neunkirchen, in Rhenish Prussia, near the Saarbrück coal-field, a large experimental apparatus in which to produce the conditions existing in a coal mine during an explosion, and in which miniature explosions could be produced. This apparatus was used to solve a number of different questions, but the particular point which interests us to-day is their experiments with explosives, which promised to combine safety with efficiency; and of these they tried a considerable number. The first of the compounds which appeared likely to satisfy the conditions required was *Sekurite*,* the invention of a local chemist of Dudweiler, near Saarbruck. This gentleman produced a small quantity of *Sekurite*, which was tested in the apparatus at Neunkirchen and in the Government mines; but he was for some reason or other unable to produce any more for some time, and we are not in possession of any reliable information respecting its qualities. It is believed, however, that from its nature

* An explosive almost identical with *Sekurite* is described by Sprengel in his pamphlet on "A new class of Explosives which are non-explosive during their fabrication, storage, and transport," London, 1873.

it will be subject to decomposition if not kept in carefully sealed vessels. Another compound called *Carbonite*, the invention of Mr. C. E. Bichel, of Schlebusch, near Cologne (see *ante*), is also reported to have fulfilled the requirements laid down by the Commissioners, and this compound is now being tested practically in some of the Westphalian collieries. *Roburite*, invented by Dr. Roth, of Berlin, appears to combine, as already stated, in a marvellous manner almost all the conditions required. As I have said, it is a substance so perfectly harmless that it can be handled with perfect impunity, and no friction or blow will ignite it. It burns quietly if placed on the fire, and retains its properties for a considerable length of time. It is not readily subject to decomposition, and it is equally harmless when frozen. It is, however, a most powerful explosive. In this respect it is about equal to dynamite for a given weight, but its specific gravity is less than that of dynamite. Its action, however, is less concentrated, resembling more high-class powders when fired by detonation than a nitro-glycerine compound. When used in small quantities it brings down the coal without shattering it. In appearance it is not unlike yellowish oatmeal or light-coloured brown sugar. The most interesting feature is the nature of the organic component which Dr. Roth has obtained, after a troublesome investigation extending over three years. Dr. Roth has endeavoured to meet the practical requirements of modern mining by the careful adaption of the qualities of Roburite to all the conditions under which explosives can be used in mines. He first directed his attention to the methyl compounds, alcohols, and glycerines, and endeavoured by different chemical processes to obtain from them explosive compounds which, when mixed with oxygen-yielding substances, would not inflame fire-damp. But all these attempts (comprising experiments with many hundred members of these groups) were unsuccessful. Several powerful explosives were produced, but they possessed, in common with dynamite and other explosives of that class, elements of danger. Dr. Roth then turned to the so-called aromatic groups which belong to the hydro-carbon family, from which the French chemist, Designolles, and Sprengel in England have produced powerful explosive agents. But their practical employment seemed too dangerous in consequence of their being so readily ignited by friction, and from their being so highly corrosive that they cannot be safely handled, and must be conveyed to the place where they are used in hermetically sealed glass bottles, and moreover, all these products (like Melinite) readily decompose. Moreover, all these stuffs have the disqualification that they would ignite fire-damp. Dr. Roth now set himself to answer this question —Is it possible to retain the power which these aromatic hydrocarbons possess, and by the introduction of some other element into their molecular relations, produce a compound which, when exploded, would produce gases which would immediately quench the flame producing ignition? In other words, is it possible to combine with the explosive the means of entirely quenching all flame at the very moment of ignition? This question he has been able to answer in the affirmative. After many fruitless experiments he at last discovered the quenching element in chlorine. Hydro-chloric-acid gas is generated on explosion, and, being incapable of supporting combustion, it prevents the spreading of the flame to any dangerous surrounding atmosphere, such as fire-damp. In fact the practical effect is very similar to the water envelope in the case of Settle's water cartridge, with the immense advantage that, being intimately mixed with the explosive compound, the safe-guarding provision can never be absent, as it would be in the case of the water cartridge if the envelope containing the water should be accidentally punctured.

The results obtained in Germany in the trials before mentioned naturally created much interest in this country, and early in the present year Mr. S. Coxon (formerly of the Usworth Collieries) made a visit to Saarbrück, to see trials made with the different compounds; and shortly after Mr. Emerson Bainbridge decided to pay a visit to Germany with the same object. He kindly invited me to accompany him, and we left England together on the 9th of June. Mr. C. Stuart-Wortley, the Under-Secretary for the Home Department, who was then busily engaged with the Coal Mines Regulation Act, was specially interested in Mr. Bainbridge's visit, and through him Mr. Bainbridge got letters of introduction from Sir James Ferguson, the Under-Secretary for Foreign Affairs, to the British Consul at Dusseldorf, and through him to almost everyone in Germany who could give us any information on the subject. On the 11th of June we reached the small mining town of Witten in Westphalia, a few miles distant from Krupp's famous works at Essen, and there

Mr. Bainbridge had an interview with Herr Von Hoffmann, the President of the Gelsenkirchen Colliery Company, perhaps the largest colliery concern in Germany. A series of interesting experiments were arranged.—Those which took place on the 11th of June being in a freestone quarry on the surface, and those on the 13th underground in the Walfisch Colliery.

The experiments in the quarry took place partly in the afternoon and partly at ten o'clock in the evening, the first series being devoted to the relative power of roburite as compared with dynamite and powder in blasting rock. The results of these experiments were immensely satisfactory, the power of roburite appearing to be if anything even greater than that of dynamite. In the case of one of the experiments, an excessive charge having been placed in a horizontal shot-hole, the rock was dislodged with such a force that portions of it were projected *en parabolic* and fell at a distance of something like 100 yards from the point of discharge. On returning to the quarry at ten p.m., a further series of experiments were conducted in complete darkness. Three shot-holes, four feet in depth, had been drilled into the solid face of the rock horizontally, and into each of them was successively placed a charge of gunpowder, dynamite, and roburite. None of the holes were stemmed. The different charges being successively fired, the flame and sparks from the gunpowder were estimated to have proceeded from the mouth of the hole about 30 feet; the dynamite about 8 feet, and the roburite produced merely a flash which could hardly be measured. It should be borne in mind that all these holes were fired without tamping or stemming, and we intended in the next experiment to place a slight amount of stemming in the hole. The rock, however, was too much damaged to allow of this. The experiment was however tried on the following Monday underground. The next experiments were with cartridges suspended in the open air, and in this series of experiments the flash of the roburite cartridge was hardly perceptible, while that of the others was very considerable. But the most interesting test of all was the last. Powder, dynamite, and roburite were successively discharged with simply two or three handfuls of sand placed over them. The quarry was illuminated by the discharge of the two former, but with roburite successive tests failed to produce any flash or flame whatever. This was most striking, the detonation of the charge being exceedingly loud, but without the faintest gleam where the charge was exploded. It must not be inferred from the act that a small quantity of flame was seen from the shot-hole, and from the suspended cartridge, that roburite does produce flame in any degree when properly used, In both these cases it was fired by tests which are never obtained in practice, and have merely a scientific interest. No explosion can take place without initial flame, but the distinctive property of roburite is that that initial flame is immediately quenched by the gas from explosion. In fact this gas extinguished the flame in a similar manner to the water envelope in the case of Settle's water cartridge.

In the underground experiments which took-place on the 18th, evidence was desired on two points. The first as to flame resulting from a blown-out shot, and the other as to the size of coal brought down by roburite as compared with powder. The experiments in the mine, in the most complete darkness, convinced us that with even a very small amount of stemming, say one inch or an inch and a half, no flame whatever appeared; and the underviewer at the colliery, who has used roburite for some months and watched it very carefully in its action, firmly asserted that on no occasion had he or any of his people seen the least vestige of flame.

Mr. Bainbridge and myself subsequently visited Saarbrück. Baurath Von Eilert, the Chief Director of the Government Mines, kindly gave us much information, which entirely confirmed the impression that we had ourselves formed.

Being thoroughly satisfied from my own observation of the value of roburite, and having the management of a somewhat gaseous colliery, I naturally felt very anxious to avail myself of it as soon as possible. I therefore applied to the Roburite Explosives Co., 59, Old Broad Street, London, and they, although they had not yet commenced the manufacture in this country, very kindly undertook to obtain a supply from Germany. This was received by the middle of October, and I at once carried out some experiments with a view to testing it in a practical manner. A series of experiments were conducted on the 18th and 19th of October, in the Parkgate Seam at Wharncliffe Silkstone Colliery, in the presence of a number of gentlemen interested in mining, and the results was to fully confirm the conclusions at which I had arrived as to its excellence as a blasting agent, and its safety.

The roburite as supplied to us is in cartridges enclosed in waterproof paper, 5 in. long and $1\frac{1}{4}$ in. diameter, each cartridge weighing (including wrapper) $3\frac{1}{2}$ ounces. The following were the experiments made at Wharncliffe Silkstone Colliery on the 19th October.

1. Shot placed close to roof in Parkgate Seam 14 feet from loose end, holed 4 ft. 2 in., under charge of roburite 90 grammes—say 5 oz. Coal brought down, but somewhat too much shattered.

2. Similar to No. 1, but charge only 60 grammes—$3\frac{1}{2}$ oz. Coal brought down in better condition.

3. Shot hole 3 ft. 6 in. deep drilled into the side of a dip bord, 45 grammes of roburite, stemmed with three inches of holing dirt, fired by electricity. The party were placed at the top of the bord, and lamps covered, so as to observe if any flame proceeded from the hole when the charge was fired. The shot blew out the stemming, but no flame was observed.

4. A charge of 90 grammes of roburite was placed 9 inches from cutting-side in leading benk. Coals holed 16ft. in length 3ft. 9in. under. The shot cut the fall square with the cutting-side, the coals falling almost in a solid block. It was estimated that the large block would weigh upwards of 11 tons.

5 and 6. These were ordinary coal shots with 60 grammes of roburite. The coal was brought down in good condition.

7. A shot was placed in the top clod over the seam in a main level for the purpose of making height. 60 grammes were used. The shot brought down the clod for 4 ft. 3 in. in length and nearly across the level, which was 8 feet wide, the sides requiring to be dressed back with the pick. In all the experiments, the gases resulting from explosion were absorbed into the surrounding atmosphere almost instantaneously, and no unpleasant effects were felt.

On the 22nd and 25th some interesting experiments were made in a specially constructed apparatus on the surface adjoining the Rockingham Gas Works, from which, through the kindness of A. M. Chambers, Esq., a supply of coal gas was obtained. A sketch of the testing arrangement is given on Plate I. It is of a rough and ready description. A is a large cask filled with water, in which is inverted a smaller cask B having a capacity of 6 cubic feet; C is a pipe conveying coal gas from the works; D an india-rubber tube conveying the gas from B into the experimental chamber E (part of an old boiler); F is a loose lid, and G a hole through which to observe the explosion. The capacity of the experimental chamber was 72 feet, so that the gas forced from the barrel B would constitute an explosive mixture of 8 per cent. of gas to 92 of air. Roburite was repeatedly fired in this apparatus without igniting the gas. Powder was also exploded, and of course gave a violent explosion and large quantities of flame.

A second set of experiments took place on the 24th October at the Monk Bretton Colliery, near Barnsley, in the presence of a large number of gentlemen, of F. N. Wardell, Esq., Chief Inspector of Mines for Yorkshire, and of several of the directors of the Roburite Company.*

The following experiments were made :—

On the Surface. 1. A cartridge of roburite was placed upon the ground and exploded.

2. A cartridge fired on the ground bedded in fine coal-dust; no flame or ignition of the coal-dust was perceptible.

3. A cartridge fired suspended in a ease into which gas was conducted and atmospheric air allowed to enter, so as to form an explosive mixture. The gas was not fired.

4. A shot fired in an apparatus shewn on Plate II., consisting of an old boiler tube, into which a current of air was introduced, by means of a hand-fan, and a stream of gas from the gasometer. When the shot was fired, the detonation was so violent that the boxing at the back end of the tube was displaced, and many persons believed the gas had been fired. This opinion being controverted by others, the boxing was re-erected and a cartridge of roburite exploded without any intermixture of gas, when the boxing was again blown away, thus shewing that its destruction was due to the violence of the discharge and not to the explosion of the gas.

5. An experiment of a similar kind was tried but without gas and with coal-dust kept in suspension by means of the fan. No flame or ignition of the coal dust took place.

6. A charge of ordinary blasting gunpowder was next fired in the tube with an explosive mixture of gas and air. The explosion was

* For this account of these experiments, I am indebted to Major-General Wardell and Mr. J. L. Marshall of Monk Bretton.

violent, and the flame of the ignited gas was belched out for a distance of perhaps 10 yards from the end of the tube. The gas also continued to burn at the end of the gas-pipe till turned off.

7. The same experiment was repeated with roburite, and as the gas was again not ignited, most of those who thought the gas had been fired in Experiment 4 admitted that they had been mistaken.

In the Pit. 1. A 2 inch hole was drilled 4 ft. 6 in. deep into coal having a face 7 yards wide, fast at both ends, and holed under for a depth of 8 ft., end on; thickness of front of coal to be blown down, 2 ft. 10 in., *plus* 9 in. of dirt. This represented a most difficult shot, having regard to the natural lines of cleavage of the coal; a "heavy job," as it was locally termed. The charge was 65 grammes of roburite, which brought down a large quantity of coal, not at all too small in size. No flame was perceptible, although all the lamps were carefully covered.

2. A 2 inch hole drilled 4 ft. 6 in. into the side of the coal about 10 in. from the top; fast ends not holed under; width of space 10 ft. This was purposely a "blow-out" shot. The result was again most satisfactory, the charge exploding in perfect darkness.

3. A "breaking-up" shot placed in the stone roof for "ripping," the hole being drilled at an angle of 35 deg. or 40 deg. This is indeed to open a cavity in the perfectly smooth roof, the ripping being continued by means of the "lip" thus formed. The charge was 105 grammes (nearly 4 oz.), and it brought down large quantities of stone.

4. A "ripping" shot in the stone roof, hole 4 ft. 6 in. deep, width of place 15 ft. with a "lip" of 2 ft. 6 in. This is a strong stone "bind," and very difficult to get down. The trial was most successful, a large heap of stone being brought down and more loosened.

5- A second "blow-out" shot, under the conditions most likely to produce an accident in a fiery mine. A 2 in. hole, 4 ft. 6 in. deep, was drilled in the face of the coal near the roof, and charged with 105 grammes of roburite. A space of 2 feet was purposely left between the charge and the tamping; the hole was then strongly tamped for a distance of nearly 2 feet. The report was very loud, and a trumpet-shaped orifice was formed at the mouth of the hole, but no flame or spark could be perceived, nor was any inconvenience caused by the fumes, even the instant after the explosion.

On the following day some further experiments were made at Silksworth Colliery, near Sunderland, the property of the Marquis of Londonderry. On arriving at the colliery, a rail was shattered by the discharge of a single cartridge, for the purpose of ascertaining whether the detonators which were in use at Silksworth were sufficiently powerful. We then went underground, and accompanied by Mr. F. W. Panton, the manager, and Mr. Palmer, underviewer, &c., proceeded inbye to a staple pit which was being sunk down a trouble about two miles from the shaft. The explosive used for this job was blasting-gelatine, without the water cartridge. On arrival we found that though a number of holes were ready, yet as they had been bored with the ordinary drill and hammer, they were triangular rather than circular, and the roburite cartridges would only enter two of them. One of these was a side shot. The stone was a strong stone bind and rather wet. The hole was dried as much as possible, and two cartridges of roburite (7 oz.) were put into the hole. It was found, however, that they could not be forced within about 9 inches from the bottom of the hole.

The battery was connected, but no explosion followed, and we assumed that owing to the wetness of the hole, the roburite had got wet and would not explode. A second charge was then fired in a top canch leading to the staple. This shot loosened a part of the rock, but did not bring down what was intended, as it was not strong enough, the shot being a very heavy one. In firing this shot it was discovered that the wires had not in the previous case been properly attached to the exploder; they were, therefore, reattached to the first shot, and on the second attempt this was successfully fired, after an interval of at least half an hour.

It was then necessary to leave the place, after arranging that the holes should be enlarged with the machine-borer. The following morning a number of shots were fired by Mr. Palmer, the underviewer, when the roburite did excellent work. Three sumping-holes were simultaneously fired, and unkeyed the bottom of the staple to a depth of about 3 ft. 6 in., the stone being shaken for a foot deeper. Other shots were also tried with great success, and a blow-out shot in very hard stone was observed at a distance of about 8 yards at right angles, and no flame was observed. Mr. Palmer expressed his opinion that the roburite was about equal to gelatine in power, and did its work equally well, while it was safer to handle and the after-gases were less objectionable.

The same evening experiments were made in the Newbottle Limestone Quarry, in the presence of Mr. Lishman, Chief Engineer to the Earl of Durham; Mr. Panton, of Silksworth; Mr. J. Wood, of North Hetton Colliery, and others. The object of these experiments was to see if any flame could be observed in the dark from blown-out shots. One or two of the experiments were disappointing, as they did not prove to be blown-out shots at all, but dislodged large masses of rock. One shot, however, bored in the floor of the quarry, was more successful, the stemming being blown out without any appearance of flame.

On Thursday, the 27th, a series of experiments were made in the yard at the Philadelphia Colliery, with an apparatus similar to the one figured in Plate I., in the presence of a number of the leading Mining Engineers of the county of Durham. Similar satisfactory results were obtained to those already mentioned, and a very favourable impression produced. The only occasion on which roburite ignited the gas was when it was buried under two or three pounds of gunpowder, when ignition took place. Dr. Roth maintained that this ignition was due to friction.

I have endeavoured to include in this paper all the experience we have as yet had with roburite, and it must be admitted that this evidence, so far as the tests have gone, seem to indicated that roburite does possess in a remarkable degree many of the qualities required for that desideratum, a *safe* explosive for use in fiery mines. As, however, the thing is quite new, it is premature to speak with absolute confidence. It has yet to sustain severer trials and undergo more crucial tests than it has yet been subjected to, and we must all hope that it will emerge successfully from those trials, for if it does we shall have in it an explosive which will be an alternative to the water cartridge, of greater power, and in some respects of greater convenience and general usefulness.

The PRESIDENT: This is a paper of such considerable length, and also contains so many scientific expressions, that it would be almost impossible to discuss it on the present occasion. Therefore, if you please, I will propose that the paper be printed and inserted in the Transactions of the Institute.

Mr. C. E. RHODES: I shall be very glad to second that, coupled with the best thanks of the Institute being accorded to Mr. Walker, and that the discussion be adjourned to a future day.

The PRESIDENT: We shall come to that after.

The motion was carried.

The PRESIDENT: I think we cannot do less than pass this resolution, that the best thanks of this Institute be given to Mr. Walker for the paper he has just read to us. I have to propose it. It does not want a seconder; everybody seems agreed about it, so it is carried by acclamation. (Applause).

Mr. G. B. WALKER: I am much obliged to you, gentlemen.

[The President was now obliged to leave the chair, and he asked Mr. A. M. Chambers to preside during the remainder of the meeting.]

DISCUSSION ON MR. G. BLAKE WALKER'S PAPER ON "SPECIAL RULES UNDER THE NEW MINES ACT."

Mr. A. M. CHAMBERS (Vice-President in the chair): The next business is the discussion on Mr. G. B. Walker's paper on Special Rules. I daresay most gentlemen present may know, perhaps a few may not, that Committees have been appointed by the South Yorkshire Coalowners and Colliery Managers, and by West Yorkshire Coalowners and Colliery Managers, and these two Committees are going to meet shortly to discuss the question of Special Rules, and if possible to get united action throughout the two districts on the subject.

Several members made enquiries as to the action which the Coalowners' Committee were taking, and whether or not the results of their deliberations would be communicated to the Institute.

The CHAIRMAN remarked that it was not possible to say what the Committee would eventually do with respect to the draft rules,

The CHAIRMAN moved that the discussion be adjourned, and said that it would be quite open when the rules were prepared for any member to read a paper on them.

Mr. SOUTHALL seconded the motion, which was carried.

MR. GASCOYNE'S PAPER ON "THE EASTERLY EXTENSION OF THE LEEDS AND NOTTINGHAM COAL-FIELD."

Mr. A. M. CHAMBERS: There is Mr. Gascoyne's paper, are any remarks to be made on that subject? If not, the meeting is closed.

MR. PATTISON exhibited at the meeting, by permission of the President, a Safety-Lamp having a "shut-off" appliance.

ICAL ENGINEERS.

GE BLAKE WAL
RBON EXPLOSIVE

OT.

From Gas Works.

D

MIDLAND INSTITUTE OF MINING, CIVIL, AND MECHANICAL ENGINEERS.

TO ILLUSTRATE MR. GEORGE BLAKE WALKER'S PAPER
ON "HYDRO-CARBON EXPLOSIVES."

Fig 1.

Scale ½ Inch to 1 Foot.

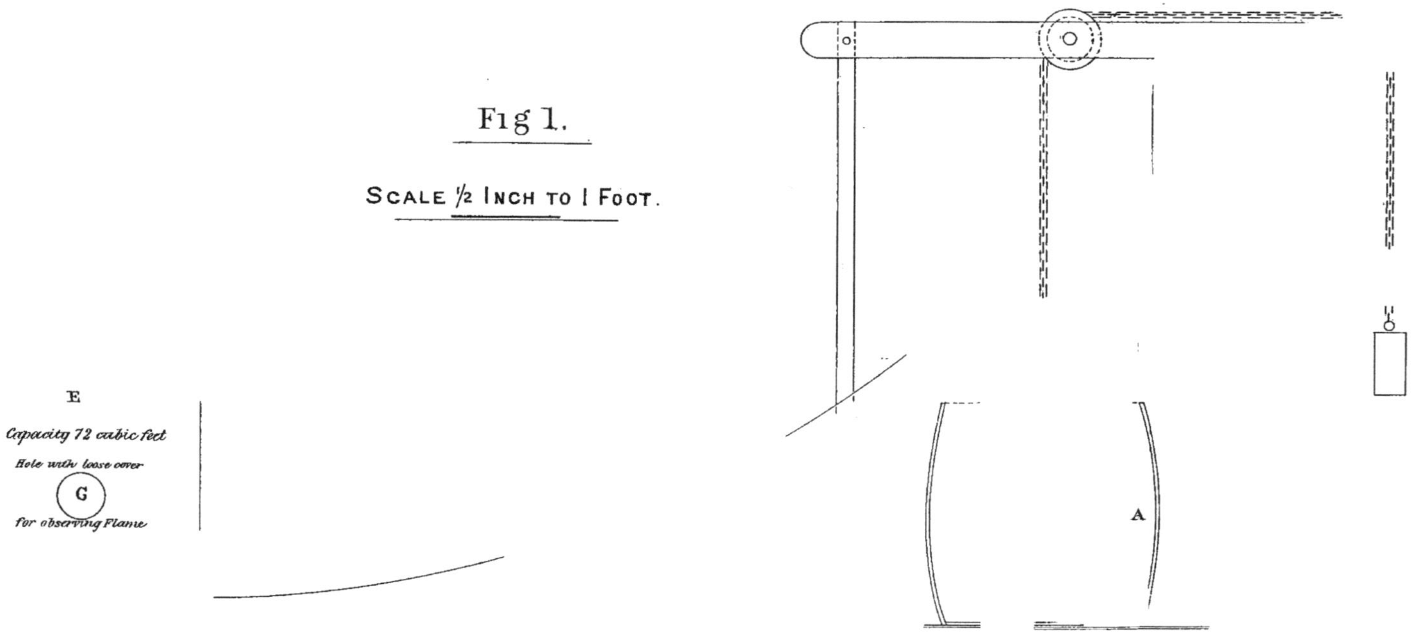

E

Capacity 72 cubic feet

Hole with loose cover
for observing Flame

G

A

)INEERS.

\KE WALK

(PLOSIVES."

o l Foo

Gas Supply

MIDLAND INSTITUTE OF MINING, CIVIL, AND MECHANICAL ENGINEERS.

TO ILLUSTRATE MR. GEORGE BLAKE WALKER'S PAPER ON "HYDRO-CARBON EXPLOSIVES."

Fig 2.

SCALE ¼ INCH TO 1 FOOT.

MIDLAND INSTITUTE OF MINING, CIVIL, AND MECHANICAL ENGINEERS.

GENERAL MEETING.

HELD AT THE INSTITUTE ROOM, BARNSLEY, ON WEDNESDAY, DECEMBER 14TH, 1887.

T. W. EMBLETON, Esq., President, in the Chair.

The minutes of the last meeting were read and confirmed.

The following gentlemen were elected members of the Institute, having been previously nominated :—

MR. HENRY MUSGRAVE, Colliery Proprietor, Havercroft Main Colliery, Wakefield.

Mr. GEORGE EDWIN JAMES MCMURTRIE, Mining Engineer, Car House Colliery, Rotherham.

THE FEDERATION OF THE MINING INSTITUTES OF GREAT BRITAIN.

The PRESIDENT: The next business is as to the Federation of Mining Institutes of Great Britain. You are aware, according to the minutes just read, that a Committee has been appointed to consider this question, and the Council have determined this Committee shall be called together within a very short time. At present we shall be very glad to hear the views of any of the members present as to the desirability or non-desirability of entering into this Federation. The first question to be determined is, whether this Institute will join this Federation or not. If any members wish to make any observations on this matter, we shall be very glad to hear what they have to say.

Mr. G. B. WALKER: We all of us feel that although we are scattered up and down the country in a number of different districts, that there is a great deal which is published in the Transactions of the different Institutes which is of interest not merely to the district where these Transactions are published, but also to ourselves here,— and that unless every Institute is to republish every interesting paper that may be read before one of them, and thus a great deal of unnecessary printing be incurred, we lose practically a great deal of very valuable information which is being brought before other Institutes. During the past week a paper, very interesting to myself, has come under my notice. It was by Mr. Frank Brain, and was read before the South Wales Institute of Mining Engineers, on "Electric Pumping." If I had not happened accidentally to see a copy of that paper, I should not have known anything about the very interesting experience which Mr. Brain laid before the South Wales Institute. In the same way a great many other interesting papers are read before the North of England Mining Institute, and unless we happen to be members of that Institute we do not see the papers that come before them. Therefore I confess it seems very desirable that we should coalesce, and thus all the Institutes should get the advantages of the papers which are published by each. But there is one disadvantage we must not overlook. If the Society is to become somewhat similar to the Institute of Civil Engineers, and have its head-quarters in London, if it is to drain away the life blood from our local centres, I am afraid we shall lose more than we shall gain. My feeling is that the plan to be adopted would be better if it were more like that of the Society of Chemical Industries, which has its centre in Manchester, but which has a number of local centres up and down the country, where the members who reside in those parts meet. I think myself that London, not being a mining locality, is not a proper centre for the head-quarters of the proposed united Institute. It is, of course, the capital of the country, and it may be the best place to choose, and there may be very great difficulty in getting other Institutes to agree to any other place. Still, the more central it is, the more accessible it is from the different mining districts, the better I think it would be. I do not know that I need make any more remarks upon the matter just now, but I think we may all venture to agree that if the details can be arranged, the idea of our obtaining all the papers which are read before the different Mining Institutes of the country from one centre, and saving ourselves expense in the publication of our own Transactions, would be a very great gain.

Mr. C. E. RHODES: While I think this a very important subject, and one which is well worthy our careful consideration, I have very little to add to what Mr. Walker has already said. I think that a federation of this description might in the long run be a means of raising the status of the Mining Engineer considerably above the level it has now reached, and I think it might be the means of enabling us to grant some such diploma as that attached to the Fellows of the Geographical Society and Civil Engineers. Therefore I think that provided the amalgamation to a certain extent of these Societies over the country in one central Society, does not swamp the endeavours of individual Societies, it must in the long run be a very good thing. The question as to what place is most desirable for the contralisation of these Societies, is a matter which the Committee appointed by the Institute to represent them in conference with the Northern Institute and others, will be able to deal. All we have to do is to pass a resolution from the Midland Institute that as a body we are in favour of the proposal of the North of England secretary, Mr. Bunning, subject to any modification which the Committee appointed by this Institute may think it desirable to make. I have much pleasure in moving " That in the opinion of this Institute it is desirable that they should if possible amalgamate, if the Committee appointed by them can recommend this to be carried out."

Mr. G. B. WALKER : I beg to second it.

The PRESIDENT : Before putting this resolution I have one or two observations to make which are almost in accord with what you have already heard. We must take care during our negotiations that we do not blot ourselves out, that in any negotiations we must keep our individuality, for I think if anything is done to militate against that it would be a great disadvantage to any of the Institutes that join this Federation. When we meet the other members of the Committees that have been appointed, this should be one of the principal things looked to. Further, I perfectly agree with what Mr. Walker has said about meeting in London. I think it would be useless to have a meeting there, because it would take us all up to London, perhaps at very inconvenient times, in order to take part in the discussion or reading of papers there. These are all details which will have to be settled by the Joint Committee, and the main question is that which has been proposed and seconded, that we shall join the Federation upon the terms proposed by Mr. Rhodes and seconded by Mr. Walker.

The resolution was carried.

DISCUSSION ON MR. G. BLAKE WALKER'S PAPER ON "HYDRO-CARBON EXPLOSIVES."

The PRESIDENT: The next business is the discussion of Mr. Walker's paper on "Hydro-Carbon Explosives."

Mr. G. B. WALKER: I have got a note here, Mr. President, which has reference to the resulting gases of roburite. I find I did not touch upon this quite so fully as I ought to have done, and I give here the results of explosions:—

MEMORANDUM UPON THE GASES EVOLVED BY THE DECOMPOSITION OF ROBURITE.

1. There are no obnoxious nitrous fumes evolved by the explosion of roburite. Like every other explosive,—nay, every other combustible substance,—it must produce a certain amount of carbonic-acid gas, but the following investigation will shew that the gases are perfectly harmless, thus bearing out the practical experience of all those who have been present at the explosion of roburite. Dr. Carl Roth gives the chemical equation of roburite as follows:—

$$\underbrace{C_6 H_3 Cl.(NO_2)_2}_{\text{(Organic)}} + \underbrace{9 NH_4 NO_3}_{\text{(Inorganic)}} = 6 CO_2 + 19 H_2 O + 2 ON + H Cl.$$

Putting the molecular weights in place of the above symbols

$$\left[\underset{\substack{\text{(Chloro-di-nitro} \\ \text{Benzol.)}}}{202\cdot 5} + \underset{\substack{\text{(Ammonium} \\ \text{Nitrate.)}}}{720} \right] = \left[\underset{\substack{\text{(Carbonic-} \\ \text{acid.)}}}{264} + \underset{\text{Water.}}{342} + \underset{\text{Nitrogen.}}{2\cdot 80} + \underset{\substack{\text{(Hydro-chloric} \\ \text{acid.)}}}{36\cdot 5} \right]$$

$$922\cdot 5. = 922\cdot 5.$$

2. A charge of 100 grammes of roburite [1543·2 grains, or nearly ¼ lb.] may be taken as the largest which would be fired in any one shot in the gallery of a mine.

Thus $922\cdot 5 : 264 = 100 : x. = x. = 28\cdot 6$ grammes carbonic-acid gas.

If the above amount to be diffused in even the most confined space, say 4 metres each way, that is 64 cubic metres of air.

Now a cubic metre of air [1000 litre = 1 cubic metre] weighs 1293 grammes

64 cubic metres = 82,752 grammes.

Into this weight of atmospheric air is dispersed an additional quantity of 28·6 grammes carbonic-acid, which is a proportion of 0·034 per cent.

The normal percentage of carbonic-acid contained in atmospheric air is 0·04.

A single workman of about 24 years of age exhales 44·7 grammes of carbonic-acid per hour.

In the above calculation, no allowance has been made for the rapid diffusion of the air, which in a few seconds will remove the surplus percentage of carbonic-acid.

In fact the effect would be practically *nil* compared with that of an ordinary London fog, since it has been computed that in the latter case the amount of carbonic-acid gas contained in the air is increased four-fold. Those who were close to the roburite shots fired in the underground experiments can testify that the air close to the seat of the shot immediately after the explosion was far better than is the case in the London Underground Railway.

Roburite gives rise to no dense nitrous and other poisonous fumes as does dynamite and gun-cotton.

Mr. C. E. RHODES: When Mr. Walker first brought before my notice this very important subject,—to my mind the most important subject that has been brought before either this or any other Institute for a very long period,—he was good enough to ask me to attend at Wharncliffe Silkstone Colliery in order to see some experiments conducted with the new explosive, roburite. At those trials I can confirm all that he has said as to the effect of the gases produced by explosion not being any more inconvenient than those produced by an ordinary gunpowder shot, and not so unpleasant as those produced by an ordinary charge of dynamite, so far as the effect could be felt by anybody standing, as I stood, immediately at the back of the shots shortly after they were fired, in the return airway, so that I got the first returns and had a full opportunity of feeling the effect. So far as my experience goes, I am of opinion that they are not in any way more deleterious than those from an ordinary powder shot. After attending Mr. Walker's experiments, which, being the first, and not being properly provided with electric batteries, were to a certain extent unsatisfactory to my mind, as the charges had to be fired by a fuse, I next visited Monk Bretton Colliery, and saw the whole of the experiments made there, which were, to my mind, very conclusive. Since then, I wrote to Mr. Walker and pointed out that I was very anxious to assure myself that this explosive was as safe as I had every reason, from the experiments I had seen, to believe it to be and as had been asserted by its inventor, and asked where I could procure a quantity of roburite. He was good enough to send a quantity down, and I have since made a number of experiments with the object of testing (possibly in some cases some may say unfairly) the explosive, to see whether by any means it could be made to ignite the surrounding atmosphere or other explosive substances contiguous to it. I will briefly read you the results of my tests, and they are such that the inventor and the advocates of roburite may say they are distinctly unfair, but they were such as to satisfy me that in this roburite we have got an explosive of a remarkable character and one that I think everybody that has had the same opportunity of testing it that I have had, will feel a very great amount of confidence in. I do not know that it is worth my while detaining you with more remarks. I will simply give you roughly the result of 70 or 80 experiments.

Mr. RHODES then read the following account of the experiments he had made:—

First, with regard to its power. No apparatus was used to definitely register the amount of work performed, the object being simply to demonstrate practically by these experiments the relative useful effect of this explosive as compared with gunpowder and gelignite, tested under similar conditions and in shots of equal weight.

A level stratum of hard bind, which formed the surface in a quarry, was selected for the experiments, part of the bind being cut away to a depth of 3 ft., thus leaving a straight vertical face across the whole width of the quarry.

For the purpose of the experiments, holes were drilled parallel with the face and at a distance of 5 ft. from it, each hole being 2 ft. 6 in. in depth and at such a distance apart as to prevent the shot in one hole shattering the strata around the next.

The first hole was charged with $\frac{1}{4}$ lb. of gunpowder, the next with $\frac{1}{4}$ lb. of gelignite, and a third with $\frac{1}{4}$ lb. of roburite, and these were fired in succession.

The roburite completely displaced about half a cubic yard of the stone, and loosened the whole of the surrounding material within a radius of a yard of the shot-hole and six inches below the bottom of the hole. The work done by the gunpowder fell far short of this;

but the force displayed by the gelignite considerably exceeded that of the roburite, displacing as nearly as we could judge 25 per cent. more of the strata.

In the above trial we could not form a sufficiently definite idea of the comparative power of roburite as compared with powder, and therefore we made another test, putting into holes similarly placed as just described, one pound of gunpowder in one and a roburite cartridge slightly reduced in the other, the result of the test being if anything in favour of the roburite.

The next tests were with regard to the safety of roburite, and it was first tried with the object of ascertaining if any flame was produced by its combustion.

Several detonators were each covered with $\frac{1}{2}$ lb. of roburite, and on each occasion a sharp flame was noticed. Whether this proceeded from the detonator or the roburite we could not say.

Having, however, found that flame was certainly generated when the roburite was fired by a detonator, the following experiments were arranged in order te determine whether the gases produced by the roburite nullified the flame produced by the roburite or detonator, or whether combustion could be communicated or not to any inflammable substances in its immediate vicinity.

Experiments.

No. of Experiment.	Nature of Experiment.	Result.
1	Loose roburite was fired in an explosive atmosphere, composed of 1 part of coal gas to 8 parts of air.	The burning roburite did not ignite the explosive atmosphere.
2	Roburite was covered with a layer of gunpowder, and fired under the same conditions as No. 1 Experiment.	The explosive mixture was not ignited.
3	Roburite was covered with a layer of gunpowder and fine coal-dust mixed together, and fired under the same conditions as Nos. 1 and 2 Experiments.	The explosive mixture was not ignited.
4	Gunpowder and fine coal-dust, as used in Experiment No. 3, were fired in the same explosive atmosphere, without any roburite.	Violent explosion, and long tongue of flame.

N.B.—All these experiments were often repeated, with the same result.

These experiments went strongly towards convincing me that an explosive gas could not be ignited by roburite; and experiments were then conducted to prove whether that effect was produced by the gases given off by the roburite transfusing with the coal gas and air and so rendering it non-inflammable, or whether the roburite flame was rendered impervious to any combustible substance other than a gas which might be contiguous at the time of the explosion.

A series of heaps of roburite were therefore arranged in a dark place and connected with the firing apparatus.

The first heap was fired in its normal state.

The second was loosely covered with gunpowder.

The third was covered with blocks of compressed gunpowder.

The fourth was covered with a mixture of coal-dust and gunpowder.

The fifth was covered with coal-dust alone.

The sixth was covered with cotton-wool saturated with benzoline; and

The seventh with tow saturated with paraffin oil.

In face, however, of the great difficulty in accurately observing the results of what took place, the following could only be recorded:

First.—A distinct flame.

Second.—The flame was not augmented in proportion to the quantity of gunpowder present, and had it not been for the great amount of fumes, the difference would hardly have been observed,—the greater part of the gunpowder being blown away apparently without being ignited.

Third.—The flame was the same as in experiment No. 1, so far as we could judge, and the impression was that the blocks of gunpowder were blown away without being ignited; and this was afterwards proved, as several broken pieces of these blocks of gunpowder were picked up.

Fourth and fifth.—There was no intense flame as would have been the result had the coal-dust been ignited. This was demonstrated over and over again, because when a mixture of coal-dust and gunpowder was fired the ignition always resulted in a long tongue of flame.

Sixth and seventh.—The saturated cotton-wool and tow were blown away without being ignited, and were afterwards picked up.

The above experiments were unsatisfactory to me for many reasons; in the first place, because it appeared that the inflammable and explosive substances surrounding the roburite were blown away by the expansion due to the explosion before the flame either of the detonator or the roburite could reach them, and with the object of proving this a heap of roburite connected by a train of roburite to a detonator was then fired, the result being that the detonator simply exploded the roburite immediately surrounding it, and the train for a distance of two feet was blown away, leaving the heap of roburite intact.

There is no doubt in my mind that if the roburite is fired in a barrel of gunpowder, stemmed tightly around, it would fire it, as was the case in the North of England; but as these experiments seem to me, in the face of the test in an inflammable mixture of gas and air, of little utility, seeing that gunpowder is an explosive and therefore contains the oxygen necessary for its own combustion and is consequently unaffected when cut off from the surrounding atmosphere or when surrounded by gases inimical to combustion, and seeing that the contention is that the ignition of roburite produces gases which render the atmosphere incapable of supporting combustion, I have therefore confined myself to experiments in an explosive mixture of gas and air, the result being that in a series of tests the roburite did not explode the surrounding mixture, but when the same mixture was tried with gunpowder a long tongue of flame was always visible.

Mr. SOUTHALL: You said in the second experiment you used a pound of powder; how much roburite did you use?

Mr. RHODES: I cannot tell you in grammes. Mr. Walker gave me a paper showing what quantity of roburite was supposed to represent a certain weight of gunpowder. I took his figures.

Mr. WALKER: This would apply to the blasting force.

Mr. SOUTHALL: Mr. Rhodes said he used a pound of powder, but did not say what quantity of roburite.

Mr. RHODES: We used the quantity of roburite supposed to be equal to a pound of powder.

The PRESIDENT: Not equal in quantity, but in explosive force?

Mr. RHODES: Yes.

Mr. WALKER: Mr. Rhodes says that the ordinary $4\frac{1}{4}$ in. cartridge was reduced. As there are four of these to a pound, there must have been about one-fifth of a pound of roburite in the charge which he says was more than equal to a pound of powder.

Mr. RHODES: That is so.

Mr. WALKER: Theoretically the equivalent force seems to be nine times that of powder, but practically it has four times the power of powder. In this case it would seem it was only one-fifth of a pound of roburite, which did more work than a pound of powder. In every case you failed to light the gas?

Mr. RHODES: Yes.

Mr. SOUTHALL: According to Mr. Rhodes' experiments, a pound of roburite was equal to five pounds of powder?

Mr. RHODES: Yes.

Mr. LONGBOTHAM: Have you made any comparison as to cost?

Mr. WALKER: We have been steadily working away with the roburite in the Rockley district of our Parkgate seam ever since the day Mr. Rhodes was with us, which will be five or six weeks ago. We have gone on steadily day by day. The men at first were a little prejudiced against it, and thought it would be a disadvantage to them, but they have gradually got over that, and now I believe I am right in saying they would rather stick to roburite than go back to powder. We find that it is very certain in action. We have never had a single missed shot except in two cases, due either to the fuse or to the detonator, because the roburite was found, when the coal was taken away, to be as good as ever, so that the missing of these shots was not due to anything in the roburite itself. The steady experience I have had of five or six weeks, makes me like the stuff better than ever. I believe it is a thoroughly practical thing, which everyone will like when they have tried it, and that it will make its way wherever powder is prohibited in the new Mines Act.

Mr. LONGBOTHAM: But as to comparison of cost?

Mr. WALKER: We have adopted the basis of three to one. Assuming that one-third by weight of roburite is used instead of a given quantity of powder, and assuming that powder is 5d. a pound, we charge 1s. 3d. for a pound of roburite, and in that charge we include the detonators. If you use six shots to the pound, and the detonator costs you (as it does at present) a half-penny,—though I believe shortly we shall have them at a farthing,—if you have six shots to the pound, you exactly cover yourself. If you use more than six detonators you lose a little, assuming the price of the stuff is 1s. a pound. We charge it 1s. 3d. as compared with 5d. for gunpowder, and then you are just where you were as regards cost.

Mr. LONGBOTHAM: Do you use the same size cartridge?

Mr. WALKER: We use 1⅜ in. cartridge in a 1½ in. hole.

Mr. LONGBOTHAM: What is the effect on the coal?

Mr. WALKER: It is my candid opinion that the coal we get down is absolutely better than when we used powder.

Mr. LONGBOTHAM: At any rate it is no worse?

Mr. WALKER: In my paper I give a comparison taken by Georgi. He says the proportion of small coal was 27·5 with powder, and 25 with roburite. That is closely in accordance with my own opinion that the roburite is slightly better in the matter of large coal than the powder.

Mr. LONGBOTHAM: Has there been any uncertainty observed which might be the result of imperfect admixture of the roburite?

Mr. WALKER: No; all ours would come in the same consignment, and no doubt was all mixed at the same time.

Mr. SOUTHALL: Is there any difficulty in keeping it?

Mr. RHODES: I may tell you what I have done. I soaked some of the roburite in a small quantity of water, which I thought would be sufficient for the roburite to take up, with the object of seeing if any liquid left was of an explosive character; my idea being that dampness of the atmosphere would have the same effect as soaking in water, only you have it quicker. The liquid left after soaking the roburite all night in a tumbler of water had no effect in the fire, but the roburite was prejudiced by being damped all night, so I should say it would be necessary to keep it in a dry place, same as gunpowder.

Mr. WALKER: It has this advantage. If after being damped you dry it again in a dry place it is as good as ever. We tried the experiment of putting roburite on a small plate and put cokes under it, and then applying a blast, heated the plate to redness, and all it did was to fizzle away, just as sulphur would.

Mr. LONGBOTHAM: Then roburite gives immunity from risk, produces the coal at the same cost or not more, and it does not make the percentage of round any less: so that you obtain absolute immunity from risk without any additional cost. Is that so?

Mr. WALKER: That is my opinion.

Mr. RHODES: I certainly think it will want experience as to the amount to put in the charges. It is so rapid in its action that a friable, disintegrated coal, unless proper precautions were taken, and a proper amount put in, might very probably be blown and broken up considerably. But I believe it is only a question of experience to ascertain the amount to be put in.

Mr. WALKER: I will illustrate that. Our fireman began by ramming the roburite tight, but after experience he found, by ramming the stemming much more lightly and not quite up to the charge, that the gases had a little elasticity in the hole and brought the coal down in better condition, and made the action a little slower than it would be if rammed up tight. So for hard stone you would put the stemming in as tight as possible, but for coal, and soft coal especially, you would relax the tightness of the stemming a bit.

Mr. WARD: I have had to do with the roburite ever since we commenced with the experiments at Wharncliffe Silkstone. I think we have had about 400 shots of it, and I have seen two-thirds of them. In the first two charges we used we brought the coal down small, but now we have got experience we bring it down larger than we did with powder. We have brought it down 12 feet long and 5 feet broad in the solid coal.

Mr. LONGBOTHAM: You have not had missed shots?

Mr. WARD: Two with the fuse. If a man is not careful he may pull out the fuse from the detonator. As to the roburite, I say the same as Mr. Walker; I believe the fault was that the cap had not gone off.

Mr. LONGBOTHAM: Have you tried it in a wet hole?

Mr. WARD: We tried one about five weeks ago, and fired it.

Mr. LONGBOTHAM: And it did its work?

Mr. WARD: Yes. I do not say that if you let it stay over night it might not get damp; but as soon as it was ready we let it off.

Mr. HEDLEY: Did Mr. Rhodes make more than one experiment with the roburite surrounded with powder?

Mr. RHODES: Five or six.

Mr. HEDLEY: And none exploded?

Mr. RHODES: No.

Mr. HEDLEY: Because it was tried at Lord Durham's collieries in a keg of powder, and the whole thing exploded.

Mr. WALKER: That was at Philadelphia.

Mr. LONGBOTHAM: Was there a large charge?

Mr. WALKER: It was the ordinary charge, but it was put in a heap of gunpowder, about 2 lbs.

Mr. RHODES: We fired a charge of roburite, with 18 inches train, with a pound of powder. The experiment was tried three times, and each time the powder was not fired, the train only being fired. Next, a charge of roburite was fired under a loose covering of gunpowder in an explosive mixture, which was not ignited. That was repeated three times. Then gunpowder and coal-dust were fired in the same mixture, and a loud explosion took place. On one occasion only had we, with the roburite, a flame which appeared to be the ignition of any surrounding atmosphere, and that particular explosion was not anything like the same magnitude as when we fired gunpowder.

Mr. HEDLEY: I believe Dr. Roth's explanation of the explosion at Philadelphia was that it was the friction consequent on the roburite explosion which caused the gunpowder to explode. The same thing should have occurred in your experiment.

Mr. RHODES: The powder was blown away in my case.

Mr. HINCHLIFFE: The keg prevented it being blown away in the Philadelphia case. A very unfair test I think that was.

Mr. RHODES: My experiments were all tried with loose powder.

Mr. WALKER: If the detonator were not properly put in it would explode the powder.

Mr. RHODES: If the detonator were at one side, or through the roburite, the detonator would fire the powder.

Mr. W. H. CHAMBERS: The flame would be from the detonators that you saw?

Mr. RHODES: Yes.

Mr. LONGBOTHAM: The detonator would fire the powder?

Mr. RHODES: Yes.

The PRESIDENT: It appears as far as the experiments have gone with roburite, they have been very successful. Don't you think it would be well if, with all new things such as this roburite is, those who are going to use it would report to the Institute all the phenomena they observe respecting it? I think that would either tend to confirm or to deny the safety of the use of roburite. It seems to be an unfortunate thing that this keg of gunpowder in Philadelphia was set fire to, but there might be causes there which will not be found in mines. It may have been, as I thought myself when I heard the account, that the detonator might have been too near the gunpowder; or, as it was said, the explosion resulted from friction when the powder had no means of escape such as it had in the experiments Mr. Rhodes has made.

Mr. W. HOOLE CHAMBERS: There is no doubt the safety of roburite is the great thing we wish to ascertain, and unless it is perfectly safe we shall be leaning on something which will not lead to what we require. As far as I have heard of the North of England experiments, the explosion may have been led to by one of two things. Either, and I should think the last is the most likely explanation, that the sudden charge of the roburite in the centre of the barrel caused friction at the sides which would be sufficient to ignite the powder and cause the explosion; or, in the other case, the detonator may, as is said, have been too near the outside of the roburite. I think the experiments of Mr. Rhodes are very important, and shew us distinctly the action roburite takes, because we have him describing, where the roburite was fired in the open air, that the flame was visible to the eye, no doubt arising from the detonator. Again roburite was described by Mr. Walker, by its chemical composition to have the means of damping that flame and preventing its communication to any gas or gunpowder outside, and when enclosing the detonator, although you have possible sight of the flame, it is not of sufficient violence to ignite the powder resting on the roburite. I think they are very interesting that way, for they show by the eye what action roburite has where flame exists. It is claimed by the inventor that the roburite extinguishes all flame inside it, and you have it there, shown by sight, how this effect is secured. The great importance of the invention is, as I have said before, its safety. If we can secure an explosive of this sort, which renders harmless one of our great enemies in the pit,—the coal-dust which is permeating the atmosphere in some mines of this district to a great extent,—we have a great advance, and one which, in dusty mines, will enable us to a great extent to cheapen the cost of getting coal, and to do a considerable amount of work necessary to be done in connection with ripping and other operations in dry and dusty mines. In many cases in this neighbourhood the firing of shots is done away with, and it would open out a great advantage if this explosive can carry out that which has been promised, and which it has, as far as we have seen, carried out in the past. I am sure the Institute will be glad to receive any information which those who are trying this can lay before it, and I shall be glad to propose that the discussion be adjourned in order that future information that comes to hand may be laid before the Institute, so that we can have the advantage of that in future.

Mr. WALKER: I just wish to say, in regard to what Mr. Chambers has said, that I think these powder tests are very unsatisfactory. The powder may be ignited, as it was at Philadelphia, from causes which we cannot exactly get hold of, and therefore might prejudice the character of the roburite, without any corresponding advantage. What we want to ascertain is, what is the action of roburite under the conditions in which we propose to use it, and this I have emphasised in my paper. It is no use relying on the hydro-chloric acid extinguishing the flame unless the charge is so fired that the hydro-chloric acid is bound to quench the flame. If you fire it in the open air and the whole thing is scattered, the oxygen of the surrounding atmosphere comes in, and if there be initial flame, as no doubt there is, the oxygen required to spread the flame is brought into the case, and new conditions arise. But if in the shot-hole the charge is able to work out its proper chemical conclusion, if the carbonic-acid and the hydro-chloric acid gas are able to envelope the flame and put it out, then the proper action of the roburite is fulfilled, and you have an explosive which is perfectly safe. If you go and try it under conditions opposed to every principle on which the inventor relies, I think you are not acting fairly towards him or towards the explosive; and it is against the interest of all who would gladly see the thing succeed. If it is a safe explosive it is a great gain to us all, and we therefore want to give the thing fairplay.

Mr. LAMPEN: I might suggest, this being an important matter, as an agent has been appointed in Wigan as general manager, we ask him to attend, and he would be able to throw some light on the subject.

Mr. RHODES: I think with all due respect, we have got a Mines Regulation Act to face which is very clear indeed on this subject of explosives. We shall have to be in a position to satisfy the Government that an explosive is absolutely safe before the restrictions which now envelope the use of gunpowder would be withdrawn in reference to it. Much as any information would be esteemed from a gentleman acting in the capacity of agent; still I think any information as to the safety of this explosive would be to a certain extent biassed, in his case; and I think it would be as well to confine ourselves to discussion upon actual experiments undertaken by members of this Institute.

Mr. LONGBOTHAM: It appears to be quite competent for any member of the Institute to ask the agent over.

Mr. JOSEPH MITCHELL: And, with the permission of the Chairman, to speak.

The PRESIDENT: I agree with what Mr. Rhodes has said. I do not see what we can learn from the agent to the Roburite Co. Cannot we learn a great deal more from the experiments Mr. Rhodes and Mr. Walker have made, and from experiments which will no doubt follow these, because, I imagine, the agent for the Roburite Co. may not know anything of the conditions under which these gentlemen have been using roburite. He may say it is not explosive, but that is merely saying so. If we have any future experiments brought before us, then that will convince all who are capable of being convinced that this roburite is an explosive that may suit all purposes of blasting in coal mines, whether they are dusty or not.

Mr. LAMPEN: I thought he might be able to throw some light on the explosion at Durham.

Mr. WALKER: Mr. Wither was not in the country when those experiments were made.

The PRESIDENT: Mr. Walker was there himself.

Mr. WALKER: Yes, I was there. Dr. Roth said before the experiment with the powder was tried, "I know it will not fire this powder." Mr. Rhodes has surprised me in having stated that he has fired it among powder with success. My own feeling is that it is not a legitimate test, and one that is very uncertain.

The PRESIDENT: It was under different conditions.

Mr. WALKER: The agent of the Nobel's Explosives Company told me they had fired the water-cartridge in powder with varying results, so it may arise from varied and different causes.

Mr. WASHINGTON seconded Mr. Chambers' proposition, and the discussion was adjourned.

MR. G. BLAKE WALKER'S PAPER ON SPECIAL RULES UNDER THE NEW MINES ACT.

The PRESIDENT: The next subject is the adjourned discussion on Mr. Walker's paper on Special Rules under the new Mines Act. Would it not be advisable Mr. Walker, with your consent, to defer any discussion upon this, because you see the Committees are all sitting for the alteration of the rules?

Mr. WALKER: The best thing would be to adjourn it until the new rules are before us.

The PRESIDENT: I propose we adjourn the discussion of this subject until the new rules are drafted and issued.

Mr. WALKER: I will second it.

The motion was carried.

A NEW SAFETY CAGE.

Mr. R. MILLER exhibited to the meeting a model of a new Safety Cage.

The meeting then ended.

MIDLAND INSTITUTE OF MINING, CIVIL, AND MECHANICAL ENGINEERS.

GENERAL MEETING.

HELD AT THE QUEEN'S HOTEL, LEEDS, ON WEDNESDAY, FEBRUARY 21ST, 1887.

T. W. EMBLETON, Esq., President, in the Chair.

The minutes of the last meeting were read and confirmed.

The following gentlemen were elected members of the Institute, having been previously nominated :—

Mr. JONATHAN WROE, Colliery Manager, Wharncliffe Silkstone Colliery, Barnsley.

Mr. JAMES WROE, Manager, Lidgett Colliery, Tankersley, Barnsley.

FAN EXPERIMENTS.

The PRESIDENT: Gentlemen, you remember some time ago a Committee was appointed for the purpose of testing fans. Since then the Secretary has received from the North of England Institute of Mining and Mechanical Engineers certain questions which the Committe have to investigate. What I would suggest is that these questions be forwarded to each member of the Committee, in order that they may commence with their work. I think it is useless waiting so long for any reply from the South Wales Institute, because that reply was expected four or five months ago and has not come yet. The Committee have made some experiments, but I do not know what the result is. However, I think it will be necessary that all the experiments should be conducted and all replies given to these questions by our Committee as far as they can, so that the experiments made may be uniform with those of the North of England Institute, and then we shall be able to compare results. There are a great many questions; it will take a long time to answer the questions, and I think it is necessary the Committee should begin in time.

Mr. GERRARD: You would refer it to that Committee, would you?

The President: Yes, as you see there are some instruments required. I imagine these instruments should be provided at the cost of the Institute, and after they have been used and the questions answered, they should be handed back to the Institute.

The President's suggestion was adopted.

The President: The next business is an addendum to the paper on "Hydro-Carbon Explosives," by Mr. G. Blake Walker.

ADDENDUM TO PAPER ON HYDRO-CARBON EXPLOSIVES.

By GEORGE BLAKE WALKER, F.G.S.

The paper which I read before the Institute on the 15th of November last, was almost entirely confined to a description of *Roburite*, but in order to complete the subject as far as possible it was my intention to obtain information respecting other flameless explosives, and bring this information before the Institute as soon as it was available.

On page 107 of Volume XI. will be found an allusion to *Securite*, when I remarked that I had been unable to obtain any reliable information respecting the qualities of this explosive. The only information I then possessed was that the late Mr. S. B. Coxon had visited Saarbrück for the purpose of testing *Securite*, and my information was that, though approaching a standard of safety, it still ignited an explosive mixture of gas under certain conditions. The inventor, Herr Schöneweg, was understood to suggest an alteration in the composition of his compound by saturating it in oxalic acid, but what the result of the modification was I was unable to learn. Herr Baurath von Eilert, the Chief Mining Engineer of the Prussian Government, told Mr. Bainbridge and myself that at the trials he had made with *Securite*, the results had been satisfactory.

I understand that patents have been granted for the protection of *Securite* in this country but have been refused in Germany, as will appear from the following translation from "The Berliner Börsen Courier," No. 231, May 9th, 1887 :—

"Among the many-named explosives of these latter days is one, the "so-called *Securite*, for which its 'inventor,' the chemist Schöneweg, "Dudweiler, claims the highest qualities and absolute novelty. Now

" it happens that a few weeks ago it came to our knowledge from
" some of the best informed scientific quarters, that *Securite* is
" nothing else than a crib from H. Sprengel's scientific treatise
" published in 1873. [*On a New Class of Explosives which are not
" explosive during their manufacture*, &c., dated London, 1873.]

" Complete accounts containing the substance of the invention of
" Sprengel were brought out by the following German scientific
" papers—*Dingler's Polytechnic Journal* (see vol. 212, page 323)
" and quite in detail by *Wagner's Annual Report* for 1874 (see
" 434).

" The decision of the German Patent Office (to which Herr
" Schöneweg applied for a patent) has been that *Securite* is disqualified
" for a patent, must therefore be considered a gratifying acknowledge-
" ment of Sprengel's merit and priority.

" This proves clearly the immense value of the searching
" preliminary tests applied in Germany as to the novelty of an
" invention; as on the other hand the annexing of intellectual
" property is in this way limited as far as possible; while on the
" other hand German capital is assuredly better protected against
" tangible loss than is the case in England and France, where the
" preliminary tests of the novelty and originality of an invention are
" nothing more than a matter of chance; and the capitalist, who in
" most cases is a 'layman,' has no guarantee that the invention in
" which he invests his money will not at the first breath of wind fall
" to pieces like a house of cards. The German patent is, therefore,
" always the surest voucher for the international value of an
" invention. Moreover what an inexhaustible spring Sprengel opened
" for gain-thirsty souls in his above mentioned treatise of the year 1873
" is proved by the circumstance that this very thing *Securite*, which Herr
" Schöneweg has patented in England (where, as we have said, there
" are no preliminary tests in the sense in which they exist in Germany),
" this self same thing was patented by the Swede, Lamm, under the
" name of *Bellite*.

" There will probably be no one who has any practical interest in
" deciding by legal process in the latter country which of the two
" inventors was the first to copy his ' invention ' from Sprengel."

On the 16th of January, a Company was brought out in London
under the name of " The Flameless Explosives Company, Limited,"

with Lord Bury as chairman. Their prospectus stated that "The "researches of Herr Hermann Schöneweg, an eminent scientist of "Dudweiler, in Germany, have been rewarded by the discovery of a "process which renders absolutely flameless, dynamite, blasting "gelatine, and other dangerous explosives, thus securing the pro- "tection of life and property in coal mines from explosions caused "while blasting." They further state that "the rending power of "*Securite* is about four times greater than that of blasting gunpowder, "though its violent action can be modified in manufacture to any "required extent. That whereas the fumes of dynamite, blasting "gelatine, gun-cotton, and gunpowder are unhealthy and dangerous, "frequently leading to loss of life, the gasses produced by the "explosion of *Securite* are comparatively harmless, so that men can "work in close headings and other enclosed spaces immediately after "a shot has been fired."

These being the statements put forth in the prospectus, I was in hopes of being able at this meeting to lay before the Institute some particulars of the composition and action of this explosive, based, if possible, on actual experiments. The Company appear, however, to be exceedingly suspicious of my *bona-fides*, and although they have promised to furnish particulars, and to give facilities for experiments, I have not yet succeeded in getting anything but promises from them. Some experiments are to take place at Glasgow shortly under the auspices of the Company, and I shall endeavour to obtain full particulars of these, and to lay them before the Institute at a future meeting. In the meantime perhaps some other members of the Institute may be able to obtain a sample of *Securite* for trial, and report the results to the Institute.

The following report by Herr Margraf, formerly a Prussian Inspector of Mines, pending experiments in this country, will no doubt be read with interest.

THE FOLLOWING IS A TRANSLATION OF THE REPORT OF HERR MARGRAF, MEMBER OF THE PRUSSIAN FIRE-DAMP COMMISSION, UPON HIS EXPERIMENTS WITH "SECURITE."

Securite which has been discovered by Mr. Schöneweg, chemist, at Dudweiler, is a granulated powder of a light yellow colour, with an odour of bitter almonds. Brought into contact with fire, it burns with a yellow flame, slowly and without exploding, melting at the same time, and causing but little smoke. The flame is extinguished the moment the contact with fire ceases. In contact with red-hot iron, *Securite* melts, emitting smoke, while bodies of a higher temperature produce the same effects as fire. It cannot be ignited by any shock or blow, to which it might be subject during transport or manipulation; for thin layers of the size of a five-mark piece spread out upon a cast-iron anvil could no*t* be made to explode, although violently struck with heavy hammers.

Securite can only be exploded with strong caps of 1 gramme fulminate of mercury. The action is similar to that of *Helhoffit*.

It has been proved by numerous tests that *Securite* is absolutely safe in the presence of fire-damp and coal dust. Covered with the most dangerous coal dust and surrounded with 10 per cent. of fire-damp, the cartridges on explosion never showed any appearance of flame, and no fire-damp explosion took place even when, instead of coal dust, the cartridges were saturated with petroleum ether. Experiments were made to ascertain how *Securite* would act in workings, if brought into direct contract with a pocket of pure fire-damp. In these the cartridge was covered with the most dangerous description of coal-dust, saturated with petroleum ether, in the presence of 10 per cent. of fire-damp. The experimental tube was then entirely covered with coal dust, and 10 per cent. of fire-damp was let into the test drive. When exploded, not a vestige of flame could be observed. [*Here follows a drawing and exact description of the experimental tube.*]

Fourteen times this experiment was repeated without shewing any different result. Later on, petroleum ether was substituted for coal dust, but here too without an appearance of flame.

These experiments prove conclusively that *Securite* is absolutely fire-damp proof, and it has also stood the test in collieries in the presence of strong fire-damp mixtures. A further experiment was made in an advanced heading in the Carlowitz Seam, in the third level of the Mine König, which gives out a large quantity of fire-damp, but which is so well ventilated by means of compressed air, that no fire-damp mixtures can be found, and dynamite shots can be fired by electricity. Although dynamite invariably ignited the fire-damp in the fissures near the explosion, no fire-damp was exploded in five drill holes fired with *Securite*. The ventilating current was afterwards stopped, and four *Securite* shots were fired near the roof into the fire-damp mixture, when no ignition followed: with dynamite a very serious explosion would have resulted.

The fall of large-piece coal resulting from the employment of *Securite* is in no way inferior to that obtained with blasting power, especially in the hard seams in the Saar district. The explosive gases are in no way injurious, and their volume is extremely small.

In wet holes "Securite" is used in waterproof cartridges.

[*Here follows a comparison between Carbonite and "Securite," much to the advantage of the latter.*]

Neunkirchen, *September*, 1886.

To this is added:

The experiments alluded to in the foregoing Report were made by me in the presence of Mr. Schöneweg during the spring and summer of 1886, and I herewith testify to the correctness of the stated results.

(*Signed*) MARGRAF,
Inspector of Royal Mines, A.D.

Grube Hostenbach, 18*th* November, 1887.

TRANSLATION OF REPORT MADE TO HERR SCHÖNEWEG BY HERR LOHMANN, MINING ASSESSOR, NEUNKIRCHEN.

I beg to inform you that on the 28th ult., the following experiments were made according to your directions:

A gelatine dynamite cartridge (65 per cent. nitroglycerine), weighing 50 grammes, was reduced to a thickness of 10 m/m, and furnished with a Bronkard cap of 1 gramme fulminate and placed in a paper shell of 30 m/m. diameter, surrounded by your preparation. Two such cartridges were then fastened together and placed openly upon a wooded support at middle height within the fire-damp testdrive, in the presence of 10 per cent. fire-damp (10 per cent. CH_4), and free coal dust, and fired by electricity without causing an explosion. A counter experiment with a cartridge of the same gelatine dynamite of 160 grammes caused an explosion in the presence of free coal dust only, without the presence of fire-damp.

According to our experience here we should have obtained the same satisfactory results by employing dynamite surrounded with your preparation.

With regard to further tests, I should recommend you, as I have already told you personally, to mix the preparation with the explosive.

With highest esteem and best regards,
Yours sincerely,
(*Signed*) LOHMANN,
Neunkirchen, *3rd October*, 1887.   *Mining Assessor.*

If it be true that Herr Schöneweg mixes with explosives oxalic acid, or surrounds them with it, he must have been repulsed by the German Patent Office on the ground that his invention was wanting in novelty, and could not, therefore, be considered as eligible for patenting under the German Government.

In regard to the thing itself, however, I may simply remark that oxalic acid is a very expensive ingredient, the price being 1s. 6d. per kilo. On combustion (that is explosion) numerous and well known and most poisonous oxides are given off, in accordance with the formula :—

$$\left. \begin{array}{c} C_6\ OOH \\ C^6\ OOH \end{array} \right\} CO_2 + CO + H_2 O$$

The combustible gases of the oxalic acid contain, as I above calculated, 31 % of poisonous carbonic oxide.

Securite, I believe, consists of di-nitro-benzol and an admixture of various kinds of saltpetre, precisely similar to those mentioned in Sprengel's publication of 1873, but whether it is entitled to be considered a new invention or not is of less consequence than its actual value as a safety explosive in mines. I therefore much regret that I have been unsuccessful in learning more about its practical qualities.

I understand that Herr Margraf has not for the last eighteen months held the Government appointment of Assessor of Mines at Saarbrück. For the past year and a half he has been manager of a private mine, and no longer holds a Government appointment.

The accompanying photographs of the flame produced by dynamite and gelatine dynamite, protected in varying degrees by Schöeneweg's preparation, will convey graphically to the eye the results claimed to have been achieved by the inventor.

The PRESIDENT: Does any member wish to make any observations upon this paper and the addendum to it? It appears to me that a contest has arisen between the manufacturers of these explosives, for instance, of *Roburite* and *Securite*. I think any decision by us should be based entirely upon experiments made by our members, just in the same way that the details of the experiments made by Mr. Walker and Mr. C. Rhodes with *Roburite* were placed before us. Then I think we shall be able to judge more precisely which explosive it would be best to employ in blasting the coal or stone in the mine. We may be quite sure of this, that if any person has an invention, he always puts it forth to the world as the best in the world; we should not trust to any reports by the makers of these explosives, but should make our own experiments. I see it indicated by Mr. Walker that he will endeavour to obtain some of the explosive that he writes about, and he will be able, no doubt, to report to the Institute what are the results obtained from the two explosives—*Roburite* and *Securite*—and we shall be able to judge exactly which is the best that should be used. It is an important matter that this should be solved by the Institute, because under the last Act there will be great difficulty in using gunpowder. Now seeing that so much has been attributed to the effect of coal dust when an explosion happens, I think the experiments to be made should be also tested in an atmosphere filled with coal dust. If you agree with me in this, would it not be better to request Mr. Walker, from this meeting, to make experiments with this new explosive, and give us a report of his experiments. It will be necessary also to ask him to commence further experiments with *Roburite*.

Mr. J. GERRARD: I think this we ought to do. We ought to convey our thanks to Mr. Walker for the further contribution that he has made on this very important subject, and ask him to continue the experiments that he is making, and to favour us with the results in a similar manner to the paper we have heard this afternoon. We are all agreed as to the importance of the subject, and ought to be grateful to Mr. Walker for the close attention that he appears to be paying to it. Probably there are other members who have made experiments. If they would give us the result of their experiments in like manner, we might get more information. *Roburite* is in itself such a much more simple explosive than the water cartridge, which does not commend itself to all of us in some particulars, that I think it would be a great advantage to the Institute if we could follow this up, and get similar practical information to that which we have had presented to us by Mr. Walker. I beg to propose that we give our best thanks to Mr. Walker for the further paper that he has sent us, and request him to keep us furnished with any further results that he may obtain.

Mr. W. HOOLE CHAMBERS: We have been very anxious to try some experiments with this *Roburite*, and ever since last meeting we have been doing what we could to secure a supply, but there seems to be some difficulty in its way. I believe it arises principally from the dangerous nature of the explosive and some misunderstanding they have with the Custom-House authorities. At any rate we have not been able to obtain any, and therefore are not able to give you particulars of experiments, as we had hoped to have done. The advantages, if they can be secured to collieries of a similar description to our own, are so great, that we should be glad to meet with something which would answer the promises these inventors give to us.

Mr. LAMPEN: As I am interested in the matter of *Roburite*, I think you will be able to get the *Roburite* in a short time from now. There was a consignment on the Thames, but the cartridges are used abroad differently to what they are in England, and the shareholders, and also the inventor, thought it would be better, before the cartridges were sent out, that they should be right, because if they were sent out to the different collieries in that form, it might damage the shareholders. Personally I think it would be well to wait a little longer; it is going to be manufactured at Wigan, and will shortly be on the market. I think I have almost the authority of the Company to say what I have said.

Mr. W. HOOLE CHAMBERS: I have great pleasure in seconding the resolution proposed by Mr. Gerrard.

The motion was carried.

Mr. J. GERRARD: Would it come under the same head if we asked other gentlemen to give us practical information on the use of other explosives—the water cartridge and others.

The PRESIDENT: I think it would not at present. The next question is, shall this paper, diagram and report from Germany, be part of our Transactions?

Mr. W. HOOLE CHAMBERS: I have great pleasure in proposing they be printed. I think the results will be useful for reference.

The PRESIDENT: There is only one thing to remark. These papers on these subjects are information for the members, and are not advertisements for *Roburite* or *Securite*. We do not regard the shareholders whether they make money or not; what we have to look to is the things they have to sell, whether they are any use to use or not.

Mr. MADDISON: I will second the motion.

The motion was carried.

MR. G. BLAKE WALKER'S PAPER ON SPECIAL RULES UNDER THE NEW MINES ACT.

The PRESIDENT: The next business is the adjourned discussion on Mr. G. Blake Walker's paper on "Special Rules under the new Mines Act." Our Secretary wrote to me respecting this, and I told him that I thought we could not discuss Mr. Walker's paper, seeing that the new rules which have been compiled by the joint Committee had not come into our hands. Mr. Walker's paper was printed before the Special Rules were drawn up. I think any discussion upon this paper will be perfectly useless at the present time; in fact I asked the Secretary not to insert this notice. However, he has done so, and for this reason—that he expected that the Special Rules would be in the hands of all the members before this time. One of the Secretaries told me the other day the rules were sent to the printer, and as soon as they were printed they would be distributed. The only discussion could be upon the rules which have been drawn up, and it would be useless entering into a discussion now about a thing of which we know nothing at all. If you think otherwise we shall be glad to hear any discussion.

Mr. W. HOOLE CHAMBERS: When Mr. Walker's paper came forward, I believe the discussion was adjourned until the Special Rules which were to be drawn up by the Committee appointed were in the hands of members. I do not think we can discuss it before we see what they are. I shall be glad to propose the discussion be further adjourned until the members have an opportunity of seeing what these rules are.

Mr. HALL: I will second that. I think it would be quite out of place to discuss this at the present time.

The discussion was adjourned accordingly.

A lamp sent by Mr. Ashworth of Derby, "Ashworth's Patent Hepplewhite Gray Deputy's Safety Lamp," was exhibited, and examined by the members present.

Mr. GREAVES proposed a vote of thanks to the President.

Mr. TEALE seconded the proposition, which was carried.

The PRESIDENT: I am much obliged to you. I am sorry we have had so little to do; I wish I had a great deal more.

BON EXPLOSIV

n of Dynam
s of Schoene

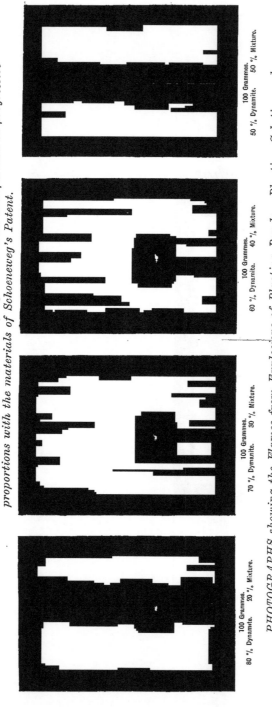

MIDLAND INSTITUTE OF MINING, CIVIL, AND
MECHANICAL ENGINEERS.

GENERAL MEETING.

HELD AT THE VICTORIA HOTEL, SHEFFIELD, ON TUESDAY, MARCH 27TH, 1888.

T. W. EMBLETON, Esq., President, in the Chair.

The minutes of the last meeting were read and confirmed. The following paper was then read:—

FOREIGN MINING RENTS AND ROYALTIES.

By H. B. NASH.

THE following Notes are extracted from the Reports by Her Majesty's Representatives abroad, being replies to a Circular Despatch sent from the Foreign Office on the 27th September, 1886, in accordance with a promise made by Lord Iddesleigh, the then Secretary of State for Foreign Affairs, in answer to a question by Mr. Philip Stanhope. These Reports were presented to both Houses of Parliament by command of Her Majesty in April, 1887.

I have made these Notes as brief as possible, so as to include only these portions relating chiefly to Coal and Ironstone Mines in the countries mentioned, thinking they would be of more general interest to members of this Institute than a long account of the laws relating to the Metaliferous Mines, which are not much worked in this district.

FRANCE.

Payment of Royalties was abolished in 1789. A law passed in that year declared all mines to be the property of the nation; the owner of land might, however, work a mine to the depth of 100 feet. In 1810 the law of 1789 was repealed, but all its chief provisions were re-enacted except the right to open shallow mines without leave of the State. That law remains unaltered.

The concession of a mine by the State, even to the owner of the land, creates a new property, independent of, and separate from, the surface, and transmissable like other property. An act of concession confers a right of perpetuity in the mine and its belongings, and gives thereunto the quality of real property.

Concessions are regulated by laws administered by a "Consul-Général des Mines," under whose authority the Engineers and Inspectors conduct their inspections. No mine can be worked without a concession, and no concession is granted except by the authority of the Council of State. The concession being obtained, the concessionaire has to pay (a) A yearly rent to the owner of the surface, seldom exceeding 10 per cent. per hectare on the whole extent of the concession ; (b) A yearly rent paid to the same owner for the surface area occupied by machinery and plant of the mine. Besides this (c) A yearly rent of 10 francs per square kilometre is paid to the State ; (d) and also a fixed annual rent of 5 per cent. on the nett produce of the mine, and (e) 10 cents per franc in addition to form a relief fund for accidents.

The total revenue from these sources to the State does not usually exceed £140,000 per annum, but it must be remembered that the area of the French coal-fields is very limited.

No person can try for minerals without previous arrangement with the owner of the land as to the indemnity to be paid.

Large coal-fields exist in France, especially in the North, in which no rent is paid to the owner of the surface: Anzin, and I believe Aniche, on the Belgian frontier, are coal-fields under these conditions. From a report written by M. Clémenceau in 1884 for presentation to the French Parliament, it appears that the coal-field of Anzin became the property of the company which now owns it in 1810. From that date the Royalty paid to the State from the net yield of the pits of Anzin, Denain, Raismes, St. Saulve, Fresnes, View Condés, Odomez, and Hasnon, forming the Anzin Colliery, appears to have been at the rate of 90,000 francs a year. In 1874 the total yield of the collieries in France was 1,992,202 tons, the Royalty paid was 208,000 francs. ; in 1883, the yield rose to 2,210,702 tons, the Royalty fell to 73,312 francs.

According to the calculation of a practical engineer the rents habitually paid in the basin of the Loire in 1863 varied according to depth between 50 c. and 70 c. per ton, which is not much less than the charge of 6d. to 10d. per ton common as Royalty in England. It is to be borne in mind that no mining rent can be charged unless it be registered in the act of concession, and the registry in that document is subject in every case to the control of the Government when granting the concession.

GERMANY.

No person may dig for minerals without a Mining Licence; any infringement of this law (even by the owners of the land) is punishable by fine or imprisonment.

A claim to a licence belongs to the person who discovered the mine in its natural state, and who thereupon makes a proper application for the concession for working the mine, in conformity with the law.

The general effect of the law is a charge of 2 per cent. on the value of the produce of the mine. The law does not recognise any right of the owner of the soil to any claim in the shape of Royalty or rent, or compensation for the minerals under his land; but the person working the mine must pay full compensation for any damage done to the land and appurtenances thereof in working the mine. Every mine-owner must pay a certain amount to the Miners' Benefit Fund, to ensure assistance in case of accident or illness, and to provide for the widows and children.

In certain special cases under the old mining laws, private owners have rights of Royalty or compensation, which still remain in force.

The taxation of mines varies in different provinces in Germany thus: (a) Some, as in Prussia, pay a tax on the gross produce, but the scale varies in different places; (b) Others pay a tax on the allotment, at so much per hectare; (c) In Alsace-Lorraine, both forms of payment are combined; (d) In Wurtemberg a general tax is imposed on mining.

The obligation to pay rents or royalties to private owners of land is, generally speaking, exceptional.

AUSTRIA—HUNGARY.

All minerals containing ores are reserved to the State, and no mining can take place except by concession in accordance with the stipulations of the mining laws. A "searching permit" for a certain district is granted for a year, and application for its renewal must be made before the expiration of that period.

When a Concession is granted, an arrangement must be made with the owner of the land, as to compensation or purchase, the value of which may be fixed by sworn experts appointed by the mining authorities, the proprietor being forced to sell his property at the price thus fixed, or at an annual rent fixed in the same manner, in which case the land reverts to the owner, after the working of the mines has ceased.

The taxes to be paid to the State are : (a) For each free digging of 424 metres diameter (no right of extracting the ore being granted), an annual tax of four florins ; (b) For the actual working of each mine of 12,545 square klafters,—one square klafter equals 3,597 square metres,—and for surface minings of 22,000 square klafters also four florins per annum ; (c) An income tax of 5 per cent. on the net income, with an addition of from 60 to 100 per cent. on the tax paid to the State as provincial and municipal taxes.

The area to be worked is limited, as regards coal to eight parallelograms, each of 12,544 square klafters, and other minerals to one half that area. Larger groups are not permitted.

ITALY.

There is no uniform law in Italy, the legislation in force in the various States previous to annexation being still maintained ; hence there is great diversity, and even the most contradictory systems are to be found in operation side by side. In the greater part of Italy, viz., in all Upper and Central Italy with the exception of the Tuscan provinces, and in Sardinia, concessions are granted by the Government in favour of the discoverer of the mines, or any person who may furnish a sufficient guarantee of his intention to undertake its working.

In the Tuscan provinces mines are the absolute property of the owners of the soil. The iron mines of the Island of Elba and of the territory of Piombino are, however, excepted. These belong to the State domain, and the State is at liberty to concede the mines to any person and upon any terms.

The following brief analysis of the provisions of the several laws and systems of legislation at present in force, is taken from the Official Report on Mineral Legislation, published in 1883. The Sardo-Lombardo Law, of 20th November, 1859, divides mines into two classes : (a) Metalliferous ores, sulphur and sulphates, coal and bitumen; (b) Peat works, quarries, and stone. Mines can only be worked by a concession from the State; a mine thus becomes a new and permanent property, which can be disposed of or transferred like any other property. Permits for exploration are granted free of charge. Concessions are subject to a fixed tax of 50 per cent. per hectare of area granted, and to the payment of income tax on the produce of the mine.

The discoverer of a mine has the preference in the granting of a concession; if he cannot carry on the works, he is entitled to compensation. Injury to other mines entails liability for damage. All operations are subject to legislative control.

The Laws of Lucca, of 3rd May, 1847, are similar to the above except in matters of procedure. The regulations respecting damage and compensation, compulsory association, and the declaration of works of ventilation or drainage are wanting.

The Laws of Napoleon, of 9th August, 1808, also distinguish mines properly so called from quarries. The provision relating to mines are analogous to those in force in Sardo-Lombard with trifling exceptions. The concession, however, is limited to 50 years, with power of renewal.

Generally, in the remaining provinces there is a fixed tax of 50 cents per hectre on the area of concession, and formerly a tax of 5 per cent. on the net produce of the mine, now converted into an income tax of 13·20 per cent. on the output of the mine as upon all other property. In some of the provinces there is a Mining Fund for accidents.

Various attempts have been made in the Italian Parliament to introduce a uniform mining law for the whole kingdom, sometimes by assimilating the Tuscan law to those in force in other provinces, and *vice versâ*, but so far without success.

BELGIUM.

By Article 552 of the Code Civil, the mineral under the soil belonged to the owner as in England, but in 1810 the law was so modified that no mining operations could be carried on without a direct concession by the Government. The concession conferred complete ownership, like other property; the machinery and plant are regarded as real property, the shares in mines and the products thereof as personal property. The mine, however, cannot be sold or divided, except by previous consent of the Government.

The owner of the soil can search for minerals, or give leave to search, but no mine can be worked without the consent of the Government. In exceptional cases Government may grant a special authority to search, without the consent of the owner, indemnity being paid for damage done to the land, but in no case can any one except the owner carry out such explorations within a hundred yards of any house or garden.

The rents and royalties payable to the State are: (a) A fixed yearly rent of 10 francs per each square kilometre of each concession; and (b) A tax not exceeding 5 per cent. per annum on the net profit.

In 1837 a new law was passed, granting some further powers to the owner of the land. This law regulates the rights of the landowners in mines, and by this law the landowners receives, like the State, a double royalty: viz., (a) An annual rent of 25 centimes per hectre of superficial area, and from 1 to 3 per cent. on the net produce of the mine.

It was also enacted by this law that the proprietor of a sufficient extent of land, and who could at the same time show that he possessed the necessary capital, was entitled to a concession if he thought fit to work the mine.

SPAIN.

Any person or company may own and work mining property with the consent of the State subject to certain conditions. All substances used generally in building operations belong to the owner of the soil. Any person can, under certain conditions, examine the surface of the land in search of minerals, and if the owner of the land refuses his consent, application may be made to the Civil Governor of the province, who may grant such permission.

A mine is reckoned to be a rectangle measuring 300 by 200 metres. and of indefinite depth. The surface remains the property of the owner of the land.

Mines of iron, coal, anthracite, lignite, carboniferous shale, salt, and some other mines of a like character, measure 500 by 300 metres.

In order to obtain the ownership of a mine, a petition must be addressed to the Civil Governor of the province, who instructs a Government Engineer to examine and report if any minerals are found to exist. If the Engineer reports there are minerals, a concession is granted. The owner has the right to have the area of his property marked out within four months after the concession.

The royalties payable to the Government are: for mines measuring 300 by 200 metres, 75 pesetas; mines measuring 500 by 330 metres, 50 pesetas, and for surface rubbish deposits, 100 pesetas for every 40,000 square metres. The earlier mining laws of Spain date from 1584; they were re-modelled in 1825, and amended in 1849 and 1852, and lastly in 1868.*

UNITED STATES.

The right of private ownership in minerals extracted from the soil is recognised in all the States and territories of the United States; all such minerals belong to the owner of the land.

All lands not sold or otherwise disposed of, are held by the Government of the United States, but when sold all rights pass without reserve to the purchaser, inclusive of the minerals, but whenever any special donation of land is made by the Federal Government to a State, it is stipulated that should any portion of such land be found to contain minerals, other lands of an agricultural value shall be selected by such State in lieu thereof.

The revised Mining Laws, passed in 1872, declares that all valuable mineral deposits in lands belonging to the Government of the United States, both surveyed and unsurveyed, are free and open to exploration and purchase by the citizens, and by foreigners who have declared their intention to become naturalized, under the regulations prescribed by law, and according to the local customs or rules of the miners in the several mining districts, so far as applicable, and not inconsistent with the laws of the United States.

Mineral lands cannot be taken up for agricultural purposes, but if in course of time minerals should be found, the Courts, as a rule, have held that the purchaser has acquired absolute title to such, and that the Government had parted with all interest therein, unless fraud or misrepresentation with regard thereto is proved to have been committed.

As regards aliens, no mining claim can be located except by United States citizens, or those who have declared their intention

---

* The figures given as the Royalties payable in Spain are those appearing in the Report, but it seems rather strange to the writer of these Notes, that the rent for the smallest area of ground should be double that of the largest area.

of becoming naturalized, but it is held that a claim jointly by a citizen and an alien conveys a perfect patent. Proofs of citizenship are required, and also a charter or certificate of incorporation in case of a company or partnership.

Owners of mines in the several mining districts make their own rules and regulations, provided that they accord with the laws of the Federal Union, and with the laws of the State or territory in which the mines are situate. In all cases the location of a mine must be distinctly marked on the ground, so that its boundaries may be readily traced. All records of mining claims must contain the name of the locator or locators, the date when made, and a description of the claim, by reference to some natural object or permanent monument which can be identified.

On all claims lodged prior to 10th May, 1872, ten dollars' worth of labour must have been expended prior to 10th June, 1874; after that date 100 dollars' worth of labour must have been expended on improvements made in each year until a patent has been issued.

Taxation on mineral lands, like that on all other property, is levied on their valuation. Corporations chartered for mining purposes are taxed on the nominal value of the shares issued. In some cases certain mines are exempt from taxation during the course of development.

The State of New York hold, in right of sovereignty, all mines of gold and silver, and such as contain proportions of the precious metals, the working of all such mines being regulated by the provisions the revised statutes of New York.

It will be noticed that in France, Germany, and part of Italy, it is compulsory upon the mine-owners to pay a certain amount annually, to form a relief fund in case of accidents, but not in Belgium.

In concluding this paper I do not propose to draw any conclusions or comparisions between the laws in force in the several countries herein mentioned; but hope that a thoughtful reading of this paper may lead to an interesting discussion on a subject which is of the greatest importance to the mining industry of this kingdom.

Foreign Term.		English Equivalents.
1 Hectare	equals	107645·83 Square Feet.
1 Square Klafter	,,	88 723 ,,
25 Pesetas	,,	One Sovereign.

The PRESIDENT: Has any member an observation to make on this paper? It appears to me that the tables are quite unnecessary, merely the value of foreign coins, which you can find in any almanack.

Mr. G. J. KELL: Are they lengthy?

Mr. J. HINCHLIFFE: In some cases they will not be much better off than we are, with paying five per cent. income tax. Some of the rents and reservations seem to get pretty nearly as high as we are.

The PRESIDENT: Yes. I think we should be better able to discuss the paper when it is printed; if anyone will move that it be printed.

Mr. G. B. WALKER: I have pleasure in proposing a vote of thanks to Mr. Nash for his paper. It is a very interesting one, and I beg to move it be printed and included in the Transactions of the Institute.

Mr. KELL: I shall be very glad to second that.

Mr. E. BAINBRIDGE: It will be a very good thing to anglicise it as far as you can. It will be a very useful paper to consider in connection with the address published by Mr. Percy lately—a very interesting address on the subject of English royalties. It would be interesting to know how these compare with the charges abroad.

The motion was carried.

DISCUSSION ON MR. G. B. WALKER'S PAPERS ON "HYDRO-CARBON EXPLOSIVES."

The PRESIDENT: The next business is the adjourned discussion on Mr. G. Blake Walker's papers on "Hydro-carbon Explosives." Mr. Chambers intimated that Mr. Bainbridge would be able to give us some further information.

Mr. BAINBRIDGE: I came up to-day through seeing the subject of the papers for discussion, and being especially interested in the matter. I have not looked so closely into the theory of explosives as Mr. Walker, and should have liked to have had more time to look through his papers, as I conducted the experiments abroad. I had been professionally engaged to report as to the value of Roburite. Mr. Walker accompanied me; we did the work together, and the results of the experiments were very satisfactory. But I had not an opportunity of trying an experiment which, as far as I know, has not been properly tried since, and that is the experiment of testing Roburite in a perfectly or intensely explosive atmosphere made up of atmospheric air and coal gas from the mine itself. The only place I have heard of where that could be done is in Staffordshire, where a drift exists in the marl, and where, by means of gas taken from a coal mine, they are able to get a mixture of that kind. Whether that may make any difference in the results which are recorded here I do not know. Mr. Walker tried some experiments at Rockingham with gas and air, and we exploded it once or twice to make sure that the explosive mixture was just what it should be. There are two explosives before us—Securite and Roburite. They have both been imported to this country and the latter is on the point of being manufactured. But although they came before the public nine months ago neither the one nor the other is yet available for practical use. I think it would be a good thing to have this discussion adjourned, as this has been the first Institute to take up the question of the so-called flameless explosives, until the time when there has been an opportunity of a number of collieries testing them carefully. There are two points to consider. There is the question of their being flameless. I think that was proved very well by my experiments in Germany. It has been said that in Lancashire they have seen flame from Roburite, and I have been told of experiments in Germany, where, at one place, Roburite has been found to explode a mixture of gas and air. Judging from my own personal experiments I neither believe the one nor the other, because I think the test must not have been made with properly made Roburite, or with Roburite without tamping. Recently a large consignment came to this country, and the Government authorities analysed it and found that it would not stand a certain heat test, which they apply to all explosives. I believe that heat test has nothing to do with the safety of the explosive, as an explosive, but having been a rule which the Government have instituted, they have made it a test as to whether an explosive should be imported into the country or not. Securite is not a patented explosive and may perhaps be purchased cheaper than the others. I do not know whether the members have read the discussion on Sir Frederick Abel's paper, but in that discussion interesting information is given not only with reference to that explosive, but also with respect to the gelatinous cased cartridge, which is being introduced into Staffordshire. Mr. Sawyer, the Inspector of Mines there, says there has not been a single case of mishap or flame in using that cartridge, but he, or one who spoke on the matter added, "whilst there has been no accident, the question of the continuous and permanent safety of the gelatinous or water cartridge depends distinctly upon the care with which it is used." As far as Roburite, Securite, and similar explosives are concerned, the question of safety is more independent of the workmen. Mr. Walker has referred to another explosive, Carbonite. It happens that I have been desired by parties who have had Carbonite brought before them to make another visit to Germany on purpose to make a report on the value of Carbonite, and I intend going over for that purpose shortly. As to Carbonite, I believe its features are as good as the other two in every respect, but I shall be glad to give more particulars of that to the Institute after my visit. I may say that, whilst I went to Northern Germany last time to make experiments in a mine and quarry, this time there has been a chamber put up for experiment at the works where Carbonite is made of the most elaborate and complete type, which will enable us to make experiments with blown out shots and explosive atmospheres in every way and in the presence of dust. It is possible they will give us an opportunity to make experiments with other explosives than Carbonite.

The PRESIDENT: I am glad you mentioned it, because I was going to suggest whether you could not try Securite and Roburite in the same way.

Mr. BAINBRIDGE: This depends on the companies who represent these explosives being willing to run the risk of a report being made against them. No doubt the three are running a great race. So far the Roburite people are handicapped in starting works in this country, and the Government are putting every possible obstacle in their way, to make it impossible that any risk can happen from any explosion occurring where it is made. I hope that experiments will be tried in the marl drift. Whilst speaking of explosives there is another thing we have to consider— the comparison between the explosive and the wedge, that is the special wedge we have tried at the Wharncliffe Silkstone Colliery. The last ten minutes that the wedge is hammered it is severe work for the collier, and any collier who can get an explosive which is safe to do the same thing, might feel more inclined to support them than to try the wedge, but the wedge has the advantage of costing nothing after the first cost. Since that I have heard that at Mr. Garforth's colliery, Altofts, he is largely employing this wedge for ripping, and recently at Birley Colliery here they have applied the wedge with success. Mr. Walker tried it recently, and had a difficulty with the men, who declined to use it, because with the old small wedge it was the habit to give something extra, and when a wedge is used which has an altogether different effect, and it has the effect of an explosive, they demand the same thing. If the new wedge could be applied without the severe shoulder labour of the last five minutes it would be an excellent substitute for any kind of explosive.

Mr. C. E. Rhodes: The wedge is no use in an elastic coal. The coal will give to any wedge you can get; the expansion is so slow the coal will spring to it.

Mr. Bainbridge: Mr. Garforth finds them useful in his coal.

Mr. Rhodes: In a friable coal where there is lots of cleats and letters-off they will do very well, but in the Barnsley coal, or the Swallow Wood coal at our pit the men cannot drive with them. I believe I am right in saying that in some cases in the collieries Mr. Kell is connected with, they have had to pay 17s. or 18s. a yard to get end headings driven, and have had practically to abandon end headings.

Mr. Kell: That is so.

Mr. Walker: You have been more exhaustive, Mr. Rhodes, than anybody else in testing Roburite, and points must have suggested themselves to you which will have to be watched. There is now an opportunity of trying Carbonite in Germany in a special chamber with every appliance possible. Was there anything in your Aldwarke experiments that you thought should be specially watched, or which you should desire to see demonstrated if you had such an apparatus in your own yard?

Mr. Rhodes: I think it is very important that you should be able to test these explosives in an enclosed gallery resembling as nearly as possible a roadway in a pit. If the tests are made in a small tube, as we have made them before, it is probable that the result in this artificial condition will be altogether different from the result which would really be obtained from a similar shot fired in a pit under the natural conditions. My opinion is that the expansion of the gases generated by the explosion is so rapid that so much of the atmosphere and coal gas is driven from the tube as to leave too small a quantity to diffuse with the inactive gases produced by the explosion in such abundance to form an explosive atmosphere, or an atmosphere capable of supporting the combustion of coal dust or the inflammable material which is introduced to ascertain whether the explosion produces flame. My contention is that the rapid diffusion of the inactive gases with the comparatively unlimited supply of air, may produce a result altogether different from tests made in the manner I have described. Again, the other extreme which we have employed to test these explosives, that is, firing them loosely in the open air, is still more unreliable than the former, because we have conclusively proved that the explosion itself and everything with it is to a great extent blown away.

Mr. BAINBRIDGE: I have said I was anxious to try the three explosives against each other, because the effects will be to show the effect of the gases given off. I am told Herr Krupp, of Essen, instead of fixing on either of the first two mentioned, has fixed on Carbonite for use in his mines or quarries. I shall be able to learn what reason there is for that.

The PRESIDENT: I am sure we shall be very glad, Mr. Bainbridge, to have a report of the experiments you make. I hope you will be able to test the three explosives, all under the same conditions. Were they so tested then a proper comparison could be made, and members would be able to select which explosive they thought was the best and safest to use.

Mr. BAINBRIDGE: My own first impression is that there will be little to choose between the three of them, but the result of my examination next month may altogether change that view.

Mr. KELL: I think it is desirable the subject should be adjourned until these gentlemen have made their visit to Germany. I hope then they will be able to furnish us with more information.

Mr. WALKER: I think this is a question which should remain open some time. It is a new subject and must come prominently forward, as new facts come to light nearly every day.

The PRESIDENT: That being the decision we had better adjourn this until we have a report from Mr. Bainbridge.

The President's suggestion was agreed to.

MIDLAND INSTITUTE OF MINING, CIVIL, AND MECHANICAL ENGINEERS.

GENERAL MEETING.

HELD AT THE BULL HOTEL, WAKEFIELD, ON FRIDAY, APRIL 27TH, 1888.

T. W. EMBLETON, Esq., President in the Chair.

The minutes of the last meeting were read and confirmed.

The PRESIDENT: The election of Mr. George Liversedge, of Sheffield, stands next on the agenda. The Council say it will be better to defer the election until Mr. Walker, who is one of the proposers, can be present. The Council think some inquiries should be instituted, and Mr. Walker will no doubt be able to give us the information we require. The next business is the discussion on Mr. Nash's paper on Foreign Mining Rents and Royalties.

'DISCUSSION ON MR. NASH'S PAPER ON "FOREIGN MINING RENTS AND ROYALTIES."

The PRESIDENT: I would ask Mr. Nash whether he has further information to give us respecting these rents and royalties. I only wish that he had written his paper more fully, instead of condensing it. If we had had the verbatim reports of the Consuls, it would perhaps have enabled us to understand it better. We shall be glad to hear any observations on the paper.

Mr. A. LUPTON: One or two observations occurred to me, Mr. President, as I was reading the paper. The paper disposes of one misapprehension that partly exists as to the mines in France belonging to the Government. They may have done so, yet, having been for the most part granted to concessionaires, who have them at a nominal price, it practically means that a property has been created, comparable to, if not the same as, property in English mines. Still, it is no more a case of coal mines in France being subject only to a royalty to the Government than in England. The royalty paid to the Government is comparable to the income tax paid in England by anyone making a profit. Anyone making profit is liable to the tax, and in France the income tax is fixed at five per cent., or a shilling in the pound. They are allowed to make sufficient deductions for depreciation the same as they are now allowed to make in English mines. At page 152 there is a figure that I do not quite understand, It says, "For the actual working of each mine of 12,545 klafters,— one square klafter equals 3,597 square metres." Is that 3·597 square metres? At page 150 you give, "(a) A yearly rent to the owner of the surface seldom exceeding 10 per cent. per hectare on the whole extent of the concession." Ten per cent. on what?

The PRESIDENT: Per hectare.

Mr. LUPTON: Yes, it says "10 per cent. per hectare," but 10 per cent. what on? It does not say 10 per cent. on the hectare, but 10 per cent. per hectare; and as printed, I quite fail to understand it.

Mr. NASH; That is exactly copied from the report. I was at a loss to understand it myself. It says in the memorandum of the Consul, "A yearly rent to the owner seldom exceeding 10 per cent. per hectare on the whole extent of the concession."

Mr. LUPTON: Her Majesty's Consul is very likely a man not connected with mines, and he has written it, and it reads very well, but one cannot tell what it means.

Mr. NASH: I have taken it as given in the report.

Mr. LUPTON: You have also a fixed annual rent of five per cent. on the produce of the mine, and then "(e) 10 cents per franc in addition to form a relief fund for accidents." Ten cents per franc,— what franc? Does it mean 10 cents per franc of previous taxation, or 10 cents per franc of wages paid? It might mean either.

Mr. JOSEPH MITCHELL: It is on the net produce of the mine.

Mr. NASH: I should think it is on the net produce of the mine.

Mr. LUPTON: But it does not say so.

Mr. NEVIN: It does say so,—ten cents per franc, additional as rent to the five per cent., to form a relief fund for accident.

Mr. LUPTON: Then with regard to American mines, you have on page 156, "Corporations chartered for mining purposes are taxed on the nominal value of the shares issued." That struck me as being not quite in accordance with what I had understood. Of course there may be different rules in different parts of the States, and there may be a difference between what you call imperial—I do not know what the corresponding phrase to "imperial" would be in America—and local taxation.

The PRESIDENT: State.

Mr. LUPTON: State, aggregate of States, and small counties. There might be differences between all these three divisions for anything I know to the contrary; but I did understand taxation was according to the market value of the shares, which would be a much more reasonable thing. A mine might be established with a great nominal capital, and have no profits, and to tax it on the nominal value, and tax the most prosperous mine on the same value, would be ruin to the less prosperous mine.

Mr. NASH: Taxation on mineral lands, like that on all other property, is levied on their valuation.

Mr. LUPTON: I allude to the next sentence.

Mr. NASH: It says in the report, "Corporations chartered for mining purposes are taxed on the nominal value of the shares issued."

The PRESIDENT: One is a tax on land, as I read it, and the other a tax on chartered corporations.

Mr. LUPTON: I was talking to a gentleman, who has to levy the tax in one mining district. It is only two or three days since I asked the question. I said, "On what basis do you levy them," and he said "On the market value of the shares." That was very distinct. Those were local taxes, school expenses and so on. With regard to the ground that is conceded, too, to the man who makes a mining claim, your paper does not draw a distinction. I do not know whether that distinction still exists or not, between the different kinds of mining claims in minerals. There is the mining claim that might be called stratified, and the mining claim that is in the nature of a vein. It was so a short time ago, and it must be so still if the law is not altered,—in some parts at any rate,—that if you set out a claim at the outcrop of a reef, you might follow that reef. Suppose your claim was a hundred yards long, you might take a hundred yards of the length of that vein, and follow it down as far as your capital would take you, any distance, to the depth of a mile if you could go. But if your claim was in the nature of a stratified deposit, then you could only mine under the precise area of ground you had marked out when you made your claim.

Mr. NASH: You are speaking of America?

Mr. LUPTON: Yes, of Colorado. That distinction has been the cause of lawsuits, where professional men have been engaged on one side to say it was a vein and you might follow is as far as it went, and on the other side to say it was of the nature of a stratified deposit which was limited by the area you had marked out; millions of money depending on the result. I think, Mr. President, these are all the observations I desire to make on this paper. I thank Mr. Nash for the care that he has taken, and if I have seemed to be critical, it has been for the sake of inducing him to increase the value of his paper.

The PRESIDENT: I suppose this paper is entirely based upon reports from Her Majesty's representatives abroad?

Mr. NASH: Yes, they are published in a lengthy report, and I extracted those referring to coal and ironstone mines.

The PRESIDENT: There are a great many technical matters the Consuls would not go into, as, for instance, that about Colorado.

Mr. NASH: As I mention in the first part of the paper, I only include the coal and ironstone mines. I think what Professor Lupton is speaking of are the gold and silver mines, which are given by themselves in the report, and which are not treated in the paper.

The PRESIDENT: There are peculiarities as to the way they are possessed, whether they possess an area extending over a considerable surface, or whether the concession or claim extended on certain lines. Of course, the Consuls would not go to inquire about these things.

Mr. NASH: No; it is a general opinion of the laws in force in the whole States, taken as a whole, not in any particular small mining district alone. It extends to the extent of the whole country to which it relates.

The PRESIDENT: I believe the mining laws in the several States are different?

Mr. NASH: Slightly different in the minor details, but the main laws seen to govern the whole lot.

Mr. LUPTON: I should like to add a word. The paper does not, so far as I see, give the amount of royalties paid in the United States, but it is rather interesting when you go there, to find the great diversity of royalty paid. Owing to the extent of the country, in some districts the coal is worth nothing, but in others, I was informed (I should like to have further information on the subject) they are paying in royalties as much as 3s. per ton of coal. One would have imagined, of course, that coal would pay a higher royalty in England than in the States, and it was interesting to be told that royalties were paid as high as 3s. per ton.

The PRESIDENT : In what State ?

Mr. LUPTON : Pennsylvania.

The PRESIDENT : A thick anthracite coal, which they work in a great measure by quarrying.

Mr. NASH : Yes ; and when the owner takes a claim, he takes the minerals and all that is in it, unless it is taken as agricultural land. When he sub-lets it they make their own terms.

The PRESIDENT : Three shillings a ton does not seem to be an extra price to pay for coal, because they have in many instances no pits to sink, no expense to incur, and they are close to the railway. If many of our coal-fields were in the same condition the rent would be a great deal more than it is at present.

Mr. GERRARD : That observation is based on the first paragraph, " The right of private ownership in minerals extracted from the soil is recognised in all the States and territories of the United States ; all such minerals belong to the owner of the land." If information could be given as to how these rights are exercised, it would be interesting.

Mr. NASH: Royalties are vested in private owners, who make their own terms.

Mr. NEVIN : For the last few months I have been on that side of the water, and so far as I could see, in most of the States where the land is owned by private owners, the arrangement is very much as it is with us. The minerals belong to the owner of the surface, and, of course, he gets what royalty he can, from ten cents per ton upwards ; in some cases ten cents, in others up to twenty-five cents, equal to about a shilling a ton. In Pennsylvania they get highest, but that is for anthracite. Out in the territories they are governed by the Federal Union,—that is, the laws are made in Washington, and all their mining laws are pretty much the same. In the bulk of the territories the land is not in private hands ; it is in the hands of the Government, and in that case you take up your claim, if it is on a vein about 500 yards in length and a certain width. But I believe, as Mr. Lupton has said, if the vein outcrops on your land you can go out of your own claim into your neighbour's claim, as long as you are following the vein. That is a fruitful cause of lawsuits. But in the settled States, as a rule, where the land is in private hands, the ownership is very much the same as it is in England.

The PRESIDENT: Is rent paid, and damage to land by mining operations also paid by the coal owner?

Mr. NEVIN: I think so, where land is damaged.

Mr. LUPTON: Do you happen to know, Mr. Nevin, of any case where the ownership of land is separated from that of the minerals? In the cases I happened to notice, the land was so cheap that it was generally owned, as a matter of convenience, by the owner of the mine.

Mr. NEVIN: Two cases I have had to do with myself, one in Ohio and the other in Alabama. In Ohio there were thirty or forty leases which were taken from the owner of the land as ours are in this country, but they were in perpetuity, not for a term of years, and paying a nominal certain rent of five dollars and a royalty varying from ten to twenty-five cents per ton. In Alabama the case was exactly similar. The metal mines out in the territories are in Government land, and you take up your claim. You pay so much for your claim, and have to do a certain amount of work in it, and it is yours, and you are only liable to the same taxes as men carrying on any other business.

The PRESIDENT: You do not find any reports about Scandinavia?

Mr. NASH: No; it is not mentioned.

The PRESIDENT: The mining laws in Scandinavia, in Norway especially, are very simple indeed. A man there has a licence from the Government, he pays no royalty for the mineral, and if he does not work the mine he retains possession so long as he fires a shot once a year in the presence of witnesses. This way of retaining possession I have seen myself. I suppose it is merely to keep other people out of the place, and to retain the right himself; but there is no royalty charged.

Mr. NASH: Have they coal there?

The PRESIDENT: No.

Mr. NASH: It is all minerals?

The PRESIDENT: During the coal famine in 1872, 1873, and 1874, the mining inspector, Mr. Daht, advertised in the English papers a coal which he had found in what he called the Virgin land. I spoke to him about this, and I said "You know there is no coal here. How much ash has it?" He replied, "Seventy or eighty per cent. of ash." Yet this so-called coal was advertised in the English papers to draw English capital there, to work what was nothing but black shale.

Mr. NASH: In comparing the average royalties in England, which Mr. Percy in his paper takes to average 8d. per ton, with the prices paid in France, Belgium, and Germany, which really are our competitive places for foreign trade, it shows in France that they pay five per cent. of the profits; Belgium five per cent. ($2\frac{1}{2}$ per cent. to the State and $2\frac{1}{2}$ per cent. to the landowner); and in Germany two per cent. of the value of the output. If, then, the average selling price was 6s. 3d. per ton, the royalty would be $1\frac{1}{2}$d. per ton in Germany. To place England on an equality with France, we should require a profit of 13s. 4d. per ton to have our royalty rents on anything like a fair basis with the French royalty rents, and about the same with Belgium; whilst with Germany we should require an average selling price of 33s. 4d. per ton. But I think royalty rents should not be fixed on the basis of profits, because the same mine worked by two different managers could be made to produce totally different results, greatly to the advantage or disadvantage of the landlord, and in the appointment of which said manager he would have no voice.

The PRESIDENT: There is no doubt in France a great deal of profit is made at some places and at others none at all, but that does not arise entirely from the way in which the place is managed, but from the way in which the coal lies. At some pits in the North of France, the same bed of coal has been sunk through three times in one shaft, so great is the difference of the rise or dip of the seam in fact the same seam being turned right over and doubled twice.

Mr. GERRARD produced a map showing how concessions are mapped out in Belgium. Generally the concessions are of convenient form, but whilst following the boundary of the province, they disregard entirely the boundary of the ownership of land, and the modern concessions are very much larger in area than those of older date.

Mr. WM. HY. CHAMBERS: It is a good deal like Derbyshire lead mining, they put a pit down where they like.

The PRESIDENT: I think our English system is better than this. I think no Englishman would like persons to go mining under his land in this way.

DISCUSSION ON MR. G. B. WALKER'S PAPERS ON "HYDRO-CARBON EXPLOSIVES."

The PRESIDENT: The next business is the adjourned discussion on Mr. G. B. Walker's papers on Hydro-Carbon Explosives. As Mr. Walker is not here, would it not be better to defer this until he comes, and also until we have the account promised by Mr. Bainbridge of his investigations respecting these explosives on his visit to Germany.

Mr. JOSEPH MITCHELL: Mr. Bainbridge asks for it to be adjourned.

The PRESIDENT: He does. I should also add something about another explosive—Tonite. When it is exploded Tonite produces a flame, but its inventor has invented a powder,—what its composition is is not known, but when that is put in with the Tonite there is no flame seen. At Rothwell Haigh and at another place several experiments have been made; I hope to give particulars of these, and then have a discussion upon all of the four explosives.

Mr. GERRARD: Does it occur to you, Mr. President, that the compound is an absolute necessity to the safety of Tonite, just as the water is to the water cartridge?

The PRESIDENT: I suppose so, but I cannot give you any particulars about it. I know when this powder is put in it destroys the flame.

Mr. GERRARD: It is open to the same objection as the water-cartridge—minus water, minus safety.

The PRESIDENT: If you put the powder in you have no flame, if you do not you have.

Mr. GERRARD: It rests with the users.

Mr. NASH: Is the powder mixed with the Tonite, or does the person using it put it in?

The PRESIDENT: It is used by the person using the Tonite.

Mr. NASH: And if he makes a mistake—does not put enough in, or any thing of that kind—it will be as explosive as before?

The PRESIDENT: Yes.

Mr. LUPTON: There has been no allusion in the discussion to the use of jelly instead of water to stop the flame of an explosion. I have seen it tried, and there did not appear to be any flame when the jelly was used.

The PRESIDENT: What kind of jelly has been used?

Mr. LUPTON: I believe it was a saponaceous jelly.

The discussion was adjourned.

FAN EXPERIMENTS.

The PRESIDENT: It appears these fan experiments require from us a maximum sum of £100 for the purchase of instruments; and we have not the money. It can only be met by sending circulars to the coal owners, the landlords that is, and the colliery owners of the district to see if we can raise it, otherwise we shall be behind hand.

Mr. JOSEPH MITCHELL: The North of England Institute and the South Wales Institute have agreed to find that sum.

The PRESIDENT: There would be no objection to members contributing something towards it.

Mr. HEDLEY moved, and Mr. MARSHALL seconded, a vote of thanks to the President.

The PRESIDENT: I am much obliged to you. I only regret that we have such a poor attendance when we come to Wakefield. Why it should be so I cannot tell, because it seems to me this is a most central place. I hope that at next meeting at Leeds, which will be the last before the annual meeting, we shall have a large gathering.

MIDLAND INSTITUTE OF MINING, CIVIL, AND MECHANICAL ENGINEERS.

GENERAL MEETING.

HELD AT THE QUEEN'S HOTEL, LEEDS, ON TUESDAY, MAY 29TH, 1888.

T. W. EMBLETON, Esq., President, in the Chair.

The minutes of the last meeting were read and confirmed.

The following gentlemen were elected members of the Institute, having been previously nominated:—

Mr. WILLIAM BLACKBURN, Colliery Manager, Astley, Woodlesford, near Leeds.

Mr. ALBION THOMAS SNELL, Electrician, Malden Crescent, Prince of Wales Road, London.

Mr. JAMES MELLORS, H. M. Inspector of Mines, Leeds.

The meeting was adjourned in consequence of a meeting of the Special Rules Committee being held at the same time.

MIDLAND INSTITUTE OF MINING, CIVIL, AND
MECHANICAL ENGINEERS.

GENERAL MEETING.

HELD AT THE QUEEN'S HOTEL, LEEDS, ON TUESDAY, JUNE 19TH, 1888.

G. BLAKE WALKER, Esq., Vice-President, in the Chair.

The minutes of the last meeting were read and confirmed.

The CHAIRMAN: I think it might be well if I were to occupy a few minutes in case Mr. Embleton should come in later, in telling you what took place in London a fortnight ago. On the 6th instant a meeting was held at the Rooms of the Institution of Civil Engineers, when two joint Committees met under the presidency of Sir Isaac Lowthian Bell, the first to consider the question of the federation of the Mining Institutes of the country, and the other to consider a joint scheme for testing fans where the circumstances seem likely to be instructive. The Institutes which met there were the North of England Institute, the Midland Institute, the Chesterfield Institute, and the two Staffordshire Institutes. The South Wales Institute attended the second Committee with regard to the fan experiments, but were not willing to enter into a scheme of federation. The Mining Institute of Scotland and the Geological Society of Manchester also declined. It was felt, nevertheless, that it might be useful to enter into a federation if a proper basis could be arranged, as the other Institutes which have so far dissented from joining might be hereafter disposed to join. I believe that a full report of what took place at both meetings will be laid before the members, and therefore it is perhaps undesirable for me to enter into any lengthy remarks in explaining what took place. I may say, however, that very great stress was laid on preserving the individuality of the Institutes, and this seemed to be the opinion of the representatives of all the Institutes who were represented. At the meeting of the

Fan Committee, I and some others were of opinion that former experiments with fans had been anything but satisfactory, and that some extension of the original idea (namely, that only those fans should be tested where there were fans of a different type on the same pit), should be agreed on. The feeling of the meeting was that at any rate that inquiry should be restricted in the first instance to those cases where there were fans of different types on the same pit, because in those cases more closely analogous results could be obtained than elsewhere. The principles on which the experiments were to be conducted were also settled and are in the report which is on the table to-day. I think the investigation promises to be of considerable value if only we are not deprived of a very full report at the end of them. The North of England, when they carried out their experiments issued nothing more than a very meagre table. I hope that in these experiments more will be given. About £50 has been received from various persons and firms towards the expenses, and we hope to get something more in yet. I think the subscriptions very handsome as far as they go.

Mr. A. LUPTON: Will the fan experiments go on irrespective of our collecting a larger figure in the Midland district.

The CHAIRMAN: I think so. That question was not raised at the meeting. I suppose the Institute has certain funds from which it could draw to make up the deficiency if we fail to reach the exact sum. I may say Mr. Carrington, in a letter addressed to the Secretary, says that he thinks it is more a question for individual members than for firms, and that he thinks individual members ought to come forward and subscribe.

Mr. LUPTON: Could you give us any idea of the sort of instruments which it is proposed to buy with the money for making the experiments?

Mr. NEVIN: It is not so much the instruments as the travelling expenses that would be heavy. I do not think the instruments will amount to much, because some of the instruments that the Institutes have would be available.

Mr. LUPTON: I thought they were to buy new instruments and apparatus?

The CHAIRMAN: At the meeting in London the number of instruments was cut down in order to reduce the expenses as much as possible, but great stress was laid on having them very accurate. Mr. Brown, the Secretary, was requested to ascertain whether the instruments could be tested at Greenwich Observatory in order to get absolute accuracy.

Mr. LUPTON: What sort of instruments would they be they would test there?

The CHAIRMAN: There are thermometers, anemometers, and aneroid barometers, but I think it would be chiefly adjusting anemometers.

Mr. JOSEPH MITCHELL: That is what is arranged.

Mr. LUPTON: Have they decided on an indicator?

The CHAIRMAN: Yes; Richard's indicator. There was great discussion as to whether it was possible to have any other means of measuring than the anemometer; whether the use of very light, carefully poised bladders would give more accurate results.

Mr. LUPTON: Because there is another kind of indicator said to be superior to Richard's.

Mr. JOSEPH MITCHELL: It will be for the Council to decide whether they would publish the whole report of the London meeting. The Council have not discussed it yet,

The following paper was then read :—

MINING IN THE MIDDLE AGES.

By GEORGE BLAKE WALKER, F.G.S.

THERE is no more ancient art than that of mining if we except that of the chase. Before men could till the soil metallurgy must have made some progress, in order that the implements necessary to tillage might be forthcoming. As a matter of fact, we find that the earliest use to which the metals were applied was in the construction of the spears, arrow heads, and swords, with which primitive man brought down his game or carried on war with his fellow-man. The history of the progress of metallurgy is the history of civilization. All the great nations of antiquity were skilled in the working and treatment of metals; and in artistic excellence their works even surpassed the greatest efforts of our own age. In the Book of Job there is a remarkable passage alluding to the evidently well-known operations of mining, which is interesting as shewing that at a period, possibly as early as 3000 B.C., there existed mines worked by means of shafts and subterranean galleries, and of considerable magnitude.

The passage is from the xxviii. chapter of the Book of Job. It is as follows, the translation being that of Professor Dawson, F.R.S. :—

> "Surely there is a vein for silver,
> And a place for the gold which men refine;
> Iron is taken from the earth,
> And copper is molten from the ore.
> To the end of darkness and to all extremes man searcheth
> For the stones of darkness in the shadow of death.
> He opens a shaft from where men dwell,
> Unsupported by the foot, they hang down and swing to and fro.*
> As for the earth, out of it cometh bread;
> And beneath it is changed as by fire.
> Its stones are the place of sapphires,
> And it hath lumps [nuggets] of gold.
> The path (thereto) the bird of prey hath not known,
> The vulture's eye hath not seen it:
> The wild beast's whelps have not trodden it,
> The lion hath not passed over it.
> Man layeth his hand on the hard rock,
> He turneth up the mountains from their roots,
> He cutteth channels [passages] in the rocks,
> His eye seeth every precious thing.
> He restraineth the streams from trickling,
> And bringeth the hidden thing to light."

It is probable that until the application of steam power to the working of mines, the systems of mining followed in all parts of the world must have been very much the same, and susceptible of very little alteration or advance. Until a comparatively recent date, nearly all mining operations were carried on by means of slave labour,—from the time of the ancient Egyptians to that of the Spanish rule in America. The conditions of mining where slaves were forced to work under the lash can never have been governed by any considerations of humanity or economy; and no doubt the cruelties perpetrated in the mines fully justified the horror which the writers of all former ages have expressed for the condition of those condemned to labour there.

In Germany in the Middle Ages, the labour of the mines was doubtless in a sense free, but the ties of the feudal system imposed so many restraints on the freedom of the labouring classes that their

* Genesius.

condition can hardly be regarded as other than a modified slavery. Still the position of the German miner was probably as good as that of any, and the mechanical devices employed for various purposes would be of the best description.

There exists a curious old book written in Latin by one George Agricola about the year 1550,* which describes very fully and graphically the operations of mining in Germany in the 16th and 17th centuries. For the reasons already stated there is no reason to believe that any great changes had recently taken place at the time the book was written, and we may, therefore, assume that the methods and appliances it describes are equally applicable to the Middle Ages as to the time when the book was actually written. The book is entitled "*De Re Metallica*," and is an exhaustive treatise, not only on the mining of metalliferous ores, but on their reduction and treatment. Upon the latter part of the subject I do not propose to enter, but I think that some account of the mining operations described by Agricola may be of interest to the members of the Institute.

There is a charming quaintness about Agricola. There is so much simplicity in his way of putting things, while all the time one cannot help seeing how thoroughly he is master of his subject, so that the perusal of his work cannot fail to give pleasure even at this day. At the commencement of his work he thus apologises for writing on so purely utilitarian a subject as mining:—

"It hath been thought of many† that mining is an employment "base and empirical, the knowledge whereof, though it be needful "for them whose lot is to labour therein, yet reacheth not into the "dignity of an art. But he who shall duly weigh with himself the "manifold and various matters whereof that knowledge consisteth, "will hardly, methinks, be persuaded to be of this mind. For the "miner who will be held perfect in his craft must be skilled to dis-

---

* The edition from which the extracts in the paper are taken was published in 1657 at Basle.

† I have rendered this passage in Sixteenth Century English, the better to reproduce the quaintness of the author's style, but in the rest of the passages I have quoted I have used more modern English as better adapted for the description of the processes and contrivances dealt with, reverting only to Elizabethan English in the last passage.

"cern those spots of the earth where ore is like to be found, whether
"it be hid in hill or plain, in mountain or valley, and prudent to
"judge concerning such spots whether or no they will yield reasonable
"profit in the working. He must understand the courses of veins,
"the natures and kinds of rocks, and of the substances contained in
"them, as mineral oils, salts, gems, ores, and metals. It is further-
"more necessary that he be thoroughly instructed in all such
"mechanical arts and contrivances as are used underground, and in
"ways whereby men make trial of the ores and prepare them for
"melting, which ways are exceedingly diverse, for one treatment
"is proper to gold, another to copper, nor do men deal with quick-
"silver as with iron and lead.

"Moreover the instructed miner can by no means lack the aid of
"those arts and sciences of whose dignity there was never any doubt
"among men. He must enquire of philosophy touching the nature
"of that matter whereof the earth consisteth, and the causes which
"have wrought to make it such as now we see it; for thus will he
"more readily discover the veins of metal and more wisely handle
"the ore when it hath been brought forth from the mine. Some-
"what he should know of medicine, seeing that they who toil in the
"earth are exposed to the attacks of divers diseases, so that he may
"either himself minister to their needs, or provide that they be dis-
"creetly ministered to by others. Nor will astronomy disdain to
"instruct him touching the quarters of the heavens, whereby he may
"judge of the course of veins. He must by all means understand
"the art of the surveyor, whereby he may rightly estimate the depth
"and dimensions of his shaft, that it may reach to the underground
"passages of the mine; and likewise the art of the arithmetician,
"whereby he may reckon the cost of his engines, and the wages of
"his people. It is well that his hand be also skilled in making
"models and drawings, that so he may both himself fashion engines,
"and expound the manner of their fashioning to others, who will
"more readily understand that which by means of a drawing is
"presented to the eye. And lastly, it will not be amiss that he
"know something of law, more especially as touching mines, lest
"through ignorance thereof he defraud others or suffer himself to be
"defrauded. Yea, being herein instructed he may oftentimes be of
"service to his more rude and simple neighbours.

"Wherefore let him who will understand the rules of that art
"which teacheth men to find metals, and the reasons of those rules,
"give careful and diligent heed to what is set down in this book and
"in our other works. But if he will enquire further of any special
"matter, let him consult such miners as are of skill and experience,
"albeit he will find few who are thoroughly instructed in the whole
"of that art which they profess. For one is able wisely to direct the
"digging of a pit, another is skilful in the washing of ore, a third in
"the melting thereof, a fourth hath the knowledge of the surveyor,
"a fifth that of the engineer, a sixth is well instructed in the laws
"which concern mining. Nor do we ourselves pretend to perfect
"understanding of the great and manifold science of finding and
"working metals, yet we trust to be found serviceable to such as
"shall be earnestly minded to give themselves to the study of that
"science. But let us now come to the matter in hand."

With this introduction our author immediately plunges into the depths of his subject at a length which proves him to be an enthusiast for his craft, and were we to follow him into all his dissertations, the bulk of the translation would probably more than equal the whole of the past Transactions of the Institute. This being by no means desirable, I shall now proceed to give certain extracts from the Fifth and Sixth Books, with the general aim of reproducing typical examples of the chief classes of methods and appliances which Agricola describes: and I do not doubt that as I do so you will be struck by the singular manner in which most of our modern appliances have been anticipated, and you will be compelled to admit that though we have more perfect instruments than our forefathers, we cannot justly lay claim to greater ingenuity than they.

Passing on to the Fifth Book, we find the preliminary operations thus described:—

"The miner's first operation on opening a deep vein is to sink his
"shaft, and place at the top of it his winding machine. These
"(Plates I., II., and III.) he protects with a shed, both to keep rain
"out of the pit and to shelter the men who work the machine from
"cold and wet, while it also serves as a place in which to keep
"barrows and the miners' tools. Close to the pit head another shed

"will be erected for the use of the manager and other officials, and
"for the storage of ore and other output of the mine.

"The shaft is (generally speaking) an excavation from 9 to 10 ft.
"in length by about 4 ft. in width. The shaft (see Plates I., II., III.)
"is made perpendicular or slanting, according to the course of the
"vein to be worked, but the mine itself is a prolonged underground
"trench [the author is speaking of a metalliferous mine] almost
"twice as high as it is wide, so as to allow the miners to pass freely
"to and fro in it and bring out loads of minerals. The passages are
"usually about 6 ft. high and 3 ft. wide, and are dug out by two
"miners, one working in the upper and the other in the lower part
"of the heading, the former in advance of the latter."

Our author next describes the best methods of placing shafts, and the driving of cross drifts, but as these only apply to metalliferous mines I omit them, as also the very interesting passage in which the course of metallic veins is noticed, and the probable richness or poverty of the lode estimated. The following passage will serve as an example of the last mentioned subject:—"There is good hope of
"finding ore in the place where an inclined plane joins a vertical
"one, for which reasons miners dig through the roof and floor of a
"main vein, and seek another vein which will join the former within
"a few yards," and so on.   .   .   .   .

"So much for indications. I now come to the modes of working,
"which are many and various, since a loose vein must be worked in
"one way and a hard one in another, and again there are degrees of
"hardness. So also the roof is sometimes soft and crumbling, and
"sometimes it is hard and strong. By a *loose* vein I understand one
"consisting of some kind of earth, or even of soft and half liquid
"materials; by a hard vein, one consisting of mineral substance and
"moderately hard rocks, as those for the most part are which fuse
"easily with fire. A still harder vein may consist of the rocks already
"mentioned but in combination with siliceous rock or stones easily
"fusible, as pyrites, or cadmium, or very hard marbles. The hardest
"of all are those which consist of these hard rocks and combinations
"of them, and the veins which are found contained in such rocks
"entirely. The roof and floor of a vein are said to be hard when
"the clefts and faults are few, harder when fewer, hardest when there
"are hardly any or none at all: for when these are entirely absent,

"the rock is almost without water, which softens it. The hardest
" rock of roof or floor is, however, rarely harder than the vein itself.
" A soft vein may be worked with the spade only. Until metal shews
" itself no distinction is made between the roof and the vein; as
" soon, however, as they come upon metal, it is necessary to go most
" carefully to work. The roof of the vein is first separately cleared
" away, then the crumbling material of the vein itself is dug out with
" the spade and thrown into baskets (or tubs) placed in readiness, so
" that none of the ore may be lost. The contents of a hard vein, on
" the other hand, must be torn away from the sides by means of
" wedges driven in by a hammer, and the hard rock above must be
" cut away by the same tools, unless it is permitted to soften them
" by the action of fire. A vein that is only moderately hard will
" yield to the pick, but the rock covering the hardest veins is usually
" attacked with stronger tools. Should a rich vein occur of a
" peculiarly hard character so as to be unworkable by means of tools,
" fire may be employed, provided the consent of the masters of neigh-
" bouring workings be first obtained. If they refuse it, the rock above
" or below (if more yielding) should first be removed, then a little
" above the vein should be placed beams (planks), so as to form a
" roof or floor, and in the next place beginning from the front and
" the upper part where the rock shews minute cracks, nicks should
" be made with the tools most suitable for the purpose. In each of
" these nicks four wooden wedge feathers should be inserted (or for
" deeper holes, it may be necessary to use four small iron feathers).
" Afterwards single wedges are introduced between two of the wooden
" wedge feathers, and these are driven in by hammers, so as to make
" the vein give forth a sharp ringing sound; but when it begins to
" break away from the rock above or below, a cracking is heard. As
" soon as this grows loud, the miners retreat rapidly; the noise
" becomes deafening and the vein-stuff falls, and in this manner they
" bring down 100 lbs. weight, more or less. If miners attempt to
" deal with a very hard rich vein of metal in any other manner, so
" much rubbish falls with the ore that it can scarcely be picked out
" afterwards. When in a very hard vein a knot is met with which
" contains no ore, the miners, if not permitted to use fire to remove
" it go round it, diverging to the right or left, for to cleave it with
" iron wedges would be extremely laborious. Sometimes as the work-

"men are toiling at their task, they cause the depths of the earth
"to re-echo with sweet songs, whereby they solace their hard and
"dangerous labour. The force of fire also, as I have said, overcomes
"the hardness of the rocks, and the mode of applying it is far from
"simple, for if a vein cannot be worked on account of the hardness
"or narrowness of the surrounding rock, a pile of dry firewood is
"made and lighted, one, if the mine or gallery be low, or two if it
"be high (and in this latter case one is placed upon the other), and
"both burn until entirely consumed. By the action of heat so
"applied, no very great portion of the vein is usually melted, but
"only certain coatings. When the rock either above or below the
"vein is sufficiently soft to be excavated, but the vein itself is too
"hard, cavities are made in the former either above or below the
"level of the principal gallery. There are different ways of proceed-
"ing. If the cavity is wide, a considerable quantity of fuel can be
"introduced into it; if narrow, the quantity will be small. If the
"cavity be low, one heap of firewood is introduced into it; if lofty,
"two,—the one placed upon the other, whereby the lower heap
"being kindled, the upper takes fire from it, and the burning mass,
"excited by the current of air through the workings, causes the lode
"to detach itself from the rock.

"While the rocks and veins to which fire has been applied are
"giving off noxious gases and smoke issues from the shaft or gallery,
"the miners and other workmen do not go down, lest the poisonous
"exhalations should produce disease or even death (as when I come
"to deal with miners' ailments I shall explain more at length). But
"in case the noxious gases and smoke should penetrate, by means of
"veins or fissures, into neighbouring mines, which do not contain
"hard veins or rocks, the mine-master ought not to permit fire to be
"used either in the shafts or galleries, lest the workmen (in the
"adjoining mine) should be suffocated. When the contents of a
"lode, or the fault-stuff, or the rock itself have been loosened by fire,
"the miners bring it down (if fractured in the upper part) by means
"of poles; or if the rock is still hard, iron bars are inserted into the
"crevices; if laterally fractured, they use hammers. The fragments
"thus loosened fall, and if there is still resistance it is overcome by
"the pick. Rock and earth, ore and ore-bearing stuff, are consigned
"to separate receptacles, and are brought out to the upper air. If

"raised from a shaft of moderate depth, a windlass is used; if "deeper, by horse-power.

"Frequently the presence of water in large quantity, and some-"times the ventilation of the mine is a difficulty in mining. Accord-"ingly both these things must be as carefully considered by the "mine-manager as the working itself. Both veins and faults (those "specially which contain no ore) bring water into the shafts and "galleries. There may be no circulation of air in a mine. In a "deep shaft this only occurs when there is no communication by "means of a drift with any other shaft; in a gallery if this be "driven far into the rock, and no shaft has been sunk deep enough "to reach it. In neither of these cases can there be motion or cir-"culation of the air, so that dense and vaporous exhalations arise "with a smell like that of a vault or underground cell, closed upon "all sides for many years; for which reason it is impossible for the "miners, even were the mine rich in gold or silver, to remain at "work long, or, if they persist, the vapour has pernicious effects, "and causes headache, especially if many are working together in "the mine with many lights, which then give out a feeble light: for "the products of the combustion of the lights, as well as the breath "of the men, make the air still worse.

"Water, if in moderate quantity, can be raised from the mine by "means of pumps of different kinds, turned or drawn by men. If "water should rush into one pit in such quantity as to greatly "impede the working, another shaft is dug at some yards distance "from the first, so that while by the one work is carried on without "hindrance, the water is drawn into the other, which, being at a "lower level, forms a sump, whence by means of these same engines, "or others worked by horses, the water is turned into the channels "of the nearest main gallery so as to flow out at the mouth. But "when all the water from the workings, not only of one vein, but of "neighbouring veins, flows together into the deepest shaft of the "mine, it will then be necessary to make the sump very large, and "from it to draw off all the water by means of pumps furnished with "buckets, or with leathern pipes, of which I shall treat more fully "in my next book.

"Air is most effectually driven through deep shafts and exten-"sive workings by means of air engines, which I shall likewise dis-

"cuss in the ensuing book. The outer air of its own accord rushes
"in to fill up any hollow space underground, and having entered it
"makes its way out again, but not always in the same manner. In
"Spring and Summer the air enters by the higher shaft, traverses
"the galleries and cross drifts, and issues forth from the lower. So
"also at that time of the year it passes through the higher gallery,
"and thence by means of a connecting shaft into the lower, whence
"it makes its exit. In Autumn and Winter, on the contrary, the
"air enters by the lower gallery or shaft, and comes forth from the
"higher. . . . .

"Different shafts require different systems of support. If the
"vein, together with its floor and roof, be hard, there is not much
"need of support, but timbers are placed at intervals, their ends
"being secured in holes, cut out of the rock, both of the roof and
"floor,* and to these cross-beams (Plate IV.) are attached on the
"lower face of the vein (floor), to which the planks and ladders
"are fastened. The portion of the shaft containing the ladders is
"generally partitioned off from the other part of the shaft, up which
"the vein-stuff is raised, and the sides are secured by planks where
"exposed to the action of water lest fragments of rock should fall
"down the shaft, and cause alarm or injury, or loss of life to miners
"and other workmen as they go up and down the ladders. For the
"same reasons the men should not carry loads up and down the
"ladders, lest pieces of rock should roll out of the baskets; and
"other precautions should be taken whereby the difficult and peril-
"ous ascent may be made as little dangerous as possible. If the
"vein should be of a loose character more will be necessary in the
"way of support. In this case the shaft must be lined throughout
"with long timbers, the thrust being supported by the cross-beams
"let into the sides. In order to make such joinings absolutely firm,
"the bark (slabs) of trees or wooden wedges are inserted and driven
"in between them and the walls of the vein, and any cavities which
"still remain should be filled up with rubble. . . . . .

"Lest there should be accidents to those bringing loads (to
"the shaft) from falling stones when they are drawn up a deep

---

* It must be remembered that the author is describing inclined shafts for metalliferous veins, hence the propriety of speaking of the sides of the shaft as the roof and floor (of the vein).

"shaft the whole space near the pit bottom, except the spot "where the ladders are) should be covered with rough planks, "supported on bars. An aperture must be left near the ground open "to all the other parts of the shaft, so that the buckets (skips) loaded "with minerals may be drawn up by the engines and again lowered "into the mine; the men sheltered by the penthouse being perfectly "safe.

"In the working of a single vein, one, two, three or more galleries "may be driven and at different depths. If the vein, with its roof "and floor, be compact and hard, no part of the gallery needs any "support except at the adit, where the rock is less compact. If, on "the other hand, it be crumbling, and the rock of floor and roof "soft, the gallery will need strong and frequent timbering." [Here follows a description of what is usually known as "coupling" timbering, the couplings being backed with planks (Plate VI.)] . . . "Finally, if the rock and earth are to be brought out in barrows, "planks joined together are laid upon the bottom members; if in "trucks, two parallel pieces of wood 9 in. × 9 in., having between "them a space in which the guide pin of the trucks can run (the "guide pins are sometimes grooved), so as to keep the trucks on the "track; and the channels by which the water runs off are made "under the joists of the flooring. It is usual also for the cross "drifts to be supported, but they do not require the lower joists nor "water channels, as rock is not carried along them for any great "distance, nor do they contain much water. When, however, the "galleries have already been driven, and even the cross drifts, if the "upper part of the vein is rich in ore (as it sometimes is for some "yards), new cross drifts are opened repeatedly above the old ones, "till they reach the poorer part of the lode. Such workings are "supported as follows: Very strong timbers are fixed at intervals "(usually a foot and a half thick) that they may be able to bear the "superincumbent weight. Into these cross workings, when the ore "has all been got, is stowed the refuse rock from the workings, by "which means both labour and expense is saved. So much for what "relates to the construction of shafts, galleries, and cross drifts.

"All that I have hitherto written relates principally to veins "(having a high degree of inclination); what I am now about "to say has reference partly to horizontal layers, and partly to "'bunches.' . . . .

"In working a flat vein both in its length and breadth a low
" gallery is driven. In this, if circumstances and the nature of the
" ground permit, another shaft is dug in order to ascertain whether
" another vein exists below that already found: for occasionally
" beneath the first are found two, three, or even more beds contain-
"ing the same ores, which are worked in a similar manner. In
" working flat beds the men usually lie down . . . and miners,
" working thus, being obliged, in order to use their tools, to keep the
" neck bent to the left, not unfrequently get it twisted. . . . . .
"Lest, however, the mountain or hill, being thus widely under-
"mined, should settle, it is desirable either to leave pillars and
" arches of the rock itself, whereby as by foundations, the pressure
"is resisted; or to contrive other means of support to effect the
" same purpose. Rock destitute of ore is packed away in the ex-
"hausted cavities." [Here follows an interesting account of the
method of working "bunch" veins by sinking vertical shafts down
upon them; and the occurrence of gold in river gravels, which the
limitations of space compel me to omit.]

The next section on surveying displays an immense amount of
ingenuity and skill, and as a comparision of ancient with modern
methods of surveying would in itself make a very interesting paper,
but it is impossible to touch upon it in the present extracts. I now
proceed at once to the description of some of the machines for various
purposes, which Agricola mentions, taking, however, only characteristic specimens in each class.

Implements.

The Sixth Book begins with a description of miners' tools (picks,
hammers, wedges, &c.), but as these are not unlike those which are
still used, I pass on to *buckets*. " Buckets are of two kinds, not
" differing in material and shape, but in size, the smaller kind being
" equal to a measure which would hold about 9 gallons, the larger
" would perhaps hold six times as much, but the capabilities of
" neither kind is fixed, but frequently varies. Both are made of
" staves of wood bound round with iron above and below; for wood,
" whether hazel or oak, being continually chafed against the planking
" would easily wear out, while the iron lasts. . . . . .
" The smaller buckets are filled by boys, the larger by men who get
" out the stuff from the pit bottom and fill it into the buckets, for

" which reason they are called 'loaders.' They attach the handle of
" the bucket to the rope by a hook, and the bucket is then drawn out
" by the machine. . . . .
" By some, *baskets* are preferred to buckets, which will hold as much
" as the buckets, or even more, as they are lighter. Some again use
" bags made of ox-hide, the iron handles of which are attached to
" the rope by means of a hook. Three are usually drawn out full at
" once, and three lowered while three are being filled. The leather
" buckets are in use at Schneeberg, the others at Freiberg.

Barrows. [It is not necessary to reproduce the description of the familiar wheelbarrow, the special point being that to save height the wheel is kept as high as possible, and the body of the barrow when being filled rests upon the ground.]

Trucks. [As the usual word in Yorkshire *corf* is a corruption of the German word *korb*, a basket, and as a *tub* is cousin germane to a bucket, I will, to avoid ambiguity, use the term *truck* in the following paragraph.]

" The truck (Plate V.) is half as large again as the barrow, being
" about 4 ft. long by 2½ ft. wide and the same height, but being
" square it is hooped with three iron bands, which are still further
" secured by strips of iron on both sides. Two iron axles are attached
" to it beneath, at the ends of which revolve small wheels of solid
" wood, which are prevented from slipping off the fixed axles by pins.
" A large peg is also fixed under the truck, and this runs in the groove
" of the tramway and keeps the wheels on the runners. These trucks
" are pushed by a man from behind, and from the barking sound
" they produce when moving are sometimes called 'dogs.' Trucks
" are used to bring loads out of the larger galleries, both because
" they can be more easily moved and more heavily loaded.

" The miners' 'troughs' are hollowed out of the trunks of single
" trees; the small size is generally 2 ft. long by 1 broad; these being
" filled with ore (especially when there is not a great quantity to be got
" out) are either carried out of the pits or drifts on men's shoulders
" or are suspended from their necks by cords. Pliny affirms that
" with the ancients everything extracted from the earth was borne
" out on the shoulders; but to us this mode of carrying loads does
" not commend itself, being both expensive and attended with greater
" fatigue to the men. The larger sized troughs are 3 ft. long, and

"earth containing ore is washed in them, to ascertain what it con-
"tains.

"*Water Tubs.* . . . Tubs for water differ from buckets intended
"to contain rock in being made narrower towards the top, lest when
"being drawn out of a deep shaft, and striking against the wood-
"work, the water should be spilt. . . . .
"What we term bags, are those water vessels which are made of two
"bulls' hides. These, the hair being rubbed off, are rendered smooth
"and white. . . . . These bags are attached to hooks
"hanging from chains, and are let down into the water. When
"filled they are drawn out by a large machine. . . . .

MACHINERY.

"So much for the miners' tools and vessels. I have now to speak
"of the *machinery*, which may be classed under three heads:—Winding
"machines, ventilating machines, and means of ascent and descent.

"*Winding Machines.* There is considerable variety in the kinds
"of winding apparatus, some of which are very complicated, and
"unknown, as I believe, to the ancients, those especially which are
"intended for drawing water out of the depths of the earth beyond
"the point to which the galleries have been driven, or minerals out
"of the shaft, which are not in communication with any drift or
"only with the longest. But as the depth of pit is by no means the
"same there is great diversity in the machinery employed for the
"purpose. Of those, used to get dry loads out of the pit, there are
"five varieties most commonly used,—the first of which is con-
"structed after this manner. A couple of beams rather longer than
"the sides of the pit's mouth are placed on either side of it. Holes
"are made in the end of each beam, to receive wedge-shaped pegs
"driven deeply into the earth to keep them firm. They are also cut
"out to receive the ends of two cross-beams laid on the right and
"left of the opening—that to the left, however, being set so far back
"as to give room for fixing the ladders. Into the centre of these
"cross-beams are fixed two upright posts or thick planks secured
"with iron rails. In strong sockets, in the hollowed ends of these
"posts, are placed the iron ends of the windlass. These ends pro-
"jecting beyond the posts are fixed into a wooden handle. The
"rope is wound round the windlass and made fast to it at the
"middle, each end bring furnished with an iron hook to the bucket,

"which can be attached by its handle. The windlass being turned
"by the winches draws up, as it revolves, the one bucket loaded,
"and sends down the other empty. The windlass is worked by two
"strong men, each of whom has a barrow beside him, into which he
"empties the bucket as it comes up. Two buckets will generally fill
"a barrow; as soon, therefore, as four have been drawn up, both
"men wheel their barrows outside the shed and shoot out their con-
"tents, so that, if the shaft be sunk to any depth, a heap is formed
"close to the pit head. If the lode is poor in ore, the earth and
"rock is thrown away without examination; if it is rich, the output
"is kept to itself, crushed, and washed. But in the case of water,
"the buckets containing it are poured out into the spouts, from
"whence it flows into the water-courses.

"Another machine (Plate VII.), used where the pit is deeper,
"differs from the first in having a fly-wheel in addition to the crank
"handle. This may be turned by one man, if the load is not to be
"drawn from too great a depth, as the momentum of the wheel
"supplies the place of the other, but if the depth is considerable,
"three men will be required, the wheel being equivalent to a fourth,
"for the windlass, once set in motion, is kept going by the rotation
"of the wheel, so that it is much easier to turn it. The wheel is
"sometimes weighted with lead, so that when turned its own weight
"may cause it to revolve more rapidly. For the same reason the
"windlass is made heavier by two, three, or four iron bars having
"their ends weighted with lead. When this kind of winding engine
"is worked by three men, four straight handles are fixed into the
"end opposite to the wheel, the other being furnished with a crank.
"This is turned by one man, the other levers by two,—the one push-
"ing and the other pulling. It is desirable, however, that the men
"set to work this machine should be powerful, as the labour is
"somewhat severe.

"A third kind of machine is worked with less fatigue even when
"heavier weights are raised by it. It is slower than other geared
"machines, but will draw from a greater depth, say 108 ft. Its
"construction is as follows:—The iron pivots of the capstan turn in
"two iron sockets, the lower one of which is fixed into a block made
"fast into the ground, the upper into a beam. The lower part of the
"capstan is surrounded by a circular platform formed of thick

"planks fitted together, while a horizontal cog-wheel is secured to
"the upper part. The latter works into a pinion which forms part of
"a horizontal roll, the pivots of which are likewise enclosed in iron
"sockets, and around which the rope is wound. Two men grasping
"a pole, secured at both ends into wooden uprights to prevent them
"from falling, drive the machine with their feet, continually pushing
"back against the radial strips nailed to the platform. As soon as
"one loaded bucket has been brought up and emptied, they reverse
"the direction of the machine in order to draw out another."

Horse Capstans. Still more powerful machines there are in which the load is raised by means of horses. Plate VIII. shews such a machine. A circular space, some 50 ft. in diameter, is covered in by a conical penthouse A. This penthouse consists of a number of stout beams B, placed at intervals, the interstices being boarded, C. The beams are firmly secured at the apex of the cone by a strong iron ring, and a strong cross-piece carries a loop or pedestal in which the top pivot of the capstan barrel F revolves. The bottom pivot revolves in an iron socket fixed to a wooden frame sunk into the ground, E. The ropes are coiled on the drum H, which is of larger diameter than the barrel, and is formed by three wheels secured to the barrel and lagged. The ropes pass over pulleys over the pit P, and are attached to the load K.

In order to obtain still greater power, a modification of the preceding machine is sometimes employed. This is shown in Plate IX. In this case the drum shaft is horizontal instead of vertical, and is placed either underground or under a platform on which the horses travel. The axis of the horse capstan is prolonged downwards, and carries close to its lower end a disc containing a number of strong wooden teeth, A. The drum shaft B is horizontal, and is supported at either end by strong frames. At one end it carries a pinion C into the recesses of which the wooden teeth of the disc A work. A disc D is acted upon by the brake H and the balanced level G, and by means of this arrangement the speed is controlled and the load stopped on arriving at the top.

The last engine for raising minerals from shafts is that shewn in Plate XVI. Here water is the motive power. A double water-wheel of large diameter is carried on the drum shaft, the floats being set in opposite directions, as shewn at G and H. The supply of water enters the cistern A by the channel B, and is admitted to either water-wheel by the launders E and F, the admission being regulated by the levers C and D, The alternate admission of the water to the different sides of the water-wheel reverses the direction of the motion of the drum shaft, and the ends of the chain are alternately raised and lowered.* The ladder shewn on the side of the illustration is intended to represent the ladder in the shaft by which the men ascend and descend. The rest of the details need no explanation.

"The minerals drawn from the shafts by means of these machines,
" or carried out of adits of the galleries, is transported to the foot of
" the mountain in several ways. It is loaded into carts or sledges,
" and so conveyed down the mountain side. Dogs are also employed
" in winter time to convey leathern bags filled with mineral, or two
" or three such bags are placed on a small sledge with high curved
" ends in front and low behind, on which a man seats himself, and,
" at some peril of his life, slides down the mountain slope, guiding
" the sledge by a pole, by means of which he contrives to check it if
" the speed becomes too great, and to recover it should it diverge
" from the right track. In the Alps of Austria and the Tyrol, the
" minerals are put into sacks of pigskin with the bristles on, and are
" dragged down by men from the highest mountains, which are in-
" accessible to horses, mules, or asses. The sacks are carried up
" empty by strong dogs used to the work and having pack saddles.
" When filled they are tied up with cords and fastened to a strong
" rope, which a man twists about his arm and body, and by him are
" dragged over the snow until a place is reached to which horses and
" mules can ascend. Here the mineral is turned out of the pigskin
" sacks and put into others of strong linen cloth, in which they are
" conveyed by the beasts of burden to the places where they are to
" be washed or smelted."

Sometimes where there is a regular slope the mineral is tipped

---

* A winding machine almost identical with the one here described, is mentioned in a paper by Mr. W. S. Gresley, in vol. vii. of *The Journal of the British Society of Mining Students*, as being still in use at the great Dalecarlia Copper Mine in Sweden. From Mr. Gresley's account it would seem that, in spite of the invention of the steam engine, water is still used in preference in Sweden for almost all purposes on account of its greater economy, in fact, speaking generally, it would seem that most of the methods described by Agricola are still practised in Sweden with excellent results.

into wooden spouts, down which it slides to the place where the mules and horses can take it.

"When again loads of ore have to be got down steep mountain "slopes, a sort of two-wheeled truck is used, having a couple of tree "trunks fastened to the back of it. These by their weight act as a "drag upon the speed. . . . .
"The drivers undertake to bring a certain number of loads, each of "which the foreman marks down on a small stick."

Pumping Machines.

"Water is either drawn or pumped out of the shafts. If the "quantity of water is not great it may either be pulled out in skips "or pumped out in tubs or buckets, but where there is much water "it must either be drawn out in skins or pumped by pistons. In "the first place I will describe the machine by which water is drawn "out in buckets. Of this there are three kinds. The first is as "follows. A square framework (Plate X.) of iron rods, K, contains "three small shafts (horizontal), which revolve in sockets or broad "rings of tempered steel, and four iron wheels, two of which are "pinions and two cog-wheels. The lowest of the three shafts, B, has "a wooden fly-wheel, C, outside the frame to equalize its rate of "motion, and inside the smaller of the two pinions, which has eight "notches. The middle cylinder does not project beyond the frame, "its length, therefore, is only $2\frac{1}{2}$ ft.; it has the smaller of the two "cog-wheels at the one end (the number of cogs is 48), and the larger "pinion (with 12 notches) at the other. The topmost cylinder has "the large cog-wheel measuring a foot from centre to circumference "and having 72 cogs. Both the cogs and the grooves or notches "into which they work are tempered steel. The topmost cylinder "projecting beyond the iron frame is so enclosed within another "cylinder as to form one cylinder with it, and passing through the "timber scaffolding above the shaft, revolves in a steel socket inserted "into a thick block of oak. Round this cylinder turns such a wheel "(or drum) as some pumping engines have, set round with curved "iron bands R, which catch in the links of the iron winding chain "and prevent the great weight from pulling it back. The links of a "chain of this sort do not resemble those of an ordinary chain, each "link is open and has two ends re-curved, so as to hook on to the

"next, which causes the chain to have the appearance of being
"double. The buckets are attached by cords at the place where one
"link fits into the other, so that a chain of 100 links would have an
"equal number of buckets. Where the shaft is inclined, the buckets
"must be provided with lids, but this is unnecessary in a perpendi-
"cular shaft. The machine is set in motion by a man turning a winch,
"which causes the lowest of the 3 shafts (in the iron frame) to revolve,
"so that its pinion, catching the cog-wheel of the middle cylinder,
"and the pinion of this again catching the large cog-wheel of the
"third cylinder, sets in motion the roll round which the wheel turns,
"to which the chain with its bucket is attached. The buckets
"descend empty to the bottom of the mine, where the water has
"been collected in a sump. Here another drum or wheel, whose
"axle turns in iron sockets, is fixed, so that the chain passes under
"it, and the buckets being filled with water are drawn up and
"empty themselves, three at a time, into a large tub placed beneath
"the upper wheel, whence the water runs off into a channel prepared
"for it.

"This sort of engine, however, is not one of the most serviceable,
"as it is both costly and draws less water and more slowly than
"others which have a greater number of wheels.

"Another machine of this sort described briefly by Vitruvius,
"draws up the buckets more quickly. It is, therefore, of greater use
"in shafts where water comes in continously and in considerable
"quantity. In this case there is no iron frame-work nor cog-wheels,
"but one large wooden wheel round a cylinder, turned by the feet of
"men (like a tread-mill). The cylinder, however, having no drum,
"is soon worked out. For the rest it resembles the engine first
"described, except in having a double chain of different pattern. The
"iron bands (for holding the chain) are in this secured to the cylinder
"itself,—some are plain, some with a treble curve, some with
"spikes. But enough of the first kind of machines for drawing water.
"I will now go on to the second, that is to the different descriptions
"of pumps. The first is as follows. A wooden scaffolding is put
"above the sump, and one or two pipes, enclosed one within the
"other, are lowered into the water, and fixed to the bottom by iron
"hold-fasts with straight spikes to keep them firm. The bottom
"part of the lower pipe is enclosed in a sort of wooden box, 2 ft. in

"height, with holes pierced in it all round, through which the water
"rushes in. In the upper part of this box, (where one pipe only is
"used), is a box of iron, brass, or copper with the bottom taken out,
"and a circular valve which shuts down, so that the water sucked
"up by the air cannot fall back again. If there are two pipes, this
"box is placed at their junction, but enclosed in the lower, while the
"upper pipe has a spout or opening into some channel of the mine.
"A man standing on the scaffolding pulls up and presses down a
"piston, which consists of a horizontal handle, a rod, a sucker or
"piston bucket, the latter being a piece of leather stitched in a
"conical shape and fastened on to the rod. Instead of this a small
"circular piece of iron or wood perforated with five or six holes,
"elongated or circular, may be fastened to the piston rod by an iron
"nail or screwed on to it. As the man at the pump draws up the
"piston rod, the water is drawn up through the holes in these suction
"valves, and runs away through the spout in the upper pipe, the
"valve in the box before mentioned opening at the same time so as
"to allow the water flowing in at the holes in the bottom of the pipe
"to be drawn up by the air; and again closing when the piston is
"pressed down."

Where the quantity of water exceeds what can economically be pumped by manual labour, horse power or water power is substituted. Plate XI. shows a two-lift pumping engine actuated by a water-wheel. On the same shaft with the water-wheel is fixed a cog-wheel C, having about 32 cogs working into a pinion E with about 8 slots, so that the ratio of gearing is about 4 to 1. On the pinion shaft are fixed two cranks F F, connected with a forked rod G G, the iron rods passing downward into two sets of pump trees as shewn, the lower pair drawing the water for perhaps 20 ft. and discharging it into a launder in which the upper set of pumps is fixed, which again discharge into a launder on the surface. The height which such a set of suction pumps would lift, would vary from 40 to 50 ft. The rods are guided by a series of rocking shafts secured to the central frame, by which the whole is supported. For greater depths the form of machine shewn in Plate XIII. is employed. This contrivance, which, after remaining forgotten for a long period, was recently made the subject of a patent under the name of "The Chain Pump," must have been a very efficient machine. Referring to the illustration, B is either a large water-wheel where water is obtainable or, as shown in the illustration, may be caused to revolve by the weight of men treading on cross-beams in the interior of the wheel, somewhat after the manner of the tread-mill. B revolves on the axle A, on which is also fixed the cog-wheel C, working into the pinion D on the axle A. The regularity of the speed of this axle is assisted by the fly-wheel F. The sheave E consists of a solid block of wood, grooved on the circumference, and divided into sections by iron bars, between which the balls G rest and are caused to revolve with the axle A. The chain to which these balls are attached is endless, and passes downward free in the shaft, and entering at the bottom of the pump trees, fits fairly close to their inner bore. As the balls ascend, the water above each ball is imprisoned and raised with the ascending chain until it is delivered at the outlet at the top of the pumps. The balls are made of leather stuffed with hair, and will vary in size according to the size of the apparatus. The winding machine shewn in Plate XIV. will, however, according to our author, draw more water even than the chain pump.

The rest of the illustration needs no explanation, unless the author intends to suggest that the creaking of the various parts has a somnolent effect, as illustrated by the dog in the foreground.

Ventilating Machines.

Ventilating machines are of three kinds. Those in which the wind is conducted by various contrivances into the mine; 2nd, where it is drawn out by means of bellows; and 3rd, where it is exhausted by means of fans.

In Fig. XIV. two shafts are shewn divided into four compartments by cross brattices. Over the top of the shaft, and corresponding with these cross brattices are fixed the vertical boards D and E. From whatever side the wind blows it is intercepted by these vertical boards, and a portion of it goes down one of the divisions of the shaft. The use of the circular board at the top of Fig. XIV. appears to be to prevent the wind from escaping upwards and to direct it downwards into the shaft. The necessity for it depends on the position of the contrivance. If it were so situated that the wind would strike it from above the top board would not be required.

In Plate XV. a large pair of bellows, such as we are familiar with for blowing an ordinary house fire, is shewn. The shaft is here covered over by a platform A. To the left hand a partition fences off the portion containing the ladders, which may be one of the downcasts, another downcast being shewn with a jack-roll over it to the right. The bellows for this are actuated by a crank and lever worked by a man at D, the air being exhausted from the main through the boxes B.

In Plate XII. there are some further developments of this process of exhaustion by bellows, and in the upper portion two bellows are shewn as actuated by a gin. On the gin shaft is a crown wheel P, working into a pinion R, on the shaft of which are two projections O, which alternately strike upon levers which work the bellows. In the lower part of the illustration a man is working the bellows by means of treadles, the air being conducted to the bellows by means of air-boxes.

The central illustration is a modification of the one above. In this case a drum H is caused to revolve by a horse which is trained to tread on a series of steps on its circumference. On the axle of this drum are projections which press down the planks over the bellows, in a manner similar to the action of the projections O O in the drawing above.

But perhaps the most interesting ventilating appliances are those in which the modern ventilating fan is anticipated, and one of these is shewn in Fig. XVII. In the wooden case A, a fan revolves, consisting of a number of straight vanes which form a series of entirely closed sections as the vanes are continued right up to the axle. The air enters at B and is discharged into the air-boxes at C. The idea of allowing the air to enter at the centre does not seem to have occurred to these old designers. The air is, it will be noticed, forced into the mine. The machine may be actuated either by hand or by gearing from a gin, but by far the most satisfactory application is that shewn in the illustration, where a stream of water is the motive power, and the speed of the fan multiplied by the interposition of gearing. In the latter case the fan requires no attention, but may be left to run day and night without expense and without an attendant. Agricola speaks of these as the most satisfactory of all the ventilating machines known to him. It is perhaps strange that he makes no mention of furnaces as means of ventilation, but in the metalliferous mines which he describes fuel would be too costly for such a method to be adopted.

Miners in all ages have had their own legendary lore, and Agricola, though his views may in many respects be called advanced, —he has small faith for example in the divining rod, though he gives an account of its employment,—does not reject the belief in—

"The demons that are found
In fire, air, flood or underground,
Whose power hath a true consent
With planet or with element."

After treating of the diseases and hardships to which miners are subject, he proceeds as follows :—" There is yet another kind of harm "and nuisance whereunto miners be liable in certain mines, albeit "in few, to wit, the presence of demons or goblins of fierce and "terrible aspect." These truculent and murderous imps are gravely placed in the list of causes which prevent mines from being sunk to a greater depth. In an appendix to his main work, which treats "Of creatures dwelling underground," our author says, "Last "among the creatures or essences (as the theologians say) which "abide in the earth, I place those goblins or demons who haunt "certain mines. Of these there are two kinds, whereof the first "includeth such as are terrible to look upon, and for the more part "hostile to miners. Such is the Anneberg fiend, who is said to have "slain more than twelve men with the poisonous breath which issued "from his gaping jaws, appearing in the likeness of a horse with "long neck and flaming eyes. And of the Schneeberg demon, who "sheweth himself covered by a black cowl, it is confidently affirmed "that he did violently hurl a miner against the roof of the mine, "greatly bruising him thereby. Psellus, who doth reckon six sorts "or classes of demons, holdeth such to be the worst. The other "kind is of milder nature. These be they whom the Germans call "*kobolds* and the Greeks *cabeiri*, and they are imitations of men. "They are often heard laughing, and seem to be busily working, "though they do nothing to purpose. They are called by some '*the* "*small folk of the mountains.*' If they be not mocked or enraged by "foul language, they will be harmless; yea, some, it is said, will "even aid the miners in their work. In mines worked out or like to "be little profitable they are not found, whereof the miner rather "rejoiceth than trembleth at their presence."

I must now take leave of our author, for the remainder of the book is more particularly devoted to the metallurgical processes by which the metals, after being extracted from the earth, are smelted and reduced to the forms in which they become articles of commerce. These subjects are not less interesting in themselves than the mining operations, but they are less so to members of this Institute. Perhaps some member of the *Iron and Steel Institute* may at some future time think it worth while to deal with the remaining portion of the work for the amusement and instruction of his fellow metallurgists.

I am indebted to Miss Hodges, of Elsecar, for the excellent reproduction of Agricola's illustrations.

Mr. R. MILLER: I beg to propose that this paper which Mr. Walker has read be printed in the Transactions of the Institute, and also that a vote of thanks be given to Mr. Walker for his paper.

Mr. NEVIN: I have great pleasure in seconding it.

Mr. LUPTON: And I have great pleasure in supporting it. I am sure we are much indebted to Mr. Walker for the research that he has made into this mining in the middle ages. I may say with regard to one of these drawings, showing the reversing winding engine driven by water power, I have seen the exact thing, almost precisely as drawn here, at work in the Hartz Mountains a few years ago. It shows that except the application of steam power, the system in application to winding has remained almost the same as in the days of Agricola, and perhaps hundreds of years before his time. Without saying anything more on the subject matter of the paper, I shall support most cordially the vote of thanks.

Mr. JOSEPH MITCHELL put the motion, which was carried unanimously.

The CHAIRMAN: Thank you, gentlemen.

DISCUSSION ON MR. G. BLAKE WALKER'S PAPERS "ON HYDRO-CARBON EXPLOSIVES."

The CHAIRMAN: I believe the next item on the programme is the discussion on "Hydro-carbon Explosives." I would like to say I think we should have had by this time a paper by Mr. Bainbridge on Carbonite, but for a bereavement which he has suffered, and which prevented him from attending to the subject. I hope we shall have it before long.

Mr. JOSEPH MITCHELL : It is promised for the August meeting.

The discussion was adjourned.

Mr. JOSEPH MITCHELL : I have to report at last that we have obtained an exchange of Transactions with the Institution of Civil Engineers, and they have presented us with 34 bound volumes. All future Proceedings will be forwarded in due course.

Mr. LUPTON : Have you got the date of these volumes ?

Mr. NEVIN : They say they are the whole of the second series.

Mr. JOSEPH MITCHELL : We have not got the date here.

Mr. NEVIN : I beg to propose that the thanks of the Institute be given to the Institution of Civil Engineers for the Transactions they have sent.

Mr. LUPTON : I second the proposition. It is a most valuable addition to our library.

The motion was carried unanimously.

Mr. JOSEPH MITCHELL : I have here a list of arrears amounting to £89. The Council have given me instructions to call the attention of members to Rule 6, and to say that it will be carried out.

Mr. LUPTON : What is Rule 6 ?

The CHAIRMAN : That subscriptions are due in advance, that if not paid within three months the member's name to be posted in the Institute room ; and if a member is six months in arrear he is supposed to be withdrawn from the Institute, and that his name be erased if he fail to pay after one month's notice from the Secretary.

Mr. LUPTON : Are you going to let them withdraw with all their arrears ?

Mr. JOSEPH MITCHELL : They are still liable.

The meeting then ended.

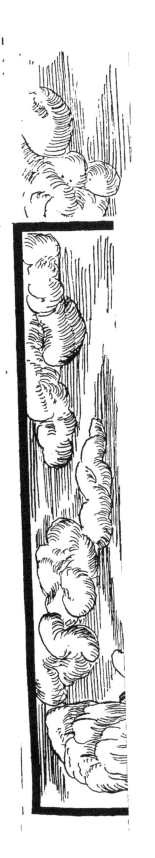

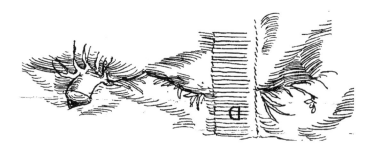

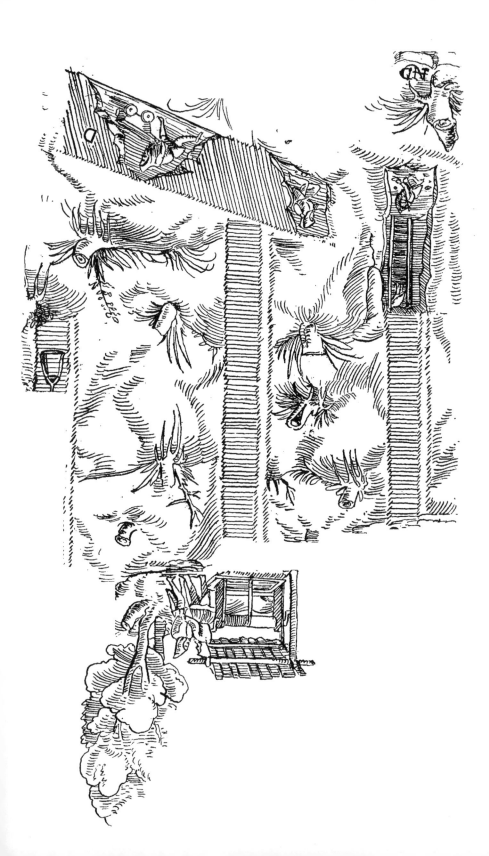

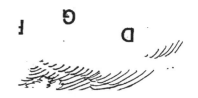

PLATE

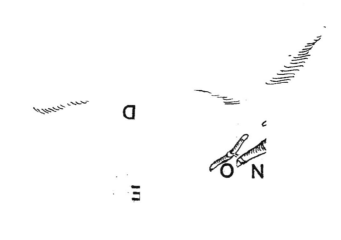

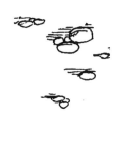

I

PLATE X.

PLATE XV.

K

N

O

A. M. CHAMBERS, J.P., Past President, Thorncliffe Iron and Collieries, Sheffield.

Secretary and Treasurer:

JOSEPH MITCHELL, Mem. Inst. C.E., F.G.S., Mining Eldon Street, Barnsley.

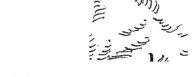

PLATE XVII.

OFFICERS—1888-89.

President:

C. E. RHODES, Aldwarke Main and Car House Collieries, Rotherham.

Vice-Presidents:

A. B. SOUTHALL, Monckton Main Colliery, Barnsley.
G. BLAKE WALKER, Wharncliffe Silkstone Colliery, Barnsley.
J. F. THOMSON, Manvers Main Colliery, Rotherham.

Council:

WM. HY. CHAMBERS, Denaby Main Colliery, Rotherham.
J. NEVIN, Dunbottle House, Mirfield, via Normanton.
JOHN GERRARD, H.M. Inspector of Mines, Wakefield.
W. E. GARFORTH, West Riding Colliery, Normanton.
G. J. KELL, Beechwood House, Kilnhurst, Rotherham.
T. W. H. MITCHELL, Mining Offices, Eldon Street, Barnsley.
J. LONGBOTHAM, Barrow Collieries, Barnsley.
C. H. COBBOLD, Dodworth, near Barnsley.

EX-OFFICIO:

T. W. EMBLETON, Past President, The Cedars, Methley, near Leeds
R. CARTER, C.E., F.G.S., Past President, Spring Bank, Harrogate.
T. CARRINGTON, C.E., F.G.S., Past President, Kiveton Park Colliery, Sheffield.
T. W. JEFFCOCK, C.E., F.G.S., J.P., Past President, Bank Street, Sheffield.
A. M. CHAMBERS, J.P., Past President, Thorncliffe Iron Works and Collieries, Sheffield.

Secretary and Treasurer:

JOSEPH MITCHELL, Mem. Inst. C.E., F.G.S., Mining Offices, Eldon Street, Barnsley.

LIFE MEMBERS.

COOPER, S. J., Mount Vernon, Barnsley.
FITZWILLIAM, THE RIGHT HON. EARL, Wentworth Woodhouse, Rotherham.
INGHAM, E. T., Blake Hall, Mirfield.

HONORARY MEMBERS.

ARMSTRONG, G. F., Yorkshire College, Leeds.
BELL, THOMAS, H.M. Inspector of Mines, Durham.
DICKINSON, J., H.M. Inspector of Mines, Manchester.
GREEN, Professor A. H., The University, Oxford.
HALL, H., H.M. Inspector of Mines, Rainhill, Prescott.
MARTIN, J. S., H.M. Inspector of Mines, Clifton.
MIALL, Professor L. C., Yorkshire College, Leeds.
MOORE, R., H.M. Inspector of Mines, Glasgow.
ROBSON, J. T., H.M. Inspector of Mines, Swansea.
RONALDSON, J. M., H.M. Inspector of Mines, 44, Athole Gardens, Glasgow.
RUSSELL, R., Sea View, St. Bees, Carnforth.
RUCKER, Professor, Errington, Clapham Park, London.
SCOTT, W. B., H.M. Inspector of Mines, Wolverhampton.
STOKES, A. H., H.M. Inspector of Mines, Derby.
THORPE, Professor T. E., Science and Art Department, South Kensington, London.
WARDELL, F. N., H.M. Inspector of Mines, Wath-on-Dearne, Rotherham.
WYNNE, THOMAS, H.M. Inspector of Mines, Gnossal, Stafford.
WILLIS, JAMES, H.M. Inspector of Mines, Newcastle-on-Tyne.

LIST OF MEMBERS.

ACKROYD, A., Morley Main Colliery, Leeds.
ADDY, J. J. Carlton Main Colliery, Barnsley.
ANDREWS, THOMAS, Wortley Iron Works, near Sheffield.
ASHTON, J. H., Waleswood Colliery, Sheffield.
BAILES, W., Cortonwood Colliery, Barnsley.
BAINBRIDGE, E., Nunnery Colliery, Sheffield.
BATTY, W., Darley Grove, Worsbro' Dale, Barnsley.
BEACHER, G. F., 101, Broomspring Lane, Sheffield.
BENNETT, J., Victoria Colliery, Wakefield.
BENNETT, J. T., Featherstone Main Colliery, Pontefract.
BLACKBURN, W., Astley, Woodlesford, near Leeds.
BLAKELEY, A. B., Soothill Wood Colliery, Batley.
BONSER, HAROLD, Newcastle-under-Lyme.
BRIERLEY, W., Roche Colliery, Batley.
BURNLEY, G. J., Birthwaite Hall, Darton, Barnsley.
CARRINGTON, T., Kiveton Park Colliery, Sheffield (PAST PRESIDENT).
CARTER, R., Spring Bank, Harrogate (PAST PRESIDENT).
CHAMBERS, A. M., Thorncliffe Collieries, Chapeltown, Sheffield (PAST PRESIDENT).
CHAMBERS, J. E., Huddersfield Road, Barnsley.
CHAMBERS, W. HOOLE, Rockingham Colliery, Barnsley.
CHAMBERS, WM. HY., Denaby Main Colliery, Rotherham (*Member of Council*).
CHILDE, H. S., Wakefield.
CLARKE, BERNARD E., Wharncliffe Silkstone Colliery, Barnsley.
CLARKE, JAMES A., Ayr Collieries, Tarbolton, N.B.
CLARKE, R. W., Wharncliffe Silkstone Colliery, Barnsley.
CLAYTON, C. D., Broxholme House, Doncaster.
COBBOLD, C. H., Dodworth, Barnsley (*Member of Council*).
COLVER, J. C., Eckington Colliery, Chesterfield.
COOPER, H., Wharncliffe Silkstone Colliery, Barnsley.
CRADOCK, G., Wakefield.
CRAIK, T., Church Street, Barnsley.

CRAVEN, JOHN, Westgate Common, Wakefield.
CRAWSHAW, C. B., The Collieries, Dewsbury.
CRESWICK, W., Sharlestone Colliery, Normanton.
CRIGHTON, JOHN, 2, Clarence Buildings, Booth St., Manchester.
EMBLETON, T. W., The Cedars, Methley, Leeds (PAST PRESIDENT).
EMBLETON, T. W., Junr., The Cedars, Methley, Leeds.
FINCKEN, C. W., Hoyland Silkstone Colliery, Barnsley.
FIRTH, J., Upper Fountain Street, Leeds.
FIRTH, W,, Water Lane, Leeds.
FOSTER, G., Osmondthorpe Colliery, Leeds.
GARFORTH, W. E., West Riding Colliery, Normanton (*Member of Council*).
GASCOYNE, ROWLAND, Mexbro', Rotherham.
GERRARD, JOHN, H.M. Inspector of Mines, Wakefield, (*Member of Council*).
GREAVES, J. O., Westgate, Wakefield.
GREENWOOD, Professor, Technical School, Sheffield.
HABERSHON, M. H., Thorncliffe Collieries, Chapeltown, Sheffield.
HAGGIE, F. W. Gateshead, Newcastle-on-Tyne.
HAGGIE, G., Sunderland.
HALL, M., Lofthouse Colliery, Wakefield.
HAMILTON, G., Bulli Coal Mining Co., Sydney, N.S. Wales.
HARGREAVES, J., Rothwell Haigh Colliery, Leeds.
HARGREAVES, W., Rothwell Haigh Colliery, Leeds.
HEDLEY, S. H., East Gawber Colliery, Barnsley.
HICKS, Professor, Firth College, Sheffield.
HIGSON, JOHN, Oak Bank, Lancaster Road, Eccles.
HINCHLIFFE, J., Bullhouse Colliery, Penistone, near Sheffield.
HODGSON, C., Rockingham Colliery, Barnsley.
HOLLIDAY, T., St. John's Colliery, Normanton,
HOWDEN, THOS., Ironfounder, Wakefield.
JACKSON, A., Howley Park Colliery, Batley.
JACKSON, W. G., Whitwood, Normanton.
JAMESON, THOMAS, Wharncliffe Silkstone Colliery, Barnsley.
JARRATT, J., Houghton Main Colliery, Barnsley.
JEFFCOCK, T. W., Bank Street, Sheffield (PAST PRESIDENT).
JEFFCOCK, C. E., Birley Colliery, Sheffield.
JEFFREY, G. F., 44a, Park Square, Leeds.
JOHNSON, J., Carlton Main Colliery, Barnsley.

MEMBERS. 205

JONES. F. J., Rother Vale Colliery, Treeton, near Rotherham.
KELL, G. J., Beechwood House, Kilnhurst, Rotherham (*Member of Council*).
LAMPEN, G. J., Kirkgate, Wakefield.
LIVERSEDGE, W. G., Norfolk Road, Sheffield.
LONGBOTHAM, JONATHAN, Barrow Collieries, Barnsley (*Member (of Council)*).
LOWRANCE, J., Peel Square, Barnsley.
LUPTON, A., 6, De Gray Road, Leeds.
MADDISON, T. R., Wakefield.
MARSHALL, J. L., Monk Bretton Colliery, Barnsley.
MARSHALL, W., Liversedge Colliery, Liversedge, via Normanton.
McMURTRIE, G. E. J., Car House Colliery, Rotherham.
MELLORS, JAMES, H.M. Inspector of Mines, Leeds.
MIDDLETON, E., Outwood House, Wakefield.
MILLER, R., Beech Grove, Locke Park, Barnsley.
MILLS, M. H., 15, Corporation Street, Chesterfield.
MIRFIELD COAL CO., Mirfield, via Normanton.
MITCALFE, F. D., Waratah Coal Co., Charlestown, N.S. Wales.
MITCHELL, JOSEPH, Mining Offices, Barnsley (*Secretary and Treasurer*).
MITCHELL, T. W. H., Mining Offices, Barnsley (*Member of Council*).
MUSGRAVE, HY., Havercroft Main Colliery, Wakefield.
NASH, H. B., Clarke's Old Silkstone Colliery, Barnsley.
NEVIN, JOHN, Dunbottle House, Mirfield, via Normanton (*Member of Council*).
NEWBOULD, T., Mineral Offices, Elsecar, near Barnsley.
NICHOLSON, M., Middleton Colliery, Leeds.
PARRY EVAN, Wharncliffe Woodmoor Colliery, Barnsley.
PATTISON, J., Morley Main Colliery, Leeds.
PEARCE, F. H., Bowling Iron Works, Bradford.
PEASEGOOD, W. G., Birley Colliery, Sheffield.
POLLARD, JOHN, Central Chambers, King Street, Wakefield.
PURCELL, S., Prince of Wales Colliery, Pontefract.
RHODES, C. E., Aldwarke Main and Car House Collieries, Rotherham (PRESIDENT).
RICHARDSON, A. M., 42, Swinegate, Leeds.
RITSON, JOSEPH, Strafford Colliery, Barnsley.

ROBINSON, J. G., Elland.
ROBERTS, SAMUEL, Park Grange, Sheffield.
ROUTLEDGE, R., Garforth Colliery, Leeds.
ROWLEY, W., 74, Albion Street, Leeds.
RYDER, W. J. H., Forth Street, Newcastle-on-Tyne.
SAINT, T. E. W., Mitchell Main Colliery, Barnsley.
SCOTT, F. W., Atlas Wire Rope Works, Reddish, Stockport.
SENIOR, A., Oak Well, Barnsley.
SHAW, JOHN, Darrington Hall, Pontefract.
SHAW, G., Wath Main Colliery, Rotherham.
SHEPHERD, F. H., South Prairie Mines, Melrose Purce Co., Washington Territory.
SHORT, W., Lambton Colliery, Newcastle, N. S. Wales.
SIMPKIN, J., Joan Royd Colliery, Heckmondwike.
SLACK, J., Rockingham Colliery, Barnsley.
SMITH, C. S., Shipley Collieries, Derby.
SMITH, H., Timber Merchant, Hull.
SMITH, SYDNEY A., Crown Buildings, Booth Street, Manchester.
SMITH, V., Strafford Colliery, Barnsley.
SNELL, ALBION T., 19, Maiden Crescent, Prince of Wales Road, London, N.W.
SOUTHALL, A. B., Monckton Main Colliery, Barnsley (VICE-PRESIDENT).
STEAR, J, Strafford Colliery, Barnsley.
STUBBS, THOS, Ibstock Colliery, Ashby-de-la-Zouch.
TEALE, W. E., Eccles, Manchester.
THIRKELL, E. W., Oaks Colliery, Barnsley.
THOMSON, J. F., Manvers Main Colliery, Rotherham (VICE-PRESIDENT).
TODD, W. G., Nunnery Colliery, Sheffield.
TYAS, A., Warren House, Sheffield Road, Barnsley.
TURNBULL, ROBERT, South Kirby Colliery, Pontefract.
WALKER, G. BLAKE., Wharncliffe Silkstone Colliery, Barnsley (VICE-PRESIDENT).
WALKER, HY. S., Great Walsingham, Norfolk.
WALLACE, J., King Street, Wigan.
WALTERS, H., Birley Colliery, Sheffield.
WARD, J., Wharncliffe Silkstone Colliery, Barnsley
WARD, W., Churwell Colliery, Leeds.

WASHINGTON, W., Mitchell Main Colliery, Barnsley.
WEEKS, J. G., Bedlington Colliery, Morpeth.
WHITE, J. F., Westgate, Wakefield.
WOODHEAD, E., Low Moor Iron Works, Bradford.
WOODHEAD, L., Beeston Colliery, Leeds.
WROE, JAMES, Lidgett Colliery, Tankersley, Barnsley.
WROE, JONATHAN, Wharncliffe Silkstone Colliery, Barnsley.

RULES.

1.—That the "SOUTH YORKSHIRE VIEWERS' ASSOCIATION" in future be called "THE MIDLAND INSTITUTE OF MINING, CIVIL, AND MECHANICAL ENGINEERS."

2.—That the objects of the Midland Institute of Mining, Civil, and Mechanical Engineers are to enable its Members to meet together at fixed periods, and to discuss the means for the ventilation of coal and other Mines, the winning and working of Collieries and Mines, the prevention of Accidents, and the advancement of the science of Mining Engineering generally.

3.—The Members of the Midland Institute of Mining, Civil, and Mechanical Engineers shall consist of Ordinary, Life, and Honorary Members.

4.—Ordinary and Life Members shall be persons educated or practising as Mining, Mechanical, or Civil Engineers, or persons having a direct interest in or the responsible management of operations connected with mining.

5.—Honorary Members shall be Mining Inspectors during the term of their office, and other persons who have distinguished themselves by their literary and scientific attainments, or who have made important communications to the society.

6,—The annual subscription for each Ordinary Member shall be £1 1s. payable in advance, and the same shall be considered due and payable on the first Tuesday in July, each year, or immediately after election. If the subscription be be not paid within *three months* the defaulter's name shall be posted up in the Meeting Room :—And any Member whose subscriptions shall be in arrear for *six months*, shall be considered as withdrawn from the Institute, and his name erased from the list of Members after one month's notice from the Secretary.

7.—Members who shall at one time make a donation of £20, or upwards, shall be Life Members.

8.—Persons desirous of becoming Members shall be proposed at a General Meeting. The nomination shall be in writing, and signed by two Ordinary Members, and shall state the name, residence, and occupation of the person proposed. The proposal shall be hung up in the Room of the Institute, at Barnsley, for one month, and the election shall take place at the next Meeting.

9.—That the Officers of the Institute shall consist of a President (who shall be a Mining, Civil, or Mechanical Engineer), three Vice-Presidents (not more than one of whom shall be a Mechanical Engineer), eight Councillors (not more than three of whom shall be Mechanical Engineers), a Treasurer and Secretary, who shall constitute a Council, for the direction and management of the affairs of the Institute. The President, Vice-Presidents, and Councillors shall be elected at the Annual Meeting, and shall be eligible for re-election with the exception of any President or Vice-President who may have held office for the three immediately preceding years, and such four Councillors as may have attended the fewest Council Meetings during the past year; but such Members shall be eligible for re-election after being one year out of office. Voting papers, with a list of Officers, shall be posted by the Secretary to all Members of the Institute at least fourteen days previous to the Annual Meeting; such Voting Papers to be by them filled up, signed, and returned under cover either personally or through the post, addressed to the Secretary, so as to be in his hands before the hour fixed for the election of Officers. The Chairman shall, in all cases of voting, appoint Scrutineers of the lists, and the scrutiny shall commence on the conclusion of the other business of the Meeting. At Meetings of the Council, five shall form a quorum, and the minutes of the Council's proceedings shall be at all times open to the inspection of the Members of the Institute.

10.—That Presidents who have become ineligible shall be *ex-officio* Members of the Council so long as they continue Members of the Institute.

11.—A General Meeting of the Institute shall be held on the first Wednesday in every month, excepting the months of January and June, at two p.m., and the Annual Meeting in the month of July shall be held in the Room of the Institute at Barnsley, at which a report of the proceedings, and an abstract of the audited accounts of the previous year shall be presented by the Council. A special meeting of the Institute may be called whenever the Council shall think fit, excepting the months excepted, and also on a requisition to the Council, signed by ten or more members. The Council shall decide where the monthly meeting shall be held. The Members in any *District* wishing the Monthly Meeting to be held in it, shall make application in writing to the Council, for permission to hold such Meeting, and shall provide a suitable room and make all necessary arrangements for that Meeting free of expense to the Institute.

12.—Every question which shall come before any Meeting of the Institute, shall be decided by the votes of the majority of the Ordinary and Life Members then present.

13.—The funds of the Society shall be deposited in the hands of the Treasurer, and shall be disbursed by him according to the directions of the Council.

14.—All papers intended to be read to the Institute shall be sent for the approval of the Council, accompanied by a short abstract of their contents.

15.—The Council shall have power to decide on the propriety of communicating to the Institute any papers which they receive, and they shall be at liberty, when they think it desirable, to direct any paper read before the Institute to be printed and transmitted to the Members. Intimation, when practicable, shall be given at the close of each General Meeting, of the subject of the paper or papers to be read, and of the questions for discussion at the next Meeting, and notice thereof shall be given by circular to each member. The reading of the papers shall not be delayed beyond such an hour as the President may think proper, and if the election of Members, or other business should not be despatched soon enough, the President may adjourn such business until after the discussion of the subject of the day.

16.—Members elected at any Meeting between the Annual Meetings shall be entitled to all the papers issued in that year, providing that subscriptions be not in arrear.

17.—The copyright of all papers communicated to, and accepted by, the Institute, shall become vested in the Institute, and such communication shall not be published, for sale or otherwise, without the permission of the Council.

18.—That each Member who may have taken part in the discussion upon any subject shall have a proof copy sent to him by the Secretary for correction; such copy to be returned to the Secretary not later than three days from the date of its receipt.

19.—The Institute is not, as a body, responsible for the facts and opinions advanced in the papers, nor in the abstracts of the conversations at the Meetings of the Institute.

20.—The Transactions of the Institute shall not be forwarded to Members whose subscriptions are in arrear.

21.—No duplicate copy of any portion of the proceedings shall be issued to any of the Members unless by written order from the Council.

22.—Each Member of the Institute shall have power to introduce a stranger to any of the General Meetings of the Institute, and shall sign in a book kept for that purpose, his own name, as well as the name and address of the person introduced, but such stranger shall not take part in any discussion or other business, unless permitted by the Meeting to do so.

23.—No alterations shall be made in any of the Laws, Rules, or Regulations of the Institute except at the Annual Meeting, and the particulars of every such alteration shall be announced at the previous General Meeting, and inserted in its minutes, and shall be exhibited in the Room of such Institute at the Meeting previous to such Annual Meeting, and such Meeting shall have power to adopt any modification of such proposed alteration or addition to the Rules.

24.—The author of each paper read before the Institute shall be allowed twelve copies of such paper (if ordered to be printed) for his own private use.

MIDLAND INSTITUTE OF MINING, CIVIL, AND MECHANICAL ENGINEERS.

ANNUAL MEETING.

HELD AT THE INSTITUTE ROOM, BARNSLEY, ON WEDESDAY, AUGUST 1ST, 1888.

T. W. EMBLETON, Esq., President, in the Chair.

The minutes of the last meeting were read and confirmed.

Mr. W. G. LIVERSEDGE, Mechanical Engineer, Sheffield, was elected a member of the Institute, having been previously nominated.

THE COUNCIL'S ANNUAL REPORT.

The SECRETARY read the Council's Report as follows:—

The Council in presenting their Annual Report for the year ending June, 1888, to the members of the Institute, have to state that the following papers have been read:—

- The President's Inaugural Address, which includes a Record of Ancient Mining, with the appliances used and treated, also of Leases, Royalties, and Payments for Coal Worked, Modes of Working, Shafts and Tubbing, with a description of the first introduction of Guides in Shafts; Pumping Appliances; Adoption of Tramways; Ventilation; Lighting of Mines; Prices per ton realized for Coal sold, &c., &c.
- Supplementary paper "On the Easterly Extension of the Leeds and Nottingham Coal-field," by Mr. ROWLAND GASCOYNE.
- This very interesting and valuable paper deals with an important subject, which the Council think should be further investigated.
- "On Considerations arising out of Sections 51, 52, 53, and 54 of the Coal Mines Regulation Act, 1887," by Mr. G. BLAKE WALKER.
- This paper was not discussed in consequence of a Committee being formed at the time, for the purpose of drawing up Special Rules for South and West Yorkshire, which Rules have since being adopted.

"On Hydro-Carbon Explosives and their Value for Mining Purposes," by Mr. G. BLAKE WALKER.

In this paper the writer reviewed the progress which had been made during the last few years in explosives extracted from hydro-carbon products, which are stated to possess certain special properties fitting them for use in mines, such as absence of flame and of deleterious gases. He also reported the results of certain experiments recently made in this country and Germany with an explosive of this class, namely, Roburite.

Addendum to paper "On Hydro-Carbon Explosives," by Mr. G. BLAKE WALKER.

"On Foreign Mining Rents and Royalties," by Mr. H. B. NASH.

This paper, which deals with the subject from a commercial point of view, is especially interesting to the lessors and lessees of coal in Yorkshire.

"On Mining in the Middle Ages," by Mr. G. BLAKE WALKER.

This was an interesting paper, and taken in connection with the President's (Mr. Embleton's) Inaugural Address, completes a valuable record of ancient and mediæval mining.

The following subjects have also been discussed :—

Mr. ROWLAND GASCOYNE's paper "On the Easterly Extension of the Leeds and Nottingham Coal-field."

Mr. B. E. CLARKE's paper "On an Arrangement for arresting the fall of Colliery Cages in cases of breakage of the rope."

The Committee's Observations on the Final Report of the Royal Commission on Accidents in Mines.

The proposed Federation of the Mining Institutes of Great Britain.

Mr. G. B. WALKER's papers "On Hydro-Carbon Explosives."

Proposed joint Fan Experiments.

Mr. H. B. NASH's paper "On Foreign Mining Rents and Royalties."

The number of members on the books of the Institute at the end of the year was 3 Life Members, 18 Honorary Members, and 150 Ordinary Members. This is a decrease of one in the number of Life Members and an increase of eight in the number of Ordinary Members.

The Council regret that many of the members are in arrear with their yearly subscriptions, but at the same time they have the satisfaction to record that this year there is a less deficiency than last, the item of arrears being £66 3s. 0d. as against £80 17s. 0d. last year.

The Accounts show a more favourable balance in the hands of the Bankers and Treasurer, but there is still a slight deficiency on account of arrears of subscriptions, and the Council earnestly hope that the members will remit their subscriptions at once.

Since last year's Report, a very important change has taken place. The Mines Regulation Bill then referred to, has since become law. In consequence of the passing of this Act, new Special Rules have been drawn up by a Committee composed principally of members of this Institute. These Rules have recently been accepted, with slight modifications, by the Government. It is reasonably hoped that this Act will settle all questions affecting mining for many years to come.

A joint Committee, composed of members from the North of England, South Wales, and this Institute, has been formed to conduct a series of experiments with ventilating fans of various types upon the same shafts under similar conditions, with a view of proving the comparative and relative useful effect produced. It is anticipated the cost of these experiments may amount to £300. To defray the amount required by this Institute, viz. £100, a special subscription list has been opened, and applications have been sent to lessors, lessees, and others interested in mining. Up to the present time the applications have been fairly responded to, but a considerable sum is still required, and, in the interest of all parties connected with mining, it is hoped the requisite amount will be raised.

The Council have pleasure in announcing that the Institution of Civil Engineers have agreed to exchange Transactions with this Institute, and have already forwarded 34 bound volumes, being the second series of their proceedings from 1879 to the present time. This contribution constitutes a very valuable addition to the Library, to which the special attention of members is called.

The Council are desirous of pointing out the fact that a further sum of money is still required for binding the Foreign Transations.

A Committee of this Institute has been appointed to confer with Committees of other Institutes of Mining Engineers in this country, with the object of forming a Federated Association to be called the Imperial Mining Institute. Since this joint Committee was appointed only one meeting has been held, and a full report of the proceedings will be laid before the members of this Institute in due course.

The Council appeal to members to take more interest in the work and proceedings of the Institute. This may be done by reading papers upon the various subjects in which they have had special experience, by attending the regular meetings, and by taking part from time to time in the discussions.

The Council are preparing a list of subjects upon which papers will be acceptable. It is hoped by such means that the usefulness of the Institute will be increased.

The Council suggest that the example of the Institution of Civil Engineers should be followed with regard to deceased members, and that after the death of any member a memoir of his professional career should be prepared and published in the Transactions of the Institute. Several esteemed members of the Institute have died without their memoirs being published, and it seems to the Council that it would be to the advantage of the Institute, and in keeping with its objects, to preserve such records in the Transactions.

The PRESIDENT: Gentlemen, you have heard the Report read; will any member propose that it be adopted, or make any observation upon it?

Mr. T. W. H. MITCHELL moved that the report be adopted.

Mr. POLLARD seconded the motion, which was carried unanimously.

ACCOUNTS.

The SECRETARY then read the annexed Statement of Accounts for the past year.

FINANCIAL STATEMENT.

THE TREASURER (JOSEPH MITCHELL) IN ACCOUNT WITH THE MIDLAND INSTITUTE OF MINING, CIVIL, AND MECHANICAL ENGINEERS.

DR.

1888.		£	s.	d.
June 30.—To Subscriptions received		176	8	0
,, Donations towards cost of proposed joint Fan Experiments		32	11	0
,, Sale of Transactions		3	3	6
,, ,, Dinner Tickets		8	5	0
,, ,, Mr. Percy's Paper		0	3	9
,, Amount received from Bank		121	18	7
,, Cheque returned		1	11	0
,, Bank Interest		0	5	1
,, Balance due to Treasurer		15	2	5
		£359	8	4

CR.

1887.		£	s.	d.
July 1.—By Balance due to Treasurer		26	10	5
1888.				
June 30.—By R. E. Griffiths		65	0	0
,, J. Fox & Son (Annual Dinner)		10	9	0
,, Reporter		6	10	0
,, Room Cleaning		2	12	0
,, Insurance Premium		0	15	9
,, Barnsley Gas Co.		0	9	5
,, John Davis & Son, Derby		0	5	6
,, Thos. Wall, Wigan		0	8	4
,, Hire of Rooms for Meetings in Leeds, Wakefield, and Sheffield		3	13	6
,, Stamps, Telegrams, Post Cards, and Sundries		15	13	1
,, Secretary's Salary		25	0	0
,, ,, Expenses		5	4	3
,, Rent of Room		15	0	0
,, Banking Co.		181	17	1
		£359	8	4

Audited and found correct,

GEO. J. KELL,
ALFRED B. SOUTHALL, } *Auditors.*

THE MIDLAND INSTITUTE OF MINING, CIVIL, AND MECHANICAL ENGINEERS.
GENERAL STATEMENT.

DR.

1888.	LIABILITIES.	£	s.	d.
June 30.—To R. E. Griffiths		67	2	0
,, Reporter		15	5	0
,, F. J. Butler		2	10	0
,, Barnsley Gas Co,		0	1	7
,, Balance being Capital		375	19	11
		£460	18	6

CR.

1888.	ASSETS.	£	s.	d.
June 30.—By Cash in Bank		78	12	6
,, Arrears of Subscriptions		66	3	0
,, Value of Transactions, 6,323 at 1s. per copy		316	3	0
		£460	18	6

Audited and found correct,

GEO. J. KELL,
ALFRED B. SOUTHALL, } Auditors.

Mr. GARFORTH: I notice in the assets you have down 6,323 copies of the Transactions at a shilling per copy, which comes to £316 3s. 0d. Every year we find the numbers increasing, and I would suggest that instead of charging members a shilling per copy you should charge a less price, say sixpence per copy, so that members who have not complete sets of the Transactions might be induced to fill up their blanks at once. I take it they would do far more good in the hands of members than locked up in a cupboard. I therefore move that the price be reduced to sixpence per copy, and that members be asked to take Transactions.

Mr. CARRINGTON: Asked what?

Mr. GARFORTH: To buy as many Transactions as they want.

The PRESIDENT: There is a note that members can obtain complete sets of the Transations at a reasonable price; it would be only, if you think proper, saying, instead of a reasonable price, sixpence each.

Mr. G. B. WALKER: If anyone wanted to buy the whole set of back numbers, sixpence might become more expensive than at present.

Mr. JOSEPH MITCHELL: We print a margin of 30 over what we require, but our numbers are variable, sometimes increasing and sometimes diminishing. When a member is in arrears for six months, or even three months, we do not send out the Transactions. If the arrears were all paid up the number of Transactions would be lessened, and we should have money in the bank instead of the numbers as an asset.

The PRESIDENT: If you have an extra number printed beyond what is absolutely required, I think it would be an advantage to the Institute if those that are in stock could be sold.

Mr. JOSEPH MITCHELL: That would have to go as a suggestion from this meeting to the Council, because they have fixed the price.

Mr. CARRINGTON: That is what I understand it to be, a suggestion to the Council.

Mr. GARFORTH: Yes; although I thought perhaps the Institute had power to settle the matter to-day.

Mr. CARRINGTON: It should be remembered that the prices fixed ought to be some considerable part altogether of the value of the the subscription; outsiders should not be able in any circumstances to get them for less than the members who pay their subscriptions.

The PRESIDENT: I propose that the Accounts be passed.

Mr. GARFORTH seconded the motion, which was carried unanimously.

ELECTION OF OFFICERS.

The PRESIDENT: It may not be in the knowledge of many members who have received the voting papers sent to each member, wherein the Council recommended that Mr. Wardell should be President for the ensuing year, that unexpectedly, although it was supposed and rather understood,—though I will not say anything further than that,—that he, if appointed, would accept the office of President, various letters have been received from him, not only by the Secretary, but by myself, altogether refusing to be elected President. He states as a reason that he has a great deal of work just now, that occupies the greater part of the day,—sometimes he has to sit up until three o'clock in the morning before he can get through his work,—and on this account he will not accept the appointment. Seeing that is the case, and that so many members were not aware of that until they entered the room, the Council think it necessary to adjourn the election of President for a month, and a voting paper will be sent out, containing the name of some one whom the Council recommend as President, and each member will then be at liberty to vote for the gentleman whose name is so given, or to cross that name out and substitute some other for it. That is the only way the Council have of getting over the difficulty, and I therefore move that course be adopted.

Mr. GARFORTH: I beg to second the motion.

The motion was carried.

Mr. GARFORTH: At the Council meeting we had, as just explained by the President, the difficulty about Mr. Wardell, but it was suggested that as Mr. Charles E. Rhodes had done so much for the interests of the Institute, especially as regards the safety lamp experiments, he was fully entitled to the office of President. Early in the afternoon we were in the difficulty of not knowing whether Mr. Rhodes would take the office or not, but as Mr. Carrington and several gentlemen have said how pleased they should be if he would accept, I am pleased to say Mr. Rhodes has consented. There may be a difficulty according to the rules in carrying his election to-day, but it has been suggested by the Secretary that it would be a very good thing to get an expression of opinion from this meeting, and with that object, but still carrying out the resolution to send slips, I beg to propose that Mr. Charles E. Rhodes be President for the ensuing twelve months.

Mr. G. B. WALKER: I have great pleasure in seconding the proposal for Mr. Rhodes' election. There can be no question amongst us that no one has deserved more than Mr. Rhodes to occupy that position.

The PRESIDENT: I think it would be a graceful tribute to Mr. Rhodes' exertions, not only with respect to lamps but other things, that he should be elected President of the Institute for the ensuing year.

The resolution was carried.

The PRESIDENT: In order to save time I think it is be necessary that the voting papers should be examined by Scrutineers, and I would ask Mr. Nevin and Mr. Saint if they would be so good as to take the voting papers, examine them, and give us the result.

Mr. W. HOOLE CHAMBERS raised a question as to the directions on the voting papers, pointing out that it was not in accordance with Rule 9.

Mr. G. B. WALKER: I will move to put the matter in order, that for this occasion only we adopt the voting paper as it stands, but in future the rule shall be strictly adhered to.

Mr. LAMPEN: I beg to second that.

Mr. W. HOOLE CHAMBERS: I think that amounts to the Council electing members of the Council. I am bound to object to that, and I shall propose "That when the voting papers are sent out for the election of President the names of the nominated Councillors also be placed upon it, and that a fresh election of the Council be made," because for myself I have only filled up this according to instructions of this paper, although I contend I was entitled to vote for eight.

Mr. GARFORTH: You vote for the other two.

Mr. W. HOOLE CHAMBERS: That is, the Council take it out of my hands and vote for me.

Mr. GARFORTH: It is only a suggestion.

The resolution was then put to the meeting and carried.

ELECTION OF OFFICERS.

The PRESIDENT: I beg to inform you of the result of the scrutiny:—

President we have none;

Vice-Presidents: Messrs. A. B. Southall, G. Blake Walker, J. F. Thompson.

Council: Messrs. W. Henry Chambers, J. Nevin, C. E. Rhodes, J. Gerrard, W. E. Garforth, G. J. Kell, T. W. H. Mitchell, J. Longbotham.

Secretary and Treasurer, Mr. Joseph Mitchell.

Mr. CARRINGTON: I think before going further a vote of thanks is due to the officers of the past year. As Mr. Embleton has just commenced a new year of office by reading the list of officers elected just now, I think we ought to pass unanimously a hearty vote of thanks to the President, Vice-Presidents, Council, Secretary and Treasurer of the Institute for their arduous labours in the past year.

Mr. NASH: I have great pleasure in seconding that.

The motion was carried.

The PRESIDENT: On behalf of myself and the other officers of the Institute, I thank you very much for the expression of kindness which you have just made towards us, and I hope that next year will be more prosperous than this.

DISCUSSION ON MR. G. B. WALKER'S PAPER ON "MINING IN THE MIDDLE AGES."

The PRESIDENT: The next business is the discussion on Mr. G. Blake Walker's paper on "Mining in the Middle Ages." I think it a very interesting paper, and it shows that although there are many professed new inventions at the present day, yet all these inventions were begun in the time of Agricola, as you see by the various drawings in the paper. In fact it proves this, that there is nothing new under the sun, that people invent things which they think are original, when, if they referred to such old books as this, they would find Agricola was many long years before them. Was the paper read from the Latin edition or the English?

Mr. G. B. WALKER: From the Latin; I do not know of an English one.

Mr. CARRINGTON: No; it is in Latin. I have seen and read the book.

Mr. G. B. WALKER: I think there is no translation in English.

The PRESIDENT: I fancy there is, but I am not quite certain about it.

Mr. GERRARD: I do not know that there is much to discuss in the paper, but I would express my own pleasure at having the opportunity of seeing such very carefully prepared extracts from a very rare book, a book which is exceedingly difficult to come across, especially with the very capital illustrations, which, of course, help us to understand what was done so very long ago. I have had the pleasure of seeing the original, and I am sure the Institute is very much indebted to Mr. Walker for enabling us to have at hand the illustrations and the pith of this very ancient and interesting old book.

Mr. G. B. WALKER: There is just one point, which is quite a subsidiary one in the paper, which is a parenthesis merely, on which I might occupy a few minutes. On page 182, Agricola mentions the fact that the miners when at their work solace their laborious task by sweet songs. In the *Nineteenth Century* for August there is an article on that subject—"On the monotous Dirges and Phrases by means of which workmen in all times have been accustomed to beguile the tedium of their labour."

Professor Attwell also writes me:

"I knew little more than the existence of George Landmann as a friend of Erasmus, who, 'almost persuaded to become a' Protestant, came to logger-heads with Luther. It is curious that such a man should have returned to Catholicism; but your closing extract shews his superstitious tendencies. His imagination seems to have been too strong for his reasoning faculties.

"Your explanations of the very curious engravings make them quite clear, even to such a reader as myself.

"I wonder whether when you were translating the passage on page 182, it occurred to you that the 'sweet songs' may have been the chants with which labourers lighten, by re-iterated cadences, monotous work. This is a subject that has never, so far as I know, been seriously treated. I always regret that when in Lausanne seven years ago I did not take the pains to note down the chants of some men who were ramming down piles. There was a remarkable development of *ten* strokes; the words *and* tunes kept up a sort of expectancy that made the intervals glide one into another. The chants, in a minor key, were very original, and must have been ancient. I can't remember the precise words, but they were something of this character :—

"Now then for *one*!
Then comes *two*!
Heigh, for the *three*!
Leading to its *four*!
Here comes *five*!
Now for good *six*!
Eight is soon here!
Now, after *nine*!
We've got to *ten*."

It was the *repetition* that gave the charm to the chant. I must have heard the series a dozen time over before the men paused. At the 'We've got to ten' there was no *drop* in the final note, but it led to the 'Now then for one.' No two chants were precisely alike."

Mr. CARRINGTON: I can confirm what Mr. Walker has described relative to the custom of miners in some countries to sing during their work as an amusement or solace of their labour. I have had a good deal of experience amongst both the Norweigan and Swedish miners, and have repeatedly heard them sing during their work certain songs, which seemed to be regularly in use amongst miners of those countries. With regard to the extract from Agricola's work *De Re Metallica*, and the allusion on page 182 in Part 96 of the Transactions as to the use of fire for driving drifts, I may say that I was concerned in the management of some argentiferous copper ore mines in Norway for many years, and that when I first went there, in 1864, I found adits, both ancient and modern, had been driven through gneiss and metamorphic schists for long distances by fire in a very simple way. I was astonished, for not only was the process new to me, but I was struck especially by the rapid way in which drifts were driven by heating the "face" or end of the drift to a red heat by timber pilled up and burnt to a strong red heat. When the timber had burnt down, water was thrown on the hot rock, which then split up into fragments, which the men easily removed. After this, a fresh pile of timber was set and fired, the process of heating and cooling being immediately repeated again. After we took the mines in hand, the driving was done by nitro-glycerine, and the work being let by contract, was of course carried on more quickly; but in a country abounding with wood, and where time or speed is not of the very greatest importance, no doubt there is something to be said, certainly as to its cheapness, for the method of driving drifts by fire. There is another matter relative to Agricola's work, which confirms what Mr. Embleton, as well as an older authority, has said as to there "being nothing new under the sun." Some years ago, on the ground that a patent-right was being infringed, an attempt was made to deter the Wingerworth Iron Company in Derbyshire from using an apparatus for sprinkling hot water upon and saturating the shale and bind adhering to argillaceous ironstone from coal measures worked at their mines, for the purpose of separating the shale and bind from the ironstone more quickly than it could be achieved by the ordinary process of letting it "weather" in the usual way. It appeared that a patent had actually been taken out for an apparatus for this purpose. However, it was accidently found from Agricola's work that a method of sprinkling ores and saturating them with water in order to clean them, instead of by the ordinary "weathering," had been used many centuries ago. On this being proved, with other points, to the parties who claimed the patent-right, they withdrew their claim.

Mr. G. B. WALKER: Have you noticed the fan, Plate 17 ? It is a singular anticipation of the Waddle.

Mr. CARRINGTON: I should have added my own expression of thanks to Mr. Walker for the trouble that he has been at in making the translation of Agricola. I know the book, I have had it in my possession, and I know it is no light task to undertake the translation of such a work, and I feel the members of the Institute are under a considerable obligation to him. It does not, perhaps, come to a practical point so much as other works, because it is an ancient record, but it is very interesting, and supplements what Mr. Embleton alluded to in his inaugural address.

The PRESIDENT: You have everything that is in use at the present time, the only improvement is that the motion is quicker, For instance, the horse working the drum is very slow compared with the engine.

Mr. G. B. WALKER: I believe there is a patent in existence which reproduces the old idea shown in Plate 13, that of the chain-pump with the balls, but in the newer patents the balls are of cast-iron.

The PRESIDENT: If you refer to Plate 16 you see there an engine worked by water, and that is so regulated that the wheel can be reversed. I once saw a similar apparatus in Wales, and I was told it was quite a new invention there, that instead of employing steam for the purpose they had water-power, and the water-wheel was so constructed that the water could be let in by sluices on either side of the wheel, so that the machine could be reversed at any time, and here we have the identical machine in No. 16. So I may go on making the same observation about every part mentioned here, and in refering to the original book you find these increased in number.

Mr. NEVIN : A similar water-wheel to that was used at the lead mines at Leadhills in Scotland a few years ago, but I believe they have replaced that by a wheel going in one direction but varying the motion of the drum by spur-wheels.

Mr. GERRARD : The fact of its having being used so recently is of interest.

The discussion then closed.

DISCUSSION ON MR. G. BLAKE WALKER'S PAPERS ON "HYDRO-CARBON EXPLOSIVES."

Mr. C. E. RHODES : I should like to make a few remarks upon Mr. Walker's paper as bearing on the subject of high explosives. Since that paper was read I have made a large number of experiments with high explosives, and to my mind it is perfectly clear that in every case, unless the explosives are properly tamped and rammed, they will almost invariably fire either an explosive mixture of fire-damp and air or an explosive substance such as coal-dust or gunpowder. These experiments I have made with an apparatus expressly put down for the purpose, and I shall be glad at any time to repeat them for the benefit of any members of the Institute who like to attend and see them carried out. I thought I would mention that, seeing that at present there is only my bare assertion that I have produced these results, and as this question is so important not only to this Institute but to the whole mining community, I think it would be very desirable that these experiments should be confirmed by others who care to see them.

Mr. G. B. WALKER : Will you explain the term "properly tamped?"

Mr. RHODES : I mean rammed and tamped in a similar way to that in which it would be used in a mine. Merely placing it in a hole and firing it it will invariably fire an explosive mixture, whether of fire-damp and air, or fire-damp, or coal-dust. The experiments have been made by firing shots in a cast-iron anvil with a hole drilled in the side, so that you have got a result the same as a blown-out shot in the pit. They have been fired in an explosive mixture travelling 12 feet per second, similar to that for testing safety lamps, and the mixture of coal-dust with an explosive mixture has been obtained by some light shutters hanging on wires, which were jerked at the time the electric spark had been applied for firing purposes. In no case could I fire the surrounding atmosphere when the explosive was properly tamped and rammed. All the experiments I made underground with blown-out shots were decidedly in opposition to the statements that have been made that there is no visible flame. I certainly got flame repeatedly five and six feet long, and about the tinge of methylated spirits, and not only was this the case with Roburite, but with Securite and Tonite.

Mr. G. B. WALKER: Have you been able to get Securite?

Mr. RHODES: Yes. I do not think there is any difference in the safety as compared with Roburite. I tested both in the presence of some gentlemen from London interested in one of them, and I do not think there is much difference between the two substances. I cannot see where the difference comes in, and the fumes after a shot is fired are almost indentical in each case. We have been using Roburite a month in stone drifting, but we do not get satisfactory results as far as power goes.

Mr. CARRINGTON: You noticed flame?

Mr. RHODES: There is a distinct flame.

Mr. CARRINGTON: I quite agree with you. I have not seen a case where there has not been flame.

Mr. RHODES: I have a letter from a gentleman connected with Roburite, who says it is light and not flame. It is too fine a distinction for me to go into. I call a light one can see 150 yards off a flame, where it is not the result of reflection. There is no doubt that it is flame, but seeing it is surrounded by a heavy gas, whether it would fire an explosive gas I do not know, but it seems to me there is a very fine line between safety and danger.

Mr. LAMPEN: Did you fire the shots by electricity?

Mr. RHODES: Yes. In order to test whether the detonators would fire the mixture, we fired 12 detonators at once and never fired the explosive mixture. This we did several times, so that there can be no question that any flame given off was the result of the explosion of Roburite.

The PRESIDENT: I am sure every gentleman present must feel very much obliged to Mr. Rhodes for the offer that he has made. It is a most important matter to investigate the behaviour of these explosives, and one that ought to be tested in every possible way, in order that we may at last arrive at some method by which those explosives may be used which will not produce flame, or be liable to ignite coal-dust or gas. I have had some experiments made with Tonite, and I suppose that those shots have been properly tamped, because in not one of them has the slightest sign of fire or reflection of any light whatever been seen. This paper, which is written by my underviewer, will be before the Institute perhaps at the next meeting. It appears from what you say that, no flame having been seen when using Tonite, the hole has been properly tamped. I am sure we are much indebted to you, Mr. Rhodes, for the invitation, and if you will name a day convenient to yourself and communicate with the Secretary, he will be able to inform the members to let them come.

Mr. CARRINGTON: What sort of tamping did you use with Roburite?

Mr. RHODES: We used unbaked brick clay exclusively for tamping in the pit.

Mr. CARRINGTON: What tamping did you use for Tonite?

The PRESIDENT: A peculiar substance that is to be used with it. It is to be rammed to the end of the hole, and there is to be a space left all round the cartridge, which is also filled with the composition. What the composition is I do not know.

Mr. RHODES: In no case have I fired the explosive mixture when the explosive has been fired as it is used in the pit. My only point is that it is a narrow margin we are relyng on, and one which ought to be considered and thoroughly investigated.

Mr. LAMPEN: I have sold a deal of Roburite, and heard several differences of opinion expressed by colliery managers with regard to it. Some say that it is everything that could be desired, and some say the same as Mr. Rhodes. I wrote to the manager and told him to send an experienced man over, and he says he will send a man over on Wednesday, the 8th, to be at the disposal of myself, which means, of course, at the disposal of any colliery proprietors in the district. That, I think, will meet the requirements of the case, because in ramming Roburite great care should be taken, and I am inclined to think whenever it has not given satisfaction it has been owing to this fault; and again in the electric apparatus, if not strong enough to fire the detonators especially manufactured for Roburite would also be the cause of failure. If you get an experienced man from Wigan he might be able to suggest things to give a different aspect to it to what it has at present.

Mr. GERRARD: With reference to the kind offer made by Mr. Rhodes to enable members of the Institute to witness experiments, I should think there cannot be two opinions about accepting it, and in addition I would suggest that a Committee be appointed at this meeting to draw up a report, so that it may appear on the Transactions of the Institute.

Mr. CARRINGTON: You would not bind them to one explosive?

Mr. GERRARD: No; it would be well for them to test all those which can be obtained. I am not speaking in the interest of anyone, but simply as to the importance of this question, which is being taken up experimentally by many collieries, and unless these facts, which Mr. Rhodes states that he has ascertained, are made known, it is possible something serious may occur. In order that it may appear on the Transactions, I would suggest that a small Committee be appointed to investigate along with Mr. Rhodes, and to report to the Institute.

The PRESIDENT: In the same way as the Committee on lamps?

Mr. GERRARD: Quite so. Failing that, perhaps Mr. Rhodes will prepare a paper, so that it could appear in the Transactions and be circulated amongst the members, so that they could have this information we have now had more fully.

Mr. RHODES: A Committee would be more satisfactory.

Mr. W. HOOLE CHAMBERS: I have great pleasure in seconding Mr. Gerrard's proposition. It appears from Mr. Rhodes' remarks that we are simply between the point of danger and safety by the fact of the shot-holes being properly tamped or not. If a shot-hole is not properly tamped in the pit it is evident you are liable to accident at any moment, therefore you are still at the mercy of the men who have to deal with and put in the shots. It is of the utmost importance this matter should be cleared up at the earliest possible moment. As **Mr.** Gerrard has said, experiments have been made in a considerable number of mines where the effect of an explosive shewing itself if a shot-hole is not properly rammed would be a source of very great danger. I think the best way to meet it is to appoint a Committee to go into experiments and made a report, in order that it may be laid before the Institute as early as possible.

Mr. BONSER : Perhaps it may not have occurred to Mr. Rhodes to see what occurred in the back of the hole. I have taken Gelignite, Gelatine-Dynamite, Tonite, and Roburite,—I have not had Securite, I could not obtain that,—and not having had an apparatus prepared with gas, and that sort of thing, to try them, I thought the next best thing would be to place the explosive to be tried in a bore-hole, filling up part of the back of the hole with fine gossamer wool, which would ignite rapidly, and has this advantage over spirits that, when blown through the atmosphere, it consumes more rapidly than when left alone. Trying one explosive after another in the same kind of hole under the same conditions, and putting fine gossamer wool in the back of the hole, I found they all behaved in the same way, the gossamer wool was consumed in each case. There is no doubt they have all an advantage over gunpowder, but there is not one of them but would ignite an explosive mixture if it came in contact with it. If the hole was well tamped you would not see flame because the tamping would not be blown, but what occurs at the back of the hole is another matter. With everyone of the explosives it would occur that, if a shot was fired in a ripping, a hole bored to a crack or fissure in the rock, or to a hugger in the coal, which was a leader from some reservoir of gas in the mine, there is no doubt that any one of these high explosives would light the gas in that fissure, which in turn would cause an explosion of the main body. It would not be safe to use any of the explosives under such circumstances unless some extinguishing agent be employed to arrest the flame. Whether Tonite will not ignite such with the agent prepared for use with it I do not know, but I think they ought all to have some extinguishing material around them. There have been two colliery explosions and one in a lead mine, arising from the explosion of high explosives used without some method of extinguishing the flame. Of those I have seen myself the only difference I could observe in the flame was,—they were all of equal size pretty nearly,—the Gelignite flame was rosy, Roburite a shade of yellow, and Tonite a shade brighter. All these high explosives should be provided with some method of protection from coming in contact with gas or fissures in which gas may exist. I do not think any of them safe to be fired in a fiery or dusty mine without such protection.

Mr. GARFORTH: In my experiments I have not found a safe explosive: they all give off flame. It was suggested the flame was due to having no resistance; I therefore had metal cases made in order to get the same resistance in a small thickness of metal as in a great thickness of rock. As a result I have seen flame in every case, both with the explosive mixtures and gunpowder.

Mr. G. B. WALKER: There is one point I should like to mention with regard to Roburite; that the very principle of the invention depends upon giving the chlorine fair play. If the explosive is so fired that the chlorine has no chance of quenching the flame which undoubtedly is in existence at the moment of detonation, then the whole principle of the explosive is as completely subverted as if you were to fire the water-cartridge without the water. What Mr. Garforth has just said rather suggests that to my mind. Do you consider in the experiments to which you just now referred that you gave the explosive every chance to act in the manner in which the inventor intended it should act, and that in spite of that the chlorine failed to quench the initial flame?

Mr. GARFORTH: I do not know what the inventor intended except to give an explosive which was entirely free from flame and which would not fire gas.

Mr. GERRARD: What was the thickness of the cases?

Mr. GARFORTH: The dimensions were $2\frac{1}{2}$ inches diameter, 9 inches long, and half inch to three-quarter inch thick.

Mr. GERRARD: What metal?

Mr. GARFORTH: Brass and gun metal. That idea was suggested by some experiments I made many years ago with compressed air. I found cylinders about the same size burst at 8,000 or 9,000 lbs. pressure per inch equal to bringing down four or five tons of coal, so that I take it that thickness of metal has an equal resistance to about four feet of coal.

Mr. G. B. WALKER: I think what Mr. Rhodes has said with respect to the experiments, that in no case has he got flame from a properly tamped shot, argues in favour of the idea that chlorine does act as a quenching agent, but it is a point which, having been disputed to-day, the Committee should look to when they make the experiments.

Mr. GARFORTH: The chlorine is not quick enough.

The PRESIDENT: This is rather an extra discussion, because the question before you is the appointment of a Committee to attend the experiments Mr. Rhodes proposes.

Mr. GERRARD: I think there will be considerable advantage in the discussion as enabling the Committee to direct attention to some important points.

Mr. GARFORTH: If you discard explosives and come to the wedge, as we did in the pit where we had the explosion, it means an extra cost of about £250 per year on 1,000 tons a day. We thought at first the four feet of strong bind could not be wedged down, but we now find we are able to do it, and keep the ripping well up to the face.

Mr. WALKER: How many places?

Mr. GARFORTH: 45 to 50 stalls.

Mr. WALKER: About a yard a week?

Mr. GARFORTH: Two yards.

Mr. RHODES: There are places where that would be very much more difficult.

Mr. GERRARD moved that the President, Messrs. Rhodes, G. B. Walker, Garforth, and Carrington be the Committee, and this was agreed to unanimously.

DISCUSSION ON MR. H. B. NASH'S PAPER ON "FOREIGN MINING RENTS AND ROYALTIES."

The PRESIDENT: The adjourned discussion on Mr. Nash's paper is the next subject.

Mr. COBBOLD: I have read Mr. Nash's and other papers on mining royalties with great interest, and I think there are several points in this, I hardly know whether to call it agitation or not, which demand a little alteration. I do not know whether it has struck the writers of these papers that whatever basis mine royalties are put upon, it will have to be a moral system. It has been said the royalties should go into the hands of the Government. If it should be so, the only system of dealing with them would be to put them in the market to the highest bidder, because everyone of us would have to demand, and for future generations should be obliged to demand, that the royalties should be put into the market, and, subject to certain conditions of stability, the highest bidder must get them. Suppose they were not dealt with in that way, and that Government was able to fix a certain low royalty which should be the price for certain minerals that were to be worked, what would be the result? It may be claimed that those who are working the coal would immediately become possessed of certain larger profits, and therefore be better off in their financial matters than at present. But I think that is a subject of very great doubt, because the fixed low royalty would immediately bring in a very severe amount of competition in obtaining these fields, and there would be a larger number of fields of coal opened out, and the competition would bring down prices, and they would be no better off than at present. But suppose it did not bring more coal into the market, that the present coal pits simply existed, and that the Government put a restriction upon the number of coal pits allowed to be worked in a district at one particular time. Then the greatest evil that exists at present would be only aggravated, for I consider at present the greatest evil which exists is that credit seems to lie in underselling your neighbour, and not as it used to be in getting a higher price than your neighbour. It seems now that if a man can come back and say that he has sold at so much less than his neighbour, he is looked upon as cleverer than the man who can say that he has sold at so much above his neighbour. So I argue that these profits, which I suppose are the object of this agitation, would disappear either in the competition or in the extra production of coal from their being no limit put upon the coal-fields. Then I see great evil in the system of doing away with landlords, for you have at present responsible possessors of the mineral properties, whereas on the other hand you have an irresponsible possessor, the Government, from whom nothing could be obtained without passing an Act of Parliament to get it. You also put into the hands of an irresponsible number of shareholders certain claims which at the present time are borne by the possessors of royalties, claims with which the directors of any company would be powerless to deal, for any shareholder can immediately object to any part of the profits being applied to any purpose outside the legitimate trade carried on under the articles of association. You would, therefore, in your articles of association, have to take up all sorts of charitable societies, and convert your financial concern into a charitable society. In that way you would knock down a certain class who, living in the district, alleviate this distress and take upon themselves a large voluntary burden, and in its place set up a class of shareholders who would be quite irresponsible, and whose pride would not be that " these collieries round here are mine, these workmen are responsible to me, these tenants are mine," but who would be living out of the district, and whose pride would be " this is my dividend, these are my consols, these are my Colonial or American shares." There is another point brought very prominently forward in these papers,—the low royalties paid in other countries. I think it has now been pointed out that in France, although the minerals belong to the Government and are worked at a very low royalty, yet by a system of taxation the royalties are raised to rather higher than we pay in this country. In Spain, it has not been contradicted that the minerals belong to the Government and are worked on a very small royalty, but then in Spain the first claimant gets this low royalty, and in very few instances works it himself, and the royalty paid in Spain on minerals, I pay it myself, is up to 40 or 50 per cent. of the gross output. So in Italy, where to make up for the Government proprietorship of the minerals there is an income tax put on of $14\frac{1}{2}$ per cent. of your profits. I think you will acknowledge, carrying this subject to the bottom and looking it fairly in the face, there is no other country where the landlord system has not existed. You will not find one where so much money and such fair profits have been made as are made in England, where the landlord system exists. The one side shews you nominally low royalties, a trade with labour remunerated by starvation wages, an irresponsible landlord, and trade bolstered up by all sorts of false props. Your own side with a freedom of contract, a well remunerated labour market. a responsible landlord, a trade with a fair average return during the past twenty years.

The PRESIDENT : Don't you think that the fault about collieries is this. If you look at the number of millions of tons raised last year as compared with 1886, it would account altogether for the low profits which have been obtained. It is the quantity which reduces profits, and every colliery owner, large or small, is endeavouring if possible to increase his quantity, because by doing so he reduces the cost per ton of all his standing charges. He forgets that raising this large quantity has the effect of reducing the selling price perhaps 6d or 1s. a ton. That seems to me to be the evil that exists at the present time.

Mr. NEVIN: A great advantage of Government having the minerals is that you get a concession of certain size and regular shape. As it is in England, if you take 300 acres of coal you may have 30 people to deal with, and the whole of these people expect a minimum rent. You have then the choice either to work the pit in an unworkmanlike manner and get a bit of coal from every royalty, or else you have to pay overgettings to some people and certain rents to the other people. Then probably you have a place where a man with a small piece of coal prevents you getting into a large royalty because he will not sell at a reasonable price.

Mr. RHODES: I do not know that I would go in for Government proprietory of the coal, but I think the Government could exercise an ameliorating influence on these bargains that have been made. There are firms in Lancashire with £200,000 minimum rents overpaid, which they can never get.

Mr. G. B. WALKER: You want a Land Court, Mr. Rhodes.

Mr. RHODES: It is perhaps what I should like.

Mr. PRESIDENT: When I mentioned standing expenses, I included all these overpaid rents and other charges.

Mr. GERRARD: Mr. Cobbold spoke from the landlord's point of view, and I quite expected that he would have been answered from various parts of the room from the lessee's point of view, because it is a most interesting and pocket-touching subject. Having lately seen a few Scotch collieries, it was brought very forcibly to our notice. There I was told they were getting an average selling price of 4s. per ton: it is costing them 4s. 3d. to raise, and they are paying royalties of as much as 1s. 4d. per ton. Such an amount of royalty rent seems out of all proportion, when probably the lessee has had to spend £100,000 in laying out the colliery. Of course there are good landlords and bad landlords. There could be cases adduced of concerns having gone down under the burden of these minimum rents, royalties, way-leaves, and so on. If it could be brought home more locally by information as to what Yorkshire royalties are, and how lessees are met, it would give a practical turn to this discussion.

The PRESIDENT: I hope Mr. Gerrard will give us such a paper.

Mr. GERRARD: No; I think I should be the last person to interfere in such a matter.

Mr. COBBOLD: I am not an advocate for high royalties, way-leaves, or minimum rents. I do know in Scotland collieries paying as high royalties as Mr. Gerrard mentions, but in the cases I have known the coal that passes through two inch bars is not paid for, so that it brings the 1s. 4d. to perhaps 6½d. or 7d. per ton. I should not like it to go out that people are paying 1s. 4d. a ton, when there is a very large rebate.

Mr. NEVIN: I think the fairest system of royalties is based on a system of prices, as in cases I know, where there is a certain basis, beyond which the royalties increased a half-penny per ton for each 6d. in price.

Mr. RHODES: That means that the land-owner is to be at the mercy of the owner of the colliery.

Mr. GARFORTH: It leads to a great deal of cheating.

The PRESIDENT: May I suggest that this is not a scientific subject, and we are becoming a revolutionary debating society.

The meeting then concluded.

The Annual Dinner was afterwards held at the King's Head Hotel.

MIDLAND INSTITUTE OF MINING, CIVIL, AND MECHANICAL ENGINEERS.

GENERAL MEETING.

HELD AT THE VICTORIA HOTEL, SHEFFIELD, ON TUESDAY, SEPTEMBER 25TH, 1888.

THOS. CARRINGTON, Esq., Past President, in the Chair.

The minutes of the Annual Meeting were read and confirmed.

ELECTION OF PRESIDENT.

The SECRETARY reported that Mr. C. E. Rhodes was unanimously elected President for the current year; also that the Council had filled up the vacancy in the Council occasioned by the election of Mr. Rhodes to be President, and appointed Mr. Cobbold.

The CHAIRMAN: Gentlemen, it unfortunately happens that there is a meeting of coalowners in the adjoining room, which we are requested to attend on urgent business. It is very unfortunate that the two meetings are called for the same time, but it is a *contretemps* we cannot help. We have just held a Council meeting, and finished the business of the Council. I am only occupying, as past President, the position of Mr. Embleton, the retiring President, for this occasion, as Mr. Embleton is unavoidably prevented being present to-day, but I am glad to say that Mr. Rhodes, the President-elect, will take his place as President for the first time to-day. But Mr. Rhodes, like myself, has to attend the meeting in the next room, and what I would suggest is that, after placing Mr. Rhodes in his position as President, you should let us go into the next room whilst the paper is being read. In asking Mr. Rhodes to take the chair as President of this Institute, I must first of all express my regret that our highly esteemed friend, Mr. Embleton, is not with us to hear from us the expression of our thanks to him for the most able and kind way in which he has discharged his duties as President during the past, the Jubilee year. I am sure we all regret that he has to retire, and we are extremely obliged to him for the kind courtesy and attention he has shown to the business and welfare of the Institute during his year of office. But whilst saying good-bye for the time being to one President, we welcome with the greatest heartiness and good feeling our friend Mr. Rhodes as our new President (hear, hear). I am sure, gentlemen, that he will do all that he can to increase the usefulness and extend the power and influence of this Institute. We all know his very great ability as a mining engineer. His great skill and great experience have been proved to us in many ways, and further I may recall to you the great kindness that the firm he represents,—because it is only through him that it has been done,—the great kindness and important help that the firm of Sir John Brown and Company have afforded us, in giving us the opportunity of carrying on, at Aldwarke Main Collieries, very valuable experiments with the object of testing and ascertaining the comparative safety of different kinds of safety lamps. I do not want to detain you with a long speech, especially as we have to go to the meeting in the next room, but I must add that in asking Mr. Rhodes to take the chair as President of the Institute, I do so with the fullest and greatest confidence that he will maintain the dignity, the usefulness, and valuable influence of our Institute, and that we may rely upon his able guidance to give us further opportunities of acquiring valuable information in the profession to which we belong. (Applause). Gentlemen, I have great pleasure in calling upon Mr. Rhodes to take the position of President of this Institute.

The PRESIDENT-ELECT (Mr. C. E. Rhodes) on taking the chair was cordially greeted. He said: Mr. Carrington and gentlemen, in taking this chair I need not say that I do so with the fullest appreciation of the honour that has been paid to a very young member of this Institute in asking him to occupy the position of President. An Institute of this character embraces men of the highest possible position in our profession, and embraces not only those, but also young aspirants to what I may call fame. Therefore, in associating those who have had the opportunity of acquiring not only a vast amount of knowledge but experience as well, the younger members, I think, cannot help but benefit. But they can only achieve this object by those who have had experience and training and have ability,—they can only, I say, receive the benefits from such an Institution as this by the older members attending the meetings and giving them the benefit of their opinions, and, if I may go so far as to say it, their advice at the same time. I trust whilst I occupy this chair that those older and more experienced members will give me their support, and endeavour so far as possible to make my year of office a practical success, and one which will conduce to the efficiency of this Institution, and, I hope, also raise the position which it occupies, not only with respect to other Institutions but also by the mining public generally. (Applause). I have to thank you most cordially, gentlemen, for the high honour you have paid me, and regret that on first taking the chair I have to ask your indulgence, along with Mr. Carrington and Mr. Mitchell, that we may attend another meeting which, it unfortunately happens, was called on the same day, the two Secretaries acting quite unknown to each other.

Mr. J. GERRARD took the chair for the transaction of the business on the agenda.

The following gentlemen were elected members of the Institute, having been previously nominated:—

Mr. THOMAS JAMESON, Mechanical Engineer, Wharncliffe Silkstone Collieries, near Barnsley.

Mr. JAMES A. CLARKE, Mining Engineer, Ayr Collieries, Tarbolton, N.B.

Mr. THOMAS ANDREWS, F.R.S., Iron Master, Ravencrag, Wortley, near Sheffield.

The following paper was then read:

ON TONITE AS AN EXPLOSIVE WHEN USED WITH A FLAME-DESTROYING COMPOUND.

BY W. HARGREAVES.

As the use of Roburite, Securite, and other explosives has been brought before the Institute, and their properties discussed, I beg to bring before the Institute another explosive called Tonite.

Tonite is made up in cartridges of such sizes as to suit the circumstances under which it is to be used. It will not explode by concussion, and a cartridge may be held in the hand and on being lighted will burn with a yellow flame like a Roman candle. During the burning of Tonite a peculiar but not an offensive odour is emitted, and when employed in blasting a considerable flame is produced.

The inventor of this explosive has discovered a compound which when put into the bore-hole with the Tonite, has the effect of destroying the flame, which always appears when Tonite is used alone.

The flame-destroying compound consists of sawdust saturated with some chemical compound, and should this mixture become dry, as it will by being kept long in a store, its active properties are renewed by adding a quantity of water, and so damping it that water can be squeezed out by pressure of the hand. The Tonite cartridge is coated with tallow, so as to prevent its being affected by the moisture of the compound.

The hole must be bored of a sufficient size to allow a space between the cartridge and the circumference of the hole. A portion of the compound is then thrown into the hole and rammed at the end until two or three inches are filled solid. The cartridge is now introduced, and a further quantity of the compound put into the hole and again rammed until the cartridge is surrounded and a few inches from the outer end covered; the remaining part of the hole is then stemmed in the usual way.

All shots are fired by electricity, and the detonator used consists of a small copper case closed at one end and filled with some easily ignited explosive which will not explode by concussion. In the open end of the case is inserted a wooden plug, through which two wires pass, and extend about $\frac{1}{8}$ of an inch into the explosive; these wires are joined together by a very small platinum wire, all enclosed within the case. The wires should be rather longer than the length of the shot-hole, so that they may reach beyond the stemming, in order to connect them to the battery cable. The battery should be placed at a safe distance from the shot, and when all is ready a current of electricity of ample power to fuse the platinum wire is passed through; this has the effect of igniting the explosive inside the detonator and so exploding the Tonite.

Notes on Work done.

Tonite is very quick in its action, and unless properly manipulated will break down the coal very much and will fracture the roof into fragments too small to be used for packing the gates. To diminish this effect, we always, when blasting the coal, fix a sprag very tightly under the hole to be fired, and remove a sprag on each side; by this plan a 2½ oz. cartridge will loose or break down 6 or 7 feet on each side of the hole with the same effect as if 6 oz. of gunpowder had been used. The hole should have a slight rise of 1 inch to 1 foot, and the end should just touch the roof of the coal.

In using Tonite for blasting the roof, to make the pack gates of sufficient height, the hole should be bored above a hard bed, so that the quick action of the Tonite does not break through it, but has the effect of making it spread more on each side of the hole, and so loose the roof that it may be easily pulled down with a pick. In stone shots we find that a 5 oz. cartridge is equal to 12 oz. of gunpowder.

The cost of drilling a shot-hole used with Tonite is the same as that with powder. The cost of the shots is as follows:—

COAL SHOTS WITH TONITE.		COAL SHOTS WITH POWDER.	
2½ oz. Tonite	3¼d.	6 oz. powder	1¼d.
Detonators	3½d.	Fuze	1d.
Flame-destroying comp^{d.}	1d.		
	7¾d.		2¼d.

STONE SHOTS WITH TONITE.		STONE SHOTS WITH POWDER.	
5 oz. Tonite	6½d.	12 oz. powder	2½d.
Detonators	3½d.	Fuze	1d.
Compound	1d.		
	11d.		3½d.

These experiments have been made in the Middleton Main Seam, Robin Hood Colliery, and a large number of shots have been fired with Tonite and the fire-destroying compound, and in no case, after careful observation, has any spark or the slightest appearance of any flame been observed. I may say that with gunpowder and the fire-destroying compound, although the flame is greatly diminished, yet it is not absolutely extinguished.

Mr. R. MILLER: I move that a hearty vote of thanks be given to Mr. Hargreaves for his paper, and that it be printed in the Transactions of the Institute.

Mr. TEALE: I have pleasure in seconding Mr. Miller's proposition. The motion was carried unanimously.

The CHAIRMAN: I take it now it will be open for any member of the Institute to ask any question, or take part in any discussion upon this paper.

Mr. MILLER: I should just like to ask, does this flame extinguisher always put the flame out.

Mr. HARGREAVES: We have fired over 2,000 shots and never seen the slightest flame.

Mr. MILLER: You would not be always that you could see it.

Mr. HARGREAVES: We have taken great care in every case where we had a fair chance.

Mr. MILLER: I should think it likely to fail sometimes, but we do not know.

Mr. HARGREAVES: If it has failed we have not seen it, and we have taken the greatest care in trying to ascertain in every way we could. We have had shots fired in all ways we could think of. We have taken blown-out shots and tried to make a flame, but we have not been able to do it; the compound has always appeared to act.

The CHAIRMAN: Mr. Hargreaves states that he has tested it by visual observation only.

Mr. MILLER: Not even in a blown-out shot?

Mr. HARGREAVES: No.

The CHAIRMAN: Where the test is by visual observation it is for each to form his own opinion, and Mr. Hargreaves states candidly that no other test has been applied than visual observation. Is the cartridge and this compound combined?

Mr. HARGREAVES: No, they are separate. The hole is bored an inch and three quarters, then we put the cartridge in and throw the compound in to fill around. After we have got what we think needful put in we leave a space at the end to stem on, and it appears to destroy the flame as far as we have been able to perceive. I may say we are now trying Roburite. We are going on for some months with Roburite to see the difference of cost and whether Roburite acts in the same form as Tonite, that is Roburite plain and simple.

The CHAIRMAN: I hope that Mr. Hargreaves will put before the Institute his experience of Roburite.

Mr. MILLER: I hope the discussion will be adjourned for the presence of Mr. G. B. Walker.

Mr. HARGREAVES: I was going to say it would be well to let the discussion wait until the general discussion takes place on papers on blasting operations.

The CHAIRMAN: I would adjourn it for several reasons. I hope the paper will shew the position of a few shots,—for instance, how the shot was placed with regard to the coal, and how with regard to ripping, and what the form of the best shot was, that the members may see for themselves, and have some further information as to the nature of the experiments.

Mr. McMURTRIE: What distance is there between the holes you put in the stone and the coal?

Mr. HARGREAVES: It is particularly in the long-face gateways we use it. We put one shot down the centre, and it is sufficient to spread to both sides to get in a very good packable stuff. If we do not get it above some strong stone, or a seam of some description, we have in the roof,—it splinters the roof into little pieces which you cannot use for packing purposes. If you have, as we have, in the Silkstone seam, an iron band,—we get it above that, and then it has the effect of toning down the power of the Tonite and breaks the rock, so that we can get it with the pick, and it is good packing. If we put it underneath the iron band it breaks in very little pieces and is fit for nothing.

Mr. MILLER: Then a great deal depends on the power of this sprag in the middle. It must be tightly put in to be effective?

Mr. HARGREAVES: It must be tightly put in.

The CHAIRMAN: Is it an ordinary sprag, some five inches in diameter?

Mr. HARGREAVES: Yes.

The CHAIRMAN: And that slight resistance causes the force to spread?

Mr. HARGREAVES: Yes.

Mr. McMURTRIE: What we have been speaking of refers to the gates, but in the coal what distance have you between the shots?

Mr. HARGREAVES: We never have but one hole put in, that is at the point opposite the gate. That is a breaking-in shot, and breaks the coal so that it will come easily, and there is not much difficulty in getting between one gate and another.

Mr. McMURTRIE: What is the distance between the gates?

Mr. HARGREAVES: Twenty-five yards.

Mr. MILLER: And you seldom require more than that shot in breaking in?

Mr. HARGREAVES: No; but if we do not put it above the ironstone band in ripping, it is difficult to get it large enough to pack with.

Mr. MILLER: Then the ironstone is really useful to reduce its power?

Mr. HARGREAVES: Yes; without it we should not get much packing.

The CHAIRMAN: Have you tried a less charge of Tonite?

Mr. HARGREAVES: We have used one or two ounce cartridges, but it did not do much good.

The CHAIRMAN: You do not give us the proportions of the extinguishing compound.

Mr. HARGREAVES: I have not noted that, but shall be pleased to do so.

Mr. McMURTRIE: Is the charge put in the roof and in the coal the same?

Mr. HARGREAVES: No; the coal shots are $2\frac{1}{2}$ oz., the roof shots 5 oz.

The CHAIRMAN: The first question that strikes us as engineers, is that of safety. As safety is of the first importance, the question that takes the first place is whether this is really an explosive which can be depended upon as a non-producer of flame; of course, so far as these experiments have been tested by visual observation, there has not been that conclusive test afforded by firing these charges in an explosive atmosphere.

Mr. HARGREAVES: No.

The CHAIRMAN: Or in a mixture of coal dust or explosive inflammable mixtures, such as other explosives have been tried in. It has been a practical series of experiments, which has been undertaken with all its risks in the mine; and Mr. Hargreaves tells us that the Middleton Main Colliery where the experiments have been made, is similar in its characteristics to the Silkstone mines, that he has fired 2,000 shots, and so far as they can see by the eye, there has been no flame produced in any single instance.

Mr. HARGREAVES: Yes.

Mr. TEALE: Does not a great deal of the safety of the shot depend on the actual surrounding of the cartridge with the material?

Suppose a shot were put in a hole and it lay free from any of this material, would not that side of the shot be likely to give off a certain amount of flame? I should imagine myself that, presuming all is done as it should be done, that it is properly stemmed and packed all round the cartridge with a material which immediately suffocates the flame, all may be well, but how are you to provide against such a use of it as might cause explosions or mistakes of that sort being made.

Mr. HARGREAVES: It is quite possible that mistakes may be made, but it seems to me that if we have got a certain quantity in, it has destroyed the flame.

Mr. MILLER: But you do not surround the cartridge with that?

Mr. HARGREAVES: Certainly we do if we possibly can.

Mr. MILLER: I thought you put it just outside?

Mr. HARGREAVES: No, we put all round if we can.

Mr. MILLER: But you cannot get it on the bottom, you cannot get round.

Mr. HARGREAVES: No; but there has always seemed to be sufficient to quench the flame; however, as the Chairman suggests, we will have it tested in gas.

The CHAIRMAN: One has been present at experiments with other explosives when one man has said that he saw a spark, or flame, and another has said that he did not. There is a want of conclusiveness with regard to what I have called visual observation.

Mr. HARGREAVES: I never allowed any but one man, the best man I have, to fire the shots. He has taken the greatest possible care, and assures me that he has not seen more than I have seen, and that we have not seen the slightest spark in any way.

Mr. TEALE: The reason I made the remark was this—the water cartridge, for example, is reliable if the water surrounds the whole cartridge, and I have seen experiments of a conclusive character with it, but wherever the cartridge is placed so that it leaves one side so that the water does not get to it, you unmistakeably have flame. The whole merit of the water cartridge lies in having it surrounded. I should think it would be an interesting experiment to test Tonite under similar conditions,—that is to say, with material surrounding the shot, and also in cases where the non-ignitable or flame-destroying material surrounds only a portion of the shot or cartridge. To put the cartridge in first, it is impossible to get it all surrounded with the flame-destroying material.

Mr. HARGREAVES: That is evidently impossible, but there seems to be sufficient to destroy the flame outside.

The CHAIRMAN: We will take the discussion as adjourned until next meeting.

Mr. HARGREAVES: I shall be glad to take up the recommendations you have made and bring the result at next meeting, and also to give particulars as to our experience with Roburite at the next meeting.

The CHAIRMAN: Then shall we take up the discussion on Mr. G. Blake Walker's papers on "Hydro-Carbon Explosives."

Mr. TEALE: I move the discussion be adjourned.

Mr. W. HOOLE CHAMBERS seconded the motion, which was carried.

Mr. W. HOOLE CHAMBERS moved a vote of thanks to Mr. Gerrard for presiding over the meeting.

Mr. TEALE seconded the motion, which was carried, and the meeting ended.

NOTES ON MATTERS OF CURRENT INTEREST.

By GEORGE BLAKE WALKER, F.G.S.

GUNPOWDER AS A SAFETY EXPLOSIVE.

Messrs. Mallard and Le Chatelier have lately been making experiments for the French Académie des Sciences on safe blasting agents. They assert, as the outcome of numerous experiments, that explosive substances fired in an atmosphere rendered dangerous by the presence of fire-damp ignite the gas only when their temperature exceeds 2,200° Centigrade or thereabouts. All the common explosives produce a higher temperature than this. It is stated that if an equal weight of carbonate or sulphate of soda, or even of ordinary coal dust, be added, dynamite and nitro-glycerine will not attain, on explosion, the temperature neecessary to ignite gas. A smaller proportion will render ordinary blasting powder safe. Such is the extraordinary statement which appeared a week or two ago in the *Colliery Guardian*. As it comes to us under the ægis of the well-known names of MM. Mallard and Le Chatelier, it cannot be summarily dismissed, but it certainly "requires confirmation."

ROBURITE.

The following experiments with Roburite at Witten in Westphalia are reported in the German mining paper *Der Compass* of Sept. 10th:

I.—230 grammes Roburite, cartridge lying free imbedded in coal dust, in the presence of 4 per cent. of fire-damp, with coal dust in suspension in the air, produced *no explosion*.

II.—575 grammes Roburite (equivalent in practice to 1,500 grammes of gunpowder), cartridge lying free embedded in coal dust, in the presence of 4 per cent. of fire-damp, with coal dust in suspension in the air, likewise produced *no explosion*.

III.—400 grammes blasting powder, cartridge lying free and conditions the same as in I. and II., produced a *violent explosion*. The flame issued far beyond the end of the gallery.

IV.—Same as No. II. *No explosion*.

V.—400 grammes blasting powder, cartridge embedded in coal dust, without gas, but with coal dust; *violent explosion*.

VI.—Repetition of preceding experiment, with the same result.

VII.—425 grammes gelatine dynamite, in the presence of disturbed coal dust and 4 per cent. of fire-damp, produced a colossal explosion with great flames and foul gases. The thick glass panes in the sides of the experimental chamber were smashed.

The organ of the Prussian Ministry of Mines (*Zeitschrift für das Berg-Hütten- und Salinenwesen im Preussischen Staate*) says in the last number:

"The use of Roburite and Carbonite, which have lately been "substituted in many coal mines in Westphalia in place of the "explosives formerly used, has been accompanied by satisfactory "results, but the trials have not so far led to a final conclusion as "to which of the two should be exclusively adopted."

AFTER-GASES OF CARBONITE.

The *Deutsche Kohlen Zeitung* of June 6th gives an account of an accident resulting from inhaling the after-gases of Carbonite at the Milseburg tunnel. Four workmen became unconscious after inhaling the fumes, one of whom died.

BICKFORD SMITH & CO.'S PATENT FUZE IGNITER.

Several cases have recently occurred in which these igniters have exploded, in one case igniting gas by which four persons were burnt. Where these igniters are in use very great care is necessary to prevent accidents—the special fuze must be used and the capsule not nipped too tightly when secured to the fuze, so as to give the gases produced on ignition sufficient vent. If, however, the igniter is put on too lightly it will be blown off. The attachment is a very delicate business.

PLOM AND D'ANDRIMONT'S CHAMBERING TOOL.

The object of this invention is to excavate at the back of shot holes a chamber, the lower part of which is filled with powder or other explosive. The advantage is supposed to consist in the concentration of the explosive at the back of the hole, and in a cavity one side of which is parallel to the face. Thus the tendency of the explosion is to force the coal not only downwards but *outwards*, and it is contended that this will render holing unnecessary. The principle will be understood from the illustration. A hole about $2\frac{1}{2}$ in.

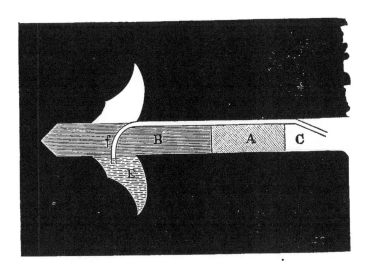

diameter is drilled in the coal by means of an ordinary machine drill. This being done, a tube having two slits near to its further end is pushed to the back of the hole. This tube contains an internal screw which carries a pair of wings. When the tube is introduced into the hole the wings are entirely within it, but on

revolving the tool they gradually expand, and cut out a circular chamber by means of their serrated edges. The dust is brought back through the tube. A charge of powder or other explosive is then introduced into the lower part of the cavity E, the wooden plug B inserted, a stemming of clay A is added, and the charge fired by means of the fuze *f*. Such is the general idea of the contrivance. The inventors propose to introduce a chamber of water by means of a special contrivance where gas is present. They assert that a saving of 10 to 50 per cent. of labour is effected, that the roof is less likely to break down, that much more large coal is made, and that headings can be driven very rapidly by means of it.

The figure given below is a reproduction of the sketch in the inventors' pamphlet, and is supposed to represent the appearance

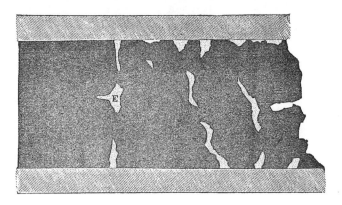

which a seam of coal would present, in section, after a shot had been fired on Plom and D'Andrimont's system.

APPLICATIONS OF ELECTRICITY TO MINING PURPOSES.

During the last twelve months the question of the application of electricity to haulage in mines has been attracting a good deal of attention. The writer is at present engaged in experiments with a locomotive actuated by means of storage batteries, which has proved itself equal to a load of 10 tons on a rising gradient of 1 in 70, the weight of the locomotive being 2 tons 5 cwts. Full particulars of these experiments will shortly be laid before the Institute.

The electric pumps at Locke & Co.'s St. John's Colliery, Normanton, continue to give excellent practical results, and it is to be hoped that a full description of them will be shortly laid before the Institute.

A SAFETY SAND CARTRIDGE.

The *Glückauf* for October 13th contains the following description of a sand cartridge invented by Herr Bergrat Jincinsky, Mährisch-Ostrau, in Austria:—

"The fact that the ignition of adjacent combustible bodies is "hindered when a flame is surrounded by fine incombustible sub-"stances (as for instance, the gauze of a safety lamp)—by means of

"which the flame, through its momentary cooling, falls below the
"temperature necessary for ignition—has led me to construct a
"dynamite cartridge which when exploded in a mixture of fire-
"damp does not ignite it.

"These cartridges consist of the usual dynamite
"cartridge a, placed as nearly central as possible in a
"paper case b, whilst the interstice c is filled with
"ordinary sharp wet fine sand up to the top d and e.
"The sand need not be dripping, but should be satu-
"rated with water; or instead of sand, foraminiferal
"earth may be used, which will absorb 30 per cent. of
"its bulk of water. The cartridge is then provided
"with an electric detonator f. The space g is filled
"with sand or foraminiferal earth, and the end of the
"cartridge envelope is bound tightly round the wires
"of the detonator. The thickness of sand surround-
"ing the dynamite should not be less than 10 mm. in
"thickness, so that if the dynamite were 18 mm. in
"diameter, the total diameter of the cartridge would
"be 38 mm.

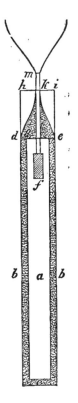

"The experiments with these cartridges have been
"most successful. In no case has a (free lying) sand
"cartridge with 200 grammes of dynamite, fired in air
"mixed with fire-damp to the extent of 9 per cent.,
"and with coal dust abundantly present, caused any
"ignition: and a cartridge having only 7 mm. of sand
"protection has given the same result.

"At the present time some further experiments are being made
"by the local committee of the Austrian Fire-damp Commission at
"Mährisch-Ostrau, with the object of deciding to what extent the
"thickness of the sand can be reduced without diminishing the
"safety of the cartridge.

"The paper envelope b is best made of glycerine paper or some
"paper which is fairly waterproof. The explanation of the safety of
"these cartridges is very simple. At the moment of the generation
"of the gas and the high temperature incident on the explosion of
"dynamite, it is involved in a granular substance the interstices of
"which are filled with water. Before this obstruction is removed,

"and the water converted into steam, the temperature is so far reduced that no ignition of the fire-damp (780° C. ignition temperature) follows.

"The advantages of these cartridges are as follows:—

"1. The non-ignition of fire-damp.
"2. The undimished effectiveness of the dynamite.
"3. The easy and cheap cost.

"The only drawback to this cartridge is that the bore-hole must be from 15 to 20 mm. larger than the ordinary bore-holes, a drawback which is not worth consideration when the safety obtained is taken into account."

Herr Bergrat Jincinsky adds: "I have described this my invention to my respected fellow engineers in the hope that it may be of service to them and be productive of the best results."

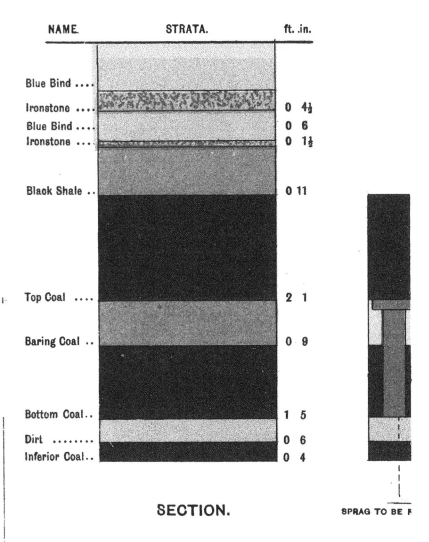

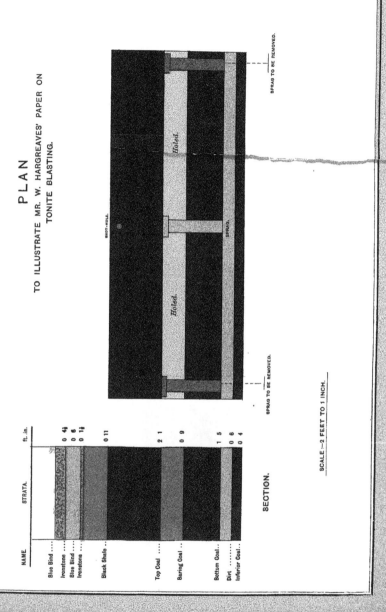

MIDLAND INSTITUTE OF MINING, CIVIL, AND MECHANICAL ENGINEERS.

GENERAL MEETING.

HELD AT THE INSTITUTE ROOM, BARNSLEY, ON WEDNESDAY, OCTOBER 31ST, 1888.

C. E. RHODES, Esq., President, in the Chair.

The minutes of the last meeting were read and confirmed.

Mr. HENRY CHAMBERS, Colliery Manager, Tinsley Park Colliery, near Rotherham, was elected a member of the Institute, having been previously nominated.

THE PRESIDENT'S INAUGURAL ADDRESS.

The PRESIDENT delivered his inaugural address as follows:—

Gentlemen,—One of the objects of our meeting to-day is to hear an inaugural address from the President.

I fear that to a great many of you presidental addresses are in the main such a well worn theme that I shall only weary you. Be that as it may, it is my duty and privilege to inaugurate the coming session by a few words, and they will be principally confined to making some suggestions as to our mode of procedure, which, if carried out, will I think aid us in our labours and add to the usefulness of this Institute.

It will be for you to say whether they meet with your approval or not; if they do, I can only say that no effort on my part shall be spared to ensure their success.

In the first place, what is the object of such an Institute as this? It appears to me to be that by our meeting together and having opportunities of discussion, not only is new light thrown upon the many subjects connected with mining, which from time to time come under our consideration, but we are also enabled to obtain assistance from the experience of others, which must help us materially in dealing with different operations and peculiar circumstances, that we, as individuals, may from time to time have to to contend with. One great advantage is thus attained, and it is obvious that the more extended the field for the dissemination of such information and practical knowledge becomes, the more valuable and useful will be the character of this Institution.

Some of its members have opportunities by travel and otherwise of acquiring a store of interesting and useful facts, which if brought more frequently before us, would be a source of instruction and interest to those who have not been so fortunate as to have had such opportunities.

There are a large number connected with our profession who, through circumstances beyond their control, are tied to one district, and in some cases to one colliery, probably for the greater part of their lives, and the natural tendency of the mind, with persons so placed, is to gradually assume a kind of hazy notion that the be-all and end-all of mining operations is to be found in the district or colliery with which they are connected.

As an instance of this, take our own district of South Yorkshire. A very few years ago there was practically only one method of working the Barnsley seam of coal,—a method eminently adapted, no doubt, to working that seam under the conditions existing fifteen or twenty years since in most of the collieries then in operation. A large number of those connected with the management of such collieries, more especially those occupying subordinate positions, had never been in other districts, and had had no experience in working coal by any other system, and the consequence was that when the coal had to be won at much greater depths, and the method which had hitherto prevailed of working and dealing with it was found to be utterly unsuited to the altered conditions which had arisen, but few of those, as I have stated, in subordinate positions were qualified by their previous experience to work the coal in a totally different way, and the result was that persons from other districts with experience of the methods required were introduced.

This is only one instance, but it serves my purpose to illustrate what I mean, and that is that if the stewards and deputies of our collieries could attend these meetings, and subjects connected more closely with the special circumstances under which they work could be brought forward and discussed, very great benefit would be conferred upon them, and this Institute would supply a want which, I am sure, has long been felt.

There are gentlemen in it who are thoroughly conversant with mining as carried out in Yorkshire, Lancashire, the thick coal of Staffordshire, anthracite and bituminous coal of Wales, and who are also conversant with mining operations in Belgium and other districts on the Continent, where difficulties utterly unknown in this country are contended with successfully, and seams which we should never attempt to work here are got with profit,—seams so dirty that no appliances in this country could be found to deal with them, yet are washed and so manipulated that they are sent into the market in a perfectly clean and saleable condition.

Now if such members would read papers upon the various subjects with which, as I have just stated, they are intimately acquainted, and if these papers could be read at times suitable and convenient for stewards, deputies, and students to attend, I feel sure that their efforts would be highly appreciated, and that they would confer a boon upon many who have not been born under such a fortunate star as themselves.

I would suggest further that more papers might be read upon topics which are familiar to some, but which to others are only familiar so far as their perhaps limited experience goes. Take haulage, for instance. I think that many papers might be read with advantage in describing various systems of haulage in use in this and other countries. This subject has I know been fully dealt with by the North of England Institute and at various times by this Institute, but the North of England Institute Transations are not within reach of all of us, and it is so long since the subject was dealt with, that I cannot help thinking that much that is new might be said upon it.

Then there is coke-making. I know no process which has more rapidly advanced during the past few years than the washing and manipulation of small coal for coke-making. Surely we can give one another the benefit of our information and experience upon this subject without injuring ourselves or our respective employers.

Ventilation, again, affords ample scope for instruction and discussion.

A mining engineer or a colliery manager to be anything should be practical, after that as much theory as you like. I believe in the old saying, "An ounce of practice is worth a ton of theory," and what I would say is that those who have had the practical experience of the many and varied subjects incidental to mining, should give others with less opportunities the benefit of it.

Interesting and valuable papers could be read upon the subject of creeps, descriptive of their origin, and the method of working which has caused them, the extent to which they have run, and the vast amount of trouble, anxiety, and expense which they have caused.

It is just possible that in describing one's experiences that we may be, to a certain extent, ventilating one's own mistakes; but even if we do admit that we have been liable to error and made mistakes, I do not think that we should be putting ourselves in a position peculiar to us. He is a wonderful man indeed who has never made a mistake, especially in mining, and those who have made the fewest are those who have seen the effect of other people's errors and have profited accordingly. Let those, therefore, who have benefitted in the past by the labour and experience of others, come forward now and assist other aspirants to fame in the mining world.

I should propose, therefore, that *six* meetings be arranged between now and next June, at Barnsley or Sheffield or some other convenient place, such meetings to be held at a time which would enable stewards, deputies, and students to attend; and that papers of the character I have indicated should be read, and I should be glad if *six* gentlemen would undertake to read papers; if they will do so a great step will, I think, be made towards a beneficial and useful session.

My proposal would take the form somewhat of a revolutionary one, inasmuch as I would give to a large number the privilege of the Institute, by extending to them the benefit of a course of lectures, without expense. At such meetings they would be allowed, of course, to take part in any discussion that might arise upon any paper, so as to encourage them to bring their own experience and powers of argument to bear upon the topic in question.

Of course I can only suggest, it will be for you to say whether you approve or not. The proposed meetings might be held on the same day of our other meetings but later on, say one to one and a half hours after our ordinary meetings, which would make the second meeting be about five or six o'clock.

As you all know, mining in this country has rapidly assumed proportions of such magnitude, and operations are carried on under such restrictions from the State, that those even in subordinate positions require qualifications entirely different from what was thought necessary twenty years ago.

If the owners are to be protected, and agents and certificated managers are to carry out the onerous and important duties they have upon their shoulders with safety and efficiency, it is necessary that those under them should have every opportunity of acquiring practical information. It is perhaps therefore with a somewhat selfish motive that I have laid the suggestions I have before you.

With respect to our other meetings, I hope that members may be induced to give papers upon one or two subjects which have not yet been touched upon.

Electricity as a motive power seems to be rapidly pushing itself to the front, and it is far from improbable, as it becomes more used and appreciated, that it will take the place of every other form of motive power now in use, more especially in mining, chiefly because of the ease with which this power can be conveyed long distances from one central station.

When one looks at the enormous strides that the science of electricity has made during the last few years, one can readily believe in almost anything taking place in respect to its use within a very short time.

If electricity is to become the important agent many anticipate for mining purposes, then in addition to the many other qualifications that a mining engineer requires, he will have to be an electrician, and it seems to me therefore that the study of electricity should form an element in the education of those who are thinking of adopting our profession, and therefore papers which could be laid before this Institute by adepts in this science would be a source of valuable information, instruction, and help.

In addition to the use of electricity for hauling and pumping, it seems to be almost a certainty that before long it will be the chief means of lighting collieries, not only from fixed lights around the pit bottom and in the main roadways, but also in the form of portable lights for the use of the workmen, as the question of a portable miners' electric safety lamp is now almost solved. I have seen one or two which nearly meet every requirement that is necessary in such a lamp except one, and that is the heavy first cost.

This question is one that I have for some time been giving my particular attention to, and I have had in use several electric safety lamps, which confirm me in the statement I have just made, and I am sanguine that before long the cost will be reduced so as to make them within reach of almost every colliery.

Another subject which I think would be of great interest to a large number of us is that of *coke-making*. I have before alluded to the manipulation of the fuel for coking, but I think papers and statistics giving the result of the working of the different kinds of coke-ovens in use would be very beneficial and instructive,—not only the working of the different types might be dealt with, but the results obtained from the same class of ovens at different places,—for instance, the percentage of coke obtained from the ordinary bee-hive ovens per ton of coal, with the weekly output per oven, at various places, might well form the subject of a paper.

These particulars vary in an astonishing degree. Some gentlemen have informed me that they get as much as 65 per cent. of coke from ordinary washed coal from bee-hive ovens, whereas others tell me they are perfectly satisfied with 50 per cent. I was informed only the other day, too, by a member of this Institute, that he was getting a yield of 12 tons of coke per week from an ordinary wagon oven, 12 ft. by 6 ft. I think. I have no experience of any such result as this, and I say therefore that much advantage would be derived by interchanging the result of our varied experiences.

There is no one connected with this Institute who has had so much experience in coke-making as Mr. A. M. Chambers, who has patented various improvements adaptable to ordinary coke-ovens, and which I am informed increase the yield of coke to a considerable extent, and at the same time improve the quality.

If that gentleman could give us the benefit of his experiments and experience on this subject, I am sure we should all feel that he was conferring a favour.

I believe, also, that he could give us some important information as to the extraction of oil and residuary products by a process which he has in operation applied to ordinary coke-ovens, and which I believe has been more successful than any of the many systems which have been from time to time brought out.

Take, again, the question of explosives. Since the passing of the Mines Regulation Act, which imposed such restrictions upon the use of powder (and rightly so), all sorts of so-called flameless explosives have been brought before the mining public, such as Tonite, Carbonite, Roburite, Securite, Bellite, &c., &c., for all of which it is asserted that they are flameless and innocuous.

Here is a topic which affords an endless field for discussion, and is of such importance not only to ourselves but to those we employ and those who employ us. Our own safety and that of those we are responsible for depends upon whether these explosives are flameless and innocuous as asserted,—or whether we may not by going solely on the *ipse dixit* of the inventors be using day by day with impunity a material which may sooner or later be the cause of some awful catastrophe.

This subject alone, if experiments could be made (and they very shortly, I hope, will) to demonstrate effectively whether there is *or is not* a flameless explosive before the mining world, could well form an important feature of the present session, and be thoroughly thrashed out.

It has so often been stated that this Institute did more to elucidate the safety lamp question than any other body in the country, not even excepting the Royal Commission on Accidents in Mines, and I therefore feel certain that if the question of explosives is taken up in the same way that that subject was, that you will have added another important work to the many you have undertaken and carried out in the interests of the mining public.

I have heard it stated that it is no use putting these explosives to trials so severe as to make it almost a certainty they should fail, because if they fail even under a trial which could never occur in actual working, a certain prejudice is created unwarrantably against the explosive so experimented upon.

My contention is, however, that such vast issues are at stake, that we should know as far as possible exactly where we are and what we are doing in this matter of explosives, and I feel sure that the gentlemen who have kindly undertaken to experiment with the various explosives now before us, will put them to such crucial tests as to leave no doubt in the minds of anyone as to their safety or not.

I hope, also, that an excursion may be arranged during the session into some other district, where objects of particular interest might attract us, and if such outing can be managed, that a thoroughly representative number may take advantage of it. I should also be glad if it is possible, if a paper could be read prior to such visit, descriptive of the plant and machinery we are going to inspect; such inspection would then be particularly interesting and useful.

Having, therefore, briefly glanced at what is proposed as our programme for the next few months, I would say that it is no use drawing up a programme however elaborate and simply leave it to your officials to carry out.

Any scheme emanating from a Council for the guidance of its members to be successful must have the full support of the whole body, and I do hope, therefore, that myself and your other officers during the coming year can count upon the hearty support of each individual member, and that our meetings will be well attended and papers plentiful.

If we can ensure that, then I feel certain that the result of our deliberations will be an advantage, and a pleasure to ourselves, and we shall have the satisfaction of feeling that we have fulfilled to some little extent what is our duty to the profession we belong to.

Considering the importance of the mining industry of this country, and that England practically owes her position in the commercial world to her mining wealth, I do not think that the status of the mining engineer or colliery manager has ever been properly recognised by other professions. I hope and believe firmly that Institutions like this can do more, with the help of its members, to let the public see how much they owe us, than can be accomplished by any other means. I hope to see the day when a mining engineer will have to pass an examination and hold a diploma exactly as is the case with a physician or a solicitor, and that no person will be allowed to practice unless he has that diploma.

The colliery manager's certificate is a step in the right direction, but I would go beyond that and have a higher grade still for those who could attain to it.

Times are changing with marvellous rapidity, and those who are not prepared to go with them will be left behind in the race; and in no case is this more true than in coal mining. Increasing difficulties will confront the mining engineer year by year. Coal will have to be won at very much greater depths, and consequently increased skill, knowledge, and experience will be necessary if the difficulties to be faced are to be successfully surmounted. I am further of opinion that the scientific training of the English mining engineer has been neglected to a very great extent in the past, and that this has been chiefly due to the enormous areas of coal which have been worked with comparative ease, and at little or no expense, and which have not necessitated the scientific knowledge and skill which will be absolutely necessary in the future.

No one can foresee and no one can predict to how great a depth coal will be won. My own opinion is that it will be got within a very few years at a depth never thought of a quarter of a century ago, and what at one time was fixed as the limit at which it could be worked, will be doubled within the terms of the lives of many who are present here to-day.

If this is to be so, the coal-fields of England have a long, long life before them, and if increased scientific skill, enterprise, and forethought be brought to bear, I do not see why we should not produce it at a cost which will still enable us to hold our own with any country in the world.

It is said that great events produce great men, and I believe that when our difficulties become greater, there will be found amongst our English mining engineers those who, by their skill and scientific training, will be quite competent to cope with them. England has never yet been behind in the race, and I am convined that she will continue to hold her proud position in the world by the help of ourselves and those who will come after us.

We are yet far from being played out, and I contend that it is our bounden duty to endeavour, as far as lies in our power, to give the benefit of our own experiences and training to those who surround us now and who will probably succeed us, and great good must necessarily be conferred upon the rising generation by having handed down to them, recorded in print, the results of the deliberations of such Institutes as ours; **and** therefore **in** conclusion, gentlemen, I ask you again to co-operate with me in making the session which we are now inaugurating, a bright spot in the annals of this Institute, and one which in addition to conducing to your satisfaction and my own, will be looked upon by similar Institutions as a credit to the Midland Institute and to the county of Yorkshire in which it is centred.

Mr. A. M. CHAMBERS : I rise to move that the best thanks of the Institute be presented to our President for the very able address that he has just read to us, and that it be printed and form part of the Transactions of the Institute. He has sketched out a line of action for the Institute which has, I am sure you will all agree, many valuable suggestions in it. I suppose it will be for the Council to take those suggestions up, and to endeavour, as far as they can see their way, to adopt them. I do not know that if we have a second class membership, it will be wise to make it a free membership. I believe most people value best what they have to pay for, and I think if to that class of members a smaller subscription, even if it were only five shillings were imposed, it would make those gentlemen feel that they had more right in the affairs of the Institute, and they would probably take a deeper interest in it. I should have thought that a subscription of half-a-guinea would have been a useful one for members of a certain class, but that is a matter of suggestion only, and one which, of course, the Council will have to consider and deal with as it thinks proper. I am sure no member of this Institute has done more for the Institute during the last few years than our President. I am pleased to see him occupying that post, and I am sure from the services he has rendered us in the past, we may be confident that in the future he will still do his best in every way to promote the interests of this Institute.

Mr. G. J. KELL : I have great pleasure indeed in seconding the vote of thanks to the President for his admirable address, and that it be printed in the Transactions.

The resolution was carried unanimously.

The PRESIDENT : I am very much obliged for your vote of thanks, and I can only say that anything I can do to further the interests of the Institute and those connected with it, any service I can render in that direction, will be most cheerfully accorded. The thanks of such a large number of gentlemen as I have from time to time been fortunate enough to obtain has been more than ample reward for any trouble I have at any time been at.

DISCUSSION ON MR. W. HARGREAVES' PAPER ON "TONITE AS AN EXPLOSIVE WHEN USED WITH A FLAME-DESTROYING COMPOUND," IN CONJUNCTION WITH MR. G. BLAKE WALKER'S PAPERS ON "HYDRO-CARBON EXPLOSIVES."

The PRESIDENT: The next discussion is on Mr. W. Hargreaves' paper on "Tonite as an Explosive when used with a Flame-destroying Compound," in conjunction with Mr. G. Blake Walker's paper on "Hydro-carbon Explosives." There is a letter from Mr. Hargreaves saying "I shall not be with you on Wednesday at the Midland Institute meeting, owing to a previous engagement."

Mr. MILLER: I do not think there can be much said about this subject as Mr. Hargreaves is not here. He promised to have some more experiments, to try it in an explosive mixture, and to give the results to this meeting.

Mr. McMURTRIE: There is a question I should like to ask Mr. Walker. Mr. Hargreaves says the cost of the detonator is $3\frac{1}{2}$d., Mr. Walker says his detonators cost a half-penny. There is a great difference in that; can Mr. Walker explain it?

Mr. JOSEPH MITCHELL: It is only fair to mention that there is a Committee appointed to deal with experiments with the various explosives, and I would suggest if there is any new matter to be brought forward which may be a guide to those who are going to experiment, now is the proper time for such remarks to be made. I would suggest the further discussion be left until after we get the report of the Experiment Committee.

Mr. G. BLAKE WALKER: With regard to the question which has been put about detonators, they vary in size and in cost, and they vary according to whether they are made in this country or in Germany. Those we are using are made by Paulus, of Vienna, and imported by the Roburite Co., and are not electric detonators. Electric detonators are very expensive on account of the attachments.

Mr. A. M. CHAMBERS: Electric detonators are 205s. a thousand, ordinary fuse detonators vary from 45s. to 56s. a thousand.

The PRESIDENT: On the subject of detonators I can perhaps communicate an important fact to the members of this Institute. We have been using Roburite for some time with detonators supplied by the Roburite Co., firing those detonators with an electric battery. Last week a shot was fired in a stone drift we are driving. An attempt was made to fire the shot, but after turning on the electric battery for some seconds, it was concluded it was a missed shot. The men, after waiting two or three seconds to put the cables on one side, walked back into the place, and as soon as they got in the shot went off, injuring one man in his eyes and severely hurting him. If it had been a point blank shot out from the face, it would probably have killed him and two or three more. The question is one of such enormous importance to everybody using detonators that it seems to me one that we ought to try to find out the reason for the occurrence in question. I do not know whether I am right in the theory I have adopted, but it is roughly this. I am of opinion the charge could not go off without the detonator being fired, because Roburite will not fire without a powerful detonator. By some means or other the wires have fired the gunpowder mixture without that mixture communicating with the fulminate of mercury, but the firing of it has ignited the paper cartridge in which it is placed, and this has burned gradually until it has fired the fulminate, acting practically like a piece of touch-paper. I have had an enlarged drawing made of one of the detonators, which I will place before the members at another meeting, and will explain exactly what I mean.

Mr. ANDREWS: How long was the current on?

The PRESIDENT: Two or three seconds.

Mr. GERRARD: I think this is a very important matter with regard to explosives—the liability to hang fire. Having made a careful investigation of the case Mr. Rhodes has named, on the spot, I must confess that to me it is very mysterious. In the absence of any other suggestion as to the cause, I am bound to think that it must be in the detonator. But the difficulty with me is that two attempts were made to fire, the first probably half an hour before the second; and if a current of electricity passed along the wires to the detonator, and a spark was produced and the detonator set in combustion, it seems difficult to me to understand that the heating of the copper cylinder of the detonator did not take place at the first attempt. It may be advanced that some foreign substance between the brimstone and the fulminate of mercury, owing to carelessness of filling the detonator, may be a reasonable explanation of it. But whether that be so or not it is a fact that a very serious accident occurred from that which we have hitherto considered it impossible to take place when using electricity, although, I believe, it has occurred on several previous occasions at the same place. I am utterly at a loss to account for this hanging fire unless it is as Mr. Rhodes has suggested, or there was some foreign substance in the detonator. I think it is a very important subject to bring forward, and that every effort should be made by experiment to arrive at a satisfactory conclusion, because we are going on with the same sort of thing and liable to serious accidents. There is another important matter which appears in the last number of the Transactions. That is in connection with Bickford Smith & Co.'s patent fuze igniter. This patent fuze igniter is put forward as being absolutely safe, absolutely flameless, and they issue certain instructions as to how it is to be used, and when their attention is drawn to a serious accident which has occurred in connection with this patent igniter they make a very shuffling explanation. The word which they use in their instructions is, I believe, that the igniter should fit "snugly" upon the fuze. My idea was that snugly meant closely, but now they say that it must not be closely, that there must be a space, and yet there is no protection; there is nothing to fix that, it is left at the will of the person using it. There should be something left to fix that place to prevent it being absolutely close, otherwise you have not a safety lighter, but something which will produce flame. I do not wish to introduce any question with regard to the use of these things in gas, because that goes without saying, but at any rate we want to produce as little flame as possible, or no flame at all. If this thing is not a safety lighter the sooner it is known and recognised the better.

Mr. G. BLAKE WALKER: At the last meeting of the Institute I produced eight of these fuze-igniters which had exploded out of two dozen. Allowing perhaps for some mismanagement in attaching them, this is a very excessive number. All that I produced were fired in the open air, and with a special view to testing how far they were reliable or not. I cannot help thinking they must have contained, this particular two dozen, more of the chemical substance on which the ignition depends than was intended. But at any rate, out of two dozen of these igniters eight exploded. A gentleman who was elected a member of this Institute at the last meeting, Mr. Clark, was burnt, and his overman was burnt, and two colliers were also burnt, from the gas which was ignited in trying a shot with one of these things in a small quantity of gas. It is rather a delicate matter to mention, but still, as at Wharncliffe Silkstone we have had something of the same kind,—not an explosion in gas, but have had these things exploding,—I think it is right to mention it, and to point out that they must be used with excessive care.

Mr. A. M. CHAMBERS: Have you had any satisfactory explanation from the maker?

Mr. G. BLAKE WALKER: No, a very unsatisfactory one indeed; Mr. Gerrard has seen it.

Mr. GERRARD: That is why I characterised it as "shuffling."

Mr. G. BLAKE WALKER: You have to draw a very fine line between just the amount of nipping to prevent the igniter being blown off the end of the fuze, and that amount which will imprison the gases and cause it to explode.

Mr. W. HOOLE CHAMBERS: My experience of the patent fuze-igniters is not very extensive. For the last week or ten days we have been using them regularly, and we have not had a single one which has not been perfectly safe up to the present time. Of course the men that use them have received instructions as to how they put them on (but they are men accustomed to fire shots), and who have not seen them in use before. I should think, altogether, we have fired 30 or 35 shots, without experiencing any accident. In some cases they have failed to ignite the fuze, but we have had no experience of them either blowing off or in any way rendering the place at all dangerous.

Mr. JONATHAN WROE: I have myself fired about a hundred shots with these caps, and found that many times they emitted flame at either end; sometimes from the outer and sometimes at the inner end.

Mr. GERRARD: Did you take account of how many produced flame?

Mr. WROE: No.

Mr. GERRARD: What was the proportion?

Mr. G. BLAKE WALKER: What do you mean by their coming out at either end of the fuze?

Mr. WROE: When you have placed the fuze in the cap, when you have pressed the cap to fire it, it has blown out at the low end.

Mr. A. M. CHAMBERS: That is, blown the cap back?

Mr. WROE: Yes.

Mr. A. M. CHAMBERS: In other words, come through the end of the cap itself?

Mr. WROE: In others I have seen it go forward right to the fuze, and produce a flame at the top of the cap.

Mr. GERRARD: Have you any idea as to how many have failed— ten or twenty?

Mr. WROE: I could not say the number, but I have seen a large quantity.

Mr. GERRARD: You know whether ten or twenty would be nearer the mark?

Mr. WROE: If you said twenty you would not be far off.

Mr. G. BLAKE WALKER: I cannot help thinking they vary a little, because in the first lot we had I never heard a word said about their being dangerous. It was in the last lot chiefly, I think, that these explosions took place.

Mr. GERRARD: Have you had any experience of hanging fire with the electric battery or detonators?

Mr. G. BLAKE WALKER: We have not been using them in that form.

SAFETY SAND CARTRIDGE.

The PRESIDENT: If there is nothing more to say on this special subject, there are in the interesting notes communicated by Mr. G. Blake Walker one or two explosives alluded to. Amongst others there is the sand cartridge which, I believe, is something quite new. Has any gentleman anything to say about that? It seems to me very much on the principle of our water cartridge but using sand instead.

Mr. GERRARD: Or to the wet moss which has been used in South Wales?

Mr. G. BLAKE WALKER: Yes.

Mr. GERRARD: It is the same thing—simply an absorbent.

Mr. G. BLAKE WALKER: I think the Committee might make a few cartridges in the way described and try them.

The PRESIDENT: I think it would be desirable if the Committee went into the merits of all these various ideas. If the mere fact of using damp sand is sufficient to prevent flame of any sort it seems to be a very simple and easy way of getting out of a great difficulty.

Mr. G. BLAKE WALKER: You see the inventor asserts that "in no case has a free lying sand cartridge with 200 grammes of dynamite fired in air mixed with fire-damp to the extent of nine per cent., and with coal dust abundantly present, caused any ignition." That is a very big statement if it is true.

The PRESIDENT: Yes.

Mr. GERRARD: Of course the great drawback to this sand cartridge, as to other cartridges, is the increased diameter of the hole and the cost of boring, always provided you can get security without that objection.

Mr. G. BLAKE WALKER: I quite agree with you. There is a very remarkable assertion, which I must say I do not credit, in the first paragraph of these notes. It comes to us with the authority of well-known names, but it amounts to this, that if you mix ordinary blasting power with coal dust, or with an equal weight of carbonate or sulphate of soda, dynamite and nitro-glycerine will not attain a temperature so high as to ignite gas. Seeing we consider stemming a hole with coal dust is a very objectionable proceeding, it looks absurd upon the face of it to think that its use will ensure safety.

Mr. GERRARD: It hardly agrees with the experiments made at Aldwarke, where you got a tremendous cloud of dust and flame from exploding a cartridge in a tub of coal dust, where there was a far greater proportion of dust than is mentioned here, where they say, "an equal weight of coal dust." A still stranger statement is that "a smaller proportion will render ordinary blasting powder safe."

The PRESIDENT: When there is nothing but coal dust, ordinary powder or dynamite will blow up and light the coal dust like powder.

Mr. G. BLAKE WALKER: The statement appeared in a paper by M. André, so that it came with some kind of authority.

EXPERIMENTS WITH ROBURITE IN GERMANY.

The PRESIDENT: It says here, No. 5, "400 grammes blasting powder, cartridge embedded in coal dust, without gas, but with coal dust; violent explosion." It does not say that the coal dust was ignited; are we to assume it was?

Mr. G. BLAKE WALKER: I translated the report just as it was. It raises the question, will coal dust explode with a slight admixture of gas?

The PRESIDENT: It will burn if not explode. Gunpowder fired amongst coal dust will light the coal dust.

Mr. G. BLAKE WALKER: When you fire it in those circumstances you have the coal dust slightly compressed, but when you have it finely divided in the air of a colliery it is more likely to explode.

Mr. W. HOOLE CHAMBERS: In his lecture at Firth College, Professor Lupton exploded dust in a gallery representing the gallery of a mine, without any gas whatever being present.

The PRESIDENT: If you blow coal dust down a tube with a candle in it you get something approaching an explosion; you get combustion so rapid as to be almost equal to an explosion.

Mr. G. BLAKE WALKER: We must distinguish between different qualities of coal dust; some give widely different results from others.

The PRESIDENT: No doubt.

PLOM AND D'ANDRIMONT'S EXCAVATOR.

Mr. G. BLAKE WALKER: There is an invention here which certainly affords subject for thought—the idea of lodging a quantity of explosive at the back of a block of coal, and so exploding it that the pressure is outwards. How the idea may work out in practice will, I should think, depend upon the nature of a particular seam, but in some coals I should not be surprised if it did answer pretty well, and obviate the necessity for holing it, and also get it in a much more lumpy state.

The PRESIDENT: The inventor alarmed me. I asked him how much powder he used, and he said "Anything over seven or eight pounds; it is a matter entirely for yourself that." I said "You will have to fire that seven or eight pounds somewhere else; it is out of the question here." It struck me that a hundred shots of eight or ten pounds each would cause a revolution in some of our collieries; but I do not see why there should not be something in it, coupled with our holing and cutting. I think it is better for powder to explode at the base of a hole or in a cutting corner; you run far less chance of a blown-out shot than you do with a straight hole.

Mr. GERRARD: I should be sorry to say anything against holing and cutting; it is one of the safeguards.

The PRESIDENT: It is an ingenious instrument; as a piece of ingenuity I was wonderfully pleased, but I do not like the idea of blowing our coal out in that sort of way.

Mr. GERRARD: We have not got to pulling it out in that way yet.

Mr. A. M. CHAMBERS: There has been a good deal of that sort of thing done in South Wales.

The PRESIDENT: They have been pretty successful in South Wales.

Mr. A. M. CHAMBERS: Before this was introduced.

The PRESIDENT: I recollect twenty-five years since in Lancashire it was a regular thing to put 2½ lbs. of powder in the first shot you fired, to give you a loose end to work from. Twenty-five years ago cutting in straight work was almost unknown in Lancashire.

Mr. GERRARD: It was an innovation. I should propose, as regards Mr. Hargreaves' paper, that it be further adjourned to give him an opportunity of being present to describe any further experiments that he has made.

Mr. A. M. CHAMBERS: Will it not be better to do as Mr. Mitchell suggests—adjourn both papers until we have the Report of the experiments Committee.

Mr. Mitchell's suggestion was accordingly agreed to.

ADJOURNED DISCUSSION ON MR. H. B. NASH'S PAPER ON
"FOREIGN MINING RENTS AND ROYALTIES."

The PRESIDENT: The next question is the adjourned discussion on Mr. H. B. Nash's paper on "Foreign Mining Rents and Royalties."

Mr. GERRARD: When last we discussed this paper Mr. Cobbold stated that a rebate was allowed on royalties in Scotland. Since then I have had a letter from one of the former Presidents of the Mining Institute of Scotland, in which he confirms the statement I made at that time.

Mr. WM. HY. CHAMBERS: I was in Scotland three weeks ago, and they were complaining then about the high royalties and the small area they were allowed to take, and there was no reduction made on any account in the way they worked the coal.

Mr. H. B. NASH: I should like to say a few words in reply to Mr. Cobbold's remarks on mining rents and royalties. I did not, either in my paper or remarks, argue for the minerals to be in the hands or under the control of Government, nor do I think the colliery owners would be in a better position were that the case than they are under the present system of landlords; but I do think there should be some person or authority who could be referred to in cases of dispute arising, as a Minister of Mines or some person appointed by Government, to whom questions of excessive royalties and way-leave rents could be referred, and whose decision should be binding upon all parties interested. I think that it would be much more satisfactory to colliery owners if the royalty rents were based upon a sliding scale system, to be regulated by the average selling price of the coal with an equitable fixed minimum, so that landlord and tenant should both suffer or benefit alike by an increase or depression in trade; whereas under the present one-sided system, the landlord has all the benefits while the colliery proprietor has to bear all the anxieties and suffer all the losses caused by the depression in trade and miners' agitations, &c., which is a most unfair division of responsibilities. I certainly do not agree with the concluding remarks of Mr. Cobbold's address, and feel sure that the coal trade in our country could not have been in a much more unsatisfactory condition as regards profits than it has been during the last ten years, with all the benefits of freedom of contract and responsible landlords of which he there speaks. With regard to Mr. Gerrard's suggestion that local information of Yorkshire royalties and how lessees are met, I think it would be making the discussion too personal to speak on this subject more than in a general manner, and might tend to create ill-feeling between the parties interested.

Mr. A. M. CHAMBERS: It appears to me an exceedingly difficult question. The fact is that conditions have entirely altered,—

the general conditions under which coal is worked, and into which the mining interest of the country have got. We have had an unprecedented depression, and whilst bound fast on one side, everything else has had to give way to the depression. In many cases that difficulty has been to a great extent met by the landlords, but it really seems as if some system of sliding scale would be a very valuable one, but I think it would have to be worked out by agreement between landlords and tenants when their time comes. I very much deprecate calling into office new bodies. We have already more official bodies than are needed in this country.

The PRESIDENT: I think the question rather a dangerous one to touch. Any interference with the rights of property is starting a train it may be very difficult to stop; and I strongly deprecate State interference. I would sooner deal with individuals by far than the State.

Mr. GERRARD: It is a very difficult question, and though a proper subject to take up, it wants careful handling. There are good and generous landlords, and hard and grasping ones, and when lessees have to deal with the latter class, then they feel it.

The discussion then closed.

DATES OF MEETINGS.

Mr. G. BLAKE WALKER: It would be a great convenience to members if the Council could decide in advance upon the dates of our meetings, and issue them upon a card so that we might know beforehand exactly when and where the meetings will be, and so enable us to make our arrangements accordingly. We sometimes get these notices three days before a meeting, and one may have made engagements which prevent his attending when he would gladly have been present. If we knew in advance when the meetings would take place it would give us a better chance of being present.

The PRESIDENT: I will arrange that with the Secretary.

The meeting then ended.

PROCEEDINGS.

WEDNESDAY, JUNE 6TH, 1888.
IN THE COUNCIL CHAMBER OF THE INSTITUTION OF CIVIL ENGINEERS, 25, GREAT GEORGE STREET, LONDON.

SIR LOWTHIAN BELL, BART., IN THE CHAIR.

FEDERATION OF MINING INSTITUTES.

PRESENT :—Sir Lowthian Bell, Bart., Messrs. J. Marley, A. L. Steavenson, M. Walton Brown, W. Cochrane, T. J. Bewick, W. Armstrong, Jun., J. Daglish, and T. Forster Brown (North of England Institute); W. H. Howard, J. Jackson, and M. H. Mills (Chesterfield); Professor Benton and Mr. Alex. Smith (South Staffordshire); Messrs. G. B. Walker, Jos. Mitchell, T. W. H. Mitchell, and A. M. Chambers (Midland); Messrs. R. Haines and J. Lucas (North Staffordshire); and Professor G. A. Lebour (Secretary).

Mr. T. FORSTER BROWN begged to propose that Sir Lowthian Bell take the chair.

Mr. W. COCHRANE seconded the resolution.

The resolution was carried unanimously.

The CHAIRMAN said he would not detain them at any great length, because he presumed that they had all made themselves acquainted with the business upon which they had met that morning. It was to discuss a project which, although set on foot mainly by a paper which had been read before the North of England Institute a few months ago, yet he believed the credit of originating the scheme itself was due to his friend and predecessor——

Mr. J. DAGLISH: It goes further back than that, Sir Lowthian. It was in the time of Mr. Forster's presidency.

The CHAIRMAN said it seemed to be buried in the mists of a remote antiquity; it had lived during all that time, and had never been called into operation. They had met there that morning in order to hear their views upon the subject, and to ascertain whether it would be favourably received, so as to justify the North of England Institute of Mining and Mechanical Engineers proceeding in their attempts to carry it into execution.

So far as he was personally concerned, he might say at once that he was in favour of the scheme, and, if for no other reason than this, that by the co-operation of all the mining engineers in the country they must, he thought, expect that they would get by a more direct road to the truth, in connection with those enquiries which it was their business to originate and discuss, than they could do single-handed. In the first place, the enquiries themselves must more or less take the colour and direction of the peculiar coal-field in which they originate, and, to correct this when necessary, it could not but be of very great advantage that the experience of one district should be compared with the experience of other districts. In addition to these, there were many questions which were almost beyond the means (he meant beyond the financial means) of a single body to investigate, but which might quite easily be brought within the powers of the union of Mining Institutes like their own. The prospectus placed in their hands very properly pointed out the example of other bodies of a cognate character, which they might themselves follow with advantage. They had the Society of Chemical Industry. Now, he believed the North of England had the credit of being one of the first, if not the first, to originate a Society of Chemical Industry. The London Society soon followed; and it saw the desirability of gathering within its fold, as it were, the Societies of Chemical Industry from the provinces which led to its establishment. The Iron and Steel Institute, of which he (the Chairman) was one of the early promoters, began at once as an Institute embracing the iron trade of every part of the country. If an instance were wanted to point out the desirability of such a mode of procedure, it was that afforded by the Iron and Steel Institute. He believed that there was no industrial Institute in the country which rose so rapidly to a position of eminence and usefulness. In speaking of the desirability of co-operation, he might mention one case in the history of that body which seemed worthy of notice, where it was thought desirable to investigate the so-called mechanical puddler, an American invention, and in order to do that in the most satisfactory way, the Iron and Steel Institute appointed four gentlemen—four, if he remembered rightly, "commissioners," as they called them—who were deputed to visit, and did visit, the United States, in order to examine the nature and success attending the use of mechanical puddling,—a matter at that time of very great importance, because, as they all knew, the labour in puddling was an extremely severe one, which it was desired to alleviate. These gentlemen went over there, and reported fully upon the mechanical puddler; and although, practically, it died a natural death in this country, it must not be inferred on that account that they disagreed with the report of the commissioners, but because the introduction of steel had in a great measure superseded puddled iron.

He would just mention another matter which was germane to their own particular profession: he meant the question of coking coal. Now there had been various kinds of coking ovens recommended, and different kinds had been tried without a proper consideration of the quality of coal to be treated. In consequence considerable sums of money had been wasted which might have been saved had the subject been examined with the care its importance deserved. He could not but think that if mining engineers in other parts had heard of the experience of the north country coke manufacturers, possibly large sums of money would have been saved.

Now, might he venture, in a company of mining engineers, to say a few words in regard to the Davy lamp? The Davy lamp was an invention made fifty years ago or more. He believed it was only very recently discovered that the presence of fine coal dust in the interior of a Davy lamp might constitute a source of danger. Then, more recently, they had been told how dangerous the presence of coal dust in the workings might be in promoting explosions, or at all events intensifying the effects of explosions. Having regard to the very small quantity of coal dust which might convert atmospheric air into a highly explosive mixture, he thought the importance of a proper investigation, which might be undertaken by the united colliery districts, could not very well be over-rated.

These were a few of the ideas which had led him to give his ready and very hearty willingness to co-operate with the other Mining Institutes of the country in securing the combination that he had endeavoured to bring before them.

He would now call upon their Secretary, Professor Lebour, to let them know what progress the movement had made in other quarters than the North of England, and then they would be better able to judge, he thought, of the probability of their carrying to a successful issue the establishment of a federation of the chief Coal Mining Institutes of this country.

Professor LEBOUR (Secretary) then read abstracts of the replies which had been received from the various Mining Institutes to the question which was addressed to them generally—" Is or is not such an arrangement as that outlined in Mr. Bunning's paper desirable ?" The Chesterfield and Midland Institute, in a letter of December 12th, 1887, answered the question in the affirmative, and appointed Messrs. J. Jackson and M. H. Mills as representatives. They had confirmed this answer by sending their representatives there that day. The Midland Institute, by their letter of January 16th, 1888, considered the scheme desirable, and named Messrs. T. W. Embleton, A. M. Chambers, T. Carrington, G. B. Walker, and Joseph Mitchell as representatives. They also had sent their representatives, and so far, therefore, confirmed their previous letter. The North Staffordshire Institute had appointed Messrs. J. Lucas, W. Y. Craig, and Richard Haines as representatives; but they did not say whether they agreed to the scheme or not. The South Staffordshire Institute considered the scheme desirable, and had appointed Messrs. W. B. Scott, J. Hughes, and Alex. Smith as representatives. Their representatives had also come to that meeting. The Mining Institute of Scotland, from their letter of December 30th, 1887, were generally of opinion that a federation for the purposes contemplated in the paper was inexpedient and unnecessary, inasmuch as the Mining Association of Great Britain already occupied the position proposed to be established. He should add, however, that in the letter in which the Secretary stated this, he added that his personal opinion was that the Mining Association of Great Britain had nothing to do with the matter at all. There was, however, an expression of opinion that an arrangement might be made for the first publication of Transactions somewhat similiar to that proposed in the paper; but no representatives had been appointed to attend this meeting. That had been confirmed by a letter received quite recently—they still wished to have nothing to do with the scheme. The South Wales Institute were not disposed at present to appoint a committee as suggested. That was in a letter dated October 27th, 1887, and that action had been confirmed. The Mining Institute of

Cornwall were of opinion that the time had not yet arrived for the Society to join the proposed federation. That also had been confirmed. The Manchester Geological Society, although its expression of opinion was informal only, was adverse to the proposal. They had also declined to send representatives to the meeting.

The CHAIRMAN said, in the meantime, he would be very glad to hear the views of any gentleman present upon the subject.

Mr. JACKSON, as representing the Chesterfield Mining Institute, the first on the paper before them, said that their Council was quite of the opinion that a federation of this sort would be desirable, and of great advantage to the mining community of great Britain. But as an Institute they did not desire to lose their individuality; they would be glad to be, as it were, a branch of a federation; but the difficulties they saw were, that they had so many members amongst them that were under-viewers or students—young men who were learning mining engineering, and also workmen in the pits—a class of men who at home were in a position to take part in discussions, and to take an active interest in the welfare of their Institute. They felt that if they became extinct, and attempted to embody themselves in one large federation, they would be doing an injury and an injustice to a large class who ably supported them; but as to the general principal of the thing they were very strongly in favour of it. They also advocated the idea of Mr. Bunning, namely, with regard to papers, that if there was a general federation they would in time be able to send the papers that were written, first, to a central committee, which would have the power of saying whether they were worthy to be read before the General Institute or not, and then confining to themselves the right of reading those papers and discussing them at home. But the better ones would go to the general committee, and, if worth anything, would be received in the Institute. Then another question came before them, namely, that of expense. They felt that if they still had to continue the same subscription, and contribute a guinea for the privilege he had mentioned, that would be detrimental to their interests. Mr. Mills and himself had come there to express these views, and to do what they could to further the objects of the meeting, so-long as it was not going to place them at a disadvantage, or to make them extinct at **Chesterfield.**

Mr. BEWICK said he would merely suggest that each gentleman, as he addressed the meeting, should say for what Institute he appeared.

Mr. MILLS said, as the other member for the Chesterfield Institute, he simply wished to say that he endorsed what Mr. Jackson had already said. There was a strong feeling in their Institution that such an Institution as the late Mr. Bunning had proposed should be established. As to the exact details of that Institution, of course neither they nor their Institute were able to say anything at present; but he was sure of this, that they had the strongest feeling that some sort of Institution such as the late Mr. Bunning had proposed should be established. He hoped that that meeting would arrange some of the details necessary.

Mr. WALKER, on behalf of the Midland Mining Institute, said he could hardly add anything to what had already been said by Mr. Jackson. They felt, as he did, that their Institute was not so important, nor composed of men (in great part, he meant) of such high status, as the North of England Mining Institute; and recognizing as they did, that theirs was a weaker Institution, and that it had a particular work to do amongst the class of men Mr. Jackson had mentioned, namely, to a large extent amongst under-managers, under-viewers, deputies, and mining students, they could not help feeling that anything that tended in any way to make their Institution less suitable to their needs would be a loss to them, and therefore, although they desired very thoroughly to support the idea of a federation of Mining Institutes, they did feel that in drawing up a scheme for that purpose the Council ought to bear in mind the nature of Institutions like theirs, whose status, perhaps, was somewhat more humble than that of the North of England Institute, from which the proposal originally came. The Council of the Midland Institute very thoroughly agreed in the general idea of the late Mr. Bunning's scheme; but still, it lacked precision. It was put forward merely in the first instance as a feeler to elicit, he presumed, the views of the different mining engineers throughout the country; and that being so, it remained, as it were, to formulate at that meeting, if the idea was gone on with, something of a more definite kind which the different Councils of the different Institutions might consider. They did not gather from the paper very clearly to what extent the subscription to their own Institution would have to be increased Their subscription was only a guinea, and that guinea was a good deal to men of the class he had alluded to,—such as under-managers, and so on,—and they could not recommend anything that very largely increased the amount of the subscription. Then again, they would like to know what was to be done in the case of those papers which were perhaps somewhat old to the profession as a whole, but which it was very desirable should be brought before local Institutions? There were certain papers which had appeared in the old numbers of the North of England Mining Institute Transactions, in connection with the special features which had been dealt with in particular collieries. He remembered a very curious question on ventilation,—he thought it was at Pontop Colliery,—which was brought forward by the late Mr. Atkinson, Government Inspector of Mines, and which he treated in such a way as brought out very clearly the principles of natural ventilation, and the influence of the different sizes of shafts, and so on. Those papers, to a certain extent, were buried in the old numbers of the Transactions of the North of England Institute; and something of a similar character might be brought forward by some one in another district and re-treated. Well, the feeling of the profession might be—" This is very old; this is something we do not very much care to be told again;" and yet, for the kind of people they had in their minds, it might be exceedingly useful. No one could read the correspondence in the weekly newspapers, such as the *Colliery Guardian* and the *Iron and Coal Trades Review*, without seeing that these papers were being constantly made the medium of correspondence of interest. Therefore, one thing he should very much like the present meeting to consider was—in how far they could permit to the different Institutions some freedom as to the selection of papers, and not impose too rigid a rule as to those which should be submitted for acceptance. Had he known that Mr. Chambers was there, he would have much preferred that that gentleman should have said what was to be said on behalf of their Institution. Later on, perhaps, Mr. Chambers might give them their views more clearly.

Mr. LUCAS said, speaking for the North Staffordshire Mining Association and Mechanical Engineers, they found themselves in a similar position to that of the former speakers, and something beyond that. Their Institution was composed partly of mechanical engineers, and it would affect those members if one part of the whole, the mechanical element, which used to combine with them, refused to subscribe. He (Mr. Lucas) had very little to say about it, because he thought the former speakers had said everything. So far as he was concerned he would simply add that personally he was strongly in favour of federation, because he saw its necessity, not only as a federation to promote their common well-being, and to exchange views, and that sort of thing, but it would be a very powerful instrument in the hands of a Mining Association as regards resolutions from such a federation, for example; if it went before the Home Secretary on matters of mining interest. In fact, on the late Mines Regulation Bill, he knew the difficulties that they had to contend against there, so that personally he was strongly in favour of such a federation being formed, but he was at a loss to define the means, and he would advocate in their Institution the formation of such a federation.

Mr. HAINES said he also represented North Staffordshire Association, and he could only endorse what Mr. Lucas, and also the representatives of the Chesterfield and the Midland Institution, had already said. The mining engineers of North Staffordshire, he might say, most heartily supported the scheme, but the colliery managers and the other members of their body who were not mining engineers pure and simple, did not see their way as a body to join it. He was sure it would have the hearty support of all those who practice as mining engineers, and he believed that advantage might accrue from it, but, as a body, he thought they had expressed very nearly the same views as those of the representatives from the Chesterfield and Midland Institute.

Mr. SMITH said, as representing the South Staffordshire Institute of Mining Engineers, he would simply express the opinion given by the Council of their Institution when this matter came before them, and that was that they thought there could scarcely be two opinions upon the subject, and certainly that such a scheme as that suggested by Mr. Bunning was very desirable. It was rather surprising to see some of the objections and answers given on the paper that they had before them. No one but the secretary of an Institution could so fully appreciate some of the troubles and difficulties set forth by Mr. Bunning in his paper. They found very often that inventors and suggestors of schemes for the improvement of mining science brought their papers to them and made them a sort of advertisement; and although in the rules of almost every Institution it was laid down distinctly that papers were the copyright of the Institution, still they constantly found that the same papers were being read throughout the country. Then again, there was no doubt whatever that if they could have Transactions of a Central Institution as suggested, they would be a very valuable addition to the mining literature of the age. Another thing was that, having a Central Institution where all the great questions would be thoroughly investigated, they would not have the difficulties they sometimes met with in the local districts, where they sometimes, he might almost say, floundered upon some questions because they were not fully conversant with the whole of the ideas involved, or where they had not the advantage of getting the pick and selection of the mining science of the day. All that would be at an end if the questions were considered by a Central Institution. A great many of the objections—well, not exactly objections, but doubts—expressed by previous speakers, were really met in Mr. Bunning's paper. In regard to the subscriptions, for instance, it was not a *sine quâ non* in accordance with the paper, as he read it, if the Institution adopted the idea, that the whole of the members should join—that the Institution should come over as a body—they might be federated, but they might have a considerable number of their members (it was distinctly stated there) who might not be members of the Central Institution. That was very well dealt with in Mr. Bunning's paper, and although he stated distinctly that they should not lose their individuality, every Institution, as the gentlemen before have expressed it, would object strongly to the scheme if there was an idea that they would lose themselves, as it were, in the Central Association. Mr. Bunning pretty clearly put it that such a thing was not desirable, because of the varying requirements of the different districts, and he (Mr. Smith) did not see that really he contemplated such a thing in his paper.

Mr. CHAMBERS said he did not think he had anything to add to what had been already said. The Midland Institute thoroughly approved of the scheme generally, though criticising some of the details. They were particularly anxious, having regard to the great number of members, as Mr. Walker had already said, that their subscriptions should not be increased. He did not entirely endorse that himself, because, of course, they were going to get additional benefits, and he thought they ought to be willing to pay something for them ; and he was quite sure a scheme of that kind would be a great benefit to the whole of the mining districts of the country. They approved of it generally, and should be happy to co-operate as regards the details of the scheme.

Mr. HOWARD: Mr. Bewick suggested that each speaker should say for what Institute he appeared, but as he had not been delegated formally by the Institute with which he was immediately connected (the Chesterfield Institution) he did not know whether his summons there came from the North or Midland Institution. He felt a little more at liberty perhaps than he otherwise would, in consequence of not having been delegated by the Chesterfield Institution, to express the opinions formed in his mind, namely, that it was scarcely federation that was practicable in their case. It was more like affiliation to a Central Institution. He thought, however they were to go about it, that would be the result, and he thought it was well worthy of the consideration of the gentlemen then present; and his hope was that, before the meeting separated, something would be formulated of that character, that the delegates could take back with them to their several Councils and put before them; and also that means should be taken to ascertain what strength there was and what probability there was of establishing a Central Institution, with the object of assisting and furthering the views and objects of the local Societies, and doing all the good that a Central Institution could do to them; not draining them, but really helping them on, and charging them no more than need be for anything that it might do for them. There would have to be something, he thought, in the way of contributions from the Societies for anything that was done for them, but it would have to take that form rather than that which had been suggested in the paper. He saw great difficulties with regard to the class of members that they and other Institutions of the same kind had ; and there was no doubt the Central Institution would be composed of what they would term the cream of the profession. The subscription itself would, no doubt, do that to a great extent, and, he thought that that being so, the thing would be worked out best upon those lines.

Mr. FORSTER BROWN said he was there not as representing the South Wales Institute of Engineers, of which he happened to be a member, but as a member of the North of England Institute. He had long held that the mining interests of the country, with which the mining profession particularly have so much to do, were of sufficient imj ortance to justify a Central Institute, which would add weight to the particular mining profession, both for legislative and other purposes; and from that point of view he had gone so far as to hold that the parent Institute (the North of England Institute) ought to take the thing up, and, whatever the consequences were, to promulgate a proper scheme. But the effect of Mr. Bunning's paper, and the opinions that had been expressed that day, showed that five-eighths of the whole of the Mining Institutions of the country were in favour of such a scheme; and it seemed to him that the next step to take was that a Committee should be appointed, and that the gentlemen representing all those Institutions which were in favour of federation should be members of that Committee, with a view to propounding some scheme of an Institution in London, leaving and still maintaining the local Institutions, but which would ultimately become the Mining Institute of England. And he had not the slightest doubt that if such an Institution was started on a sound basis all those dissentients would join in due time.

Mr. A. L. STEAVENSON said the great doubt in his mind was whether the thing was financially possible. That it would be advisable in the interests of the profession of mining engineers there could be no doubt; but, looking at Institutions of a similar character, so far as he could read the matter, the Mechanical Engineers spent £4,000 a year in doing what they practically proposed to do with the Imperial Institute; and he did not see how it was possible with an additional subscription of one guinea to meet all the contemplated expense. If it would not do so, it was possible that members such as they had now in the various Institutes would be able to contribute to the funds of the Council. Again, he would refer to the possibility of there being drawn away from the local Institutes a good many of the members of their Institute belonging to South Wales and Yorkshire. He thought those members would entirely give up their connection with the Institute at Newcastle. That course might tend to spoil their local Institutions, although it might be beneficial to the Central Institution. There was just one other point, and that was as to the decisions upon the papers which were to be printed by the Central or Imperial Institute. Mr. Bunning, he thought, in his paper suggested that a meeting of the Council should be held at certain times to select the papers that were to be printed; but he considered it was very essential that they should be decided upon at the time they were in print; because if the type had to be taken down and renewed it would add materially to the cost of producing the Transactions; so that that was a difficulty which would be met, he thought, by a very competent secretary. The central secretary in London would be better able to select the papers that he thought suitable, and would do it better than a committee; he would sit more as an impartial arbitrator in the matter, and decide what he thought best for the Institute. He could then at once give orders for the type and the plates to be put in hand and printed at a much less expense than if first done by the local centres, or if done for the local centres and afterwards reprinted for the Central Institute.

Mr. CHAMBERS said he did not know whether it was taken for granted that because Mr. Bunning had suggested that the Central Institute should be formed in London, those who approved of the scheme generally agreed to that part of it.

The CHAIRMAN: Certainly not.

Mr. CHAMBERS thought that was assumed by one or two speakers.

Professor BENTON said he had nothing to add to what they had already heard. He saw in Mr. Bunning's paper the germs of an excellent scheme, and he was waiting, personally, with great impatience to see its full development.

Mr. COCHRANE said there was one remark he should like to make. It was thrown out by one of the members that the federation should be a body dealing with parliamentary matters and resolutions being passed in that direction had been mentioned. He took it that their impression would be that such a federation would be for scientific purposes only. If those Institutions were to federate with any idea whatever of troubling themselves about the outlying matters which were indicated in the speech which fell from one gentleman, he thought it would be a great mistake. That was his impression at the present moment. He was quite capable of being impressed otherwise afterwards, but he thought the object of their Institute—at any rate, in the North of England—had been so strictly confined to the scientific and purely mining part of engineering, that he should be sorry to see any Association connected with that body that was otherwise intended. He also saw the great difficulty of dealing with an entire Institute and saying that that Institute was to pass over to the federation as an Institute, thereby forcing all its members to become chargeable therewith; and he was certain that Mr. Steavenson's prophecy would turn out to be accurate, namely, a considerably increased expense; nor did he quite agree with other speakers who said the cream of the profession would go up there—by which, he was certain, they meant not the cream intellectually, but the cream so far as their pockets were concerned. It would be a great mistake if the federation aimed at that, or thought that that was the way in which a federation was going to be supported. The scheme certainly ought to encourage not members as a body but each individual of his own voluntary action to come out from the existing Institutes and join the federation under arrangements made by each local Institute as regarded each member, having due regard to their status in that body, whether under-viewers, mining engineers, rich men or poor men, and it should be left voluntarily to each local society to make all such arrangements as to distribution of Proceedings (which, after all, was the great thing) and the attendance upon the joint meetings as they liked; but to force any member of a local Institute into any higher subscription than what he now pays, would, to his mind, be a very great mistake. One other thing was, the subject of the use of papers which already were almost taken as text books, and particularly in the North of England Institute of Mining Engineers. That was a very important point. He thought it was Mr. Jackson who mentioned that. If the affiliation was to be perfect and thorough, one of the items they would have to consider would be a retrospective affiliation as well as a forward one. They had, as every other Institute had, a very strong idea upon the subject of the copyright and value of their Proceedings. Many of the Proceedings of the North of England Institute were at that moment absolutely out of print. He did not say that that Institute would be prepared to give up the question of copyright; but, certainly, he thought, if the affiliation was to be perfect, that it would be a very desirable thing to consider the question of allowing an absolute reprint of those papers in other local Transactions; therefore, if they really intended to marry, they must go in for better or for worse, and they would have to make such a consideration as that retrospective as well as general in the future.

Mr. MARLEY said he should have been glad if their ex-President (Mr. Daglish) had given them his views first, as being to a certain extent, prior to Mr. Bunning, the father of the idea which was contained in the paper; but as Mr. Daglish had not done so, he would state very shortly some of his views upon the question. He might say, to begin with, that it was desirable that a federation of some kind should be carried out. Mr. Bunning in his paper might have pitched the case a little too high, although it was probably well to aim high so as to get a medium result. There was no doubt that the question of finance would be a very important one, and as was suggested to Mr. Daglish four years ago, when they first entered upon the subject, equally important would it be that the local Institutes should not lose their individuality. These were some of the points, after hearing the speakers that morning, to be agreed upon. The other matters would, to a very great extent, become details; for instance, it seemed that they were sufficiently unanimous so far, and that there was a sufficient adhesion to show that joint publication should be carried out; but that the members of a whole Institute should be transferred to the federated one did not seem desirable. The respective societies would probably federate for the purpose of publication, leaving it to their members' option, at some small individual increased fee, to become members of the federated society or not; and that would to a very great extent probably facilitate the question of finance—that was to say, they would save money in all probability by joint publication of selected parts of their respective Proceedings; and then the other expense would be met by special fees at the option of the individual members. As regarded the parliamentary question he was glad that Mr. Cochrane had touched upon that, because, although their Scotch friends said that the Mining Association of Great Britain would meet all that was required, that was not the case. Parliamentary matters should certainly be outside the scope of such an Institute, and the federation and the respective Institutions should not meddle with anything but what was scientific and for the saving of life—in fact that which was laid down by their first President as the principles upon which the North of England Institute was formed. These were the principal points which, subject to details, he thought they were in a position to carry out.

Mr. STEAVENSON said there was one point which might be attended to at once, whether they went on with the Institute or not; that was to provide the members of every Association (there were nine of them) with a copy of the index of their annual Proceedings, so that they might know exactly where to find any paper which had been published during the last year in any one of the societies' Transactions. It would at once keep them up to the mark as to what had been done all over the country. That might be forwarded to every member of every Institute.

Mr. DAGLISH, in answer to the Chairman, said he had really very little to say because, he thought, every point that could bear upon the question had already been touched upon. He made a few notes when they commenced, and, with their permission, he would just draw attention to the scheme which seemed to suggest itself to him, and several others with whom he had been in communication, as being the most practical, and that was to confine the federation very greatly to simply publishing joint Transactions, and, if that were so, it must be more economical than the present system. It could not lead to a greater expenditure, but would lead to a less expenditure. It would not be necessary to publish the whole of the nine volumes which were now published by the various Mining Institutes, for the reason which had already been given by several gentlemen, namely, that many of the papers were already duplicated, others were papers of only temporary interest, others of a purely local interest which it would not be necessarry to publish in the more expensive form. In the French Mining Association (La Société de l'Industrie Minérale) they publish two sets of Transactions. They publish very elaborate and beautifully got up Transactions of their more important papers; but they also publish, in a very cheap pamphlet form, the papers of a more temporary or unimportant character. If such a scheme could be adopted, it would suggest itself that there should be a Committee of selection, and that possibly each Institute would do that for itself. There might be some control over that by a joint publishing Committee; and it would be only selected papers that would be published in the larger form, leaving out, for the time being, the consideration of publishing the papers of mere temporary interest in a cheaper form. In that way, he thought, they would get all the important and interesting papers read at each Instititute which at present they did not see. In addition to the joint publication of Transactions might be added the privilege of membership of each other's Association, so that if a paper was read at any particular Association any member could attend, if he had a special interest in that subject, and take part in the discussion, not as a matter of favour, but as a matter of right. A second question of very great importance, which had already been touched upon, was that of investigations for special objects. Many of them had been repeated at nearly every Institute. Almost every Institute had had Committees upon Safety Lamps, and upon Fans, and now upon Explosives. The North of England Institute had just appointed a Committee upon Explosives, and he thought the Midland Institute had recently conducted investigations on the same subject. They had now formed amongst three of the Institutes (and they would again meet that day on the matter) a Joint Committee for Fan investigation, and, he ventured to think, it would be attended with very excellent results. Touching on a matter mentioned by Mr. Cochrane, and supported, he thought, by Mr. Marley, he would venture to say, in reference to parliamentary questions, that many of them bore upon scientific subjects, which were not dealt with in the least by the Mining Associations. In the last Mines Act, the question of the safety lamp was imported into the Act; and also the important question of the use of the fire-damp indicator, which was a purely scientific question, and one which ought to be specially worked out by a scientific Institute; but at present the Government, if they wished for reliable information upon scientific questions bearing on mining, had not any one body to whom they could refer for information. If they sought this from any one of the Mining Institutes, there would probably, and naturally, be a certain jealousy on the part of the others, besides which each district had not exactly the same peculiarities of condition. The circumstances and requirements of each district were rather different, and it would hardly do for any one Institute, in any one district, to take by itself any leading action; but if there was a general body, with whom the Government might communicate upon scientific subjects, he thought it would be a benefit, not only to the coal trade at large, but especially to mining engineers, whose character and responsibilities were so greatly affected by those legislative Acts. There was mention made by Mr. Jackson of the importance of not interfering with the individuality of each Institute at present. He thought that seemed to be the unanimous opinion. That was more needed in Mining Institutes than in any of the Engineering Institutes, because many of those gentlemen referred to who were members of the former could not go up to meetings of a central body. They could not attend any meetings unless held in their own locality; and it was of the first importance that those gentlemen should retain their membership. He thought that it would be the opinion of almost everyone who had considered the subject, that it would be fatal to in any way interfere with the individuality of the present Institutes. If each Institute, however, did not join as a body but individually, he did not say that the scheme could not be worked out, but there would be some difficulty in supplying each member with a copy of the Transactions if they were published by a central body to which they did not belong. He would just add that the points which seemed to him to be of chief importance were—1st, combination for publishing only; 2nd, combination for experimental research; 3rd, the importance of a united body to whom the Government might apply in case of requiring reliable information on scientific questions affecting mines, and obtain the opinion of those who had given time and attention to the study of those subjects as a body, rather than as individuals.

Mr. BEWICK said, in answer to the Chairman, that he did not think he had anything to add to what had been so well said by others. He certainly should very much like to see the project carried out; but he could not but also see that there were very grave difficulties in the way of it. Probably if a Committee were appointed that day to consider the whole bearings of the case, all those difficulties might be overcome. It was only as they appeared, perhaps, on the first blush of the thing, and they might be got over. He must say, taking the scheme as a whole, he was quite in favour of it. He did not know whether there was anyone there from those who had said "nay" to the project, such as the Mining Institute of Scotland, or the Cornish Institute, or the Geological Society of Manchester; if so, perhaps they would express their opinion. They were the people they would like to hear from.

Mr. COCHRANE asked if there was anybody present who had made a calculation of, or had the slightest idea as to the cost? Was there any impression on anybody's mind on that subject—the cost of the federation?

Mr. STEAVENSON said his impression was that it would cost £3 3s. each member, and that it would have a bad effect on local Institutes.

Mr. DAGLISH thought 10s. a member would be sufficient merely for publishing, and the present system of publishing cost more than that. Therefore, combination for that purpose would be attended with no extra cost whatever.

Mr. COCHRANE: Would they excuse him rising again? With regard to the question of publishing which Mr. Daglish had thrown out, which was no doubt true, it seemed to him that considerable expense would be saved in that direction, and that they should still have every paper,—not merely the chief, but the temporary ones spoken of,—which would be a great advantage, from all their Societies. Suppose they became affiliated in a small manner, and with a Secretary there, for the object of being a body that could be referred to, as had been indicated; each Society might guarantee to take from each other Society its Transactions to the full extent of the membership of the Society. It seemed to him that 10s. each,—there were 2,200 members at that moment,—would be sufficient. Those Transactions could certainly be supplied at 10s. each; that was £1,000 or £1,100. That was simply a trifle as compared with what was now spent, and that would furnish as many copies as would be required. He agreed with Mr. Steavenson that something like £3,000 or £4,000 would be about the cost of the Society.

Mr. DAGLISH: If they established a centre in London?

Mr. COCHRANE: Yes; £3,000 or £4,000 per annum. The other Societies being affiliated might say, for instance, "We will undertake to take 120 copies of Transactions of each one of the other Societies, and so on. If they were to do that, the cost of the printing would be much cheapened, one's productions would be open to everybody, and everybody would have a copy of everbody else's Proceedings.

Mr. DAGLISH said that was still making the Mining Institute publish nine volumes every year; whereas the Mechanical Engineers. with all their large numbers of able members, only published one. It was clear they were all publishing papers that ought not to be published, at least in such an expensive form.

Mr. WALKER said that, with regard to that idea, the difficulty now arising with respect to printing seemed to be that they had to print at a number of different places. He supposed that much could be arranged, that one printer should be appointed to print for all the Institutes, and that the size of the Transactions should at any rate be uniform. Then, in any case, the whole of the papers that were read before any Institute, and the discussions upon those papers, would be put into type and printed, but the actual number printed would depend upon the demand for them. For instance, supposing that a certain person read a paper before the North of England Institute, and that Institute required a thousand copies for its own members,—he said that simply to take round numbers,—a thousand copies at least of that paper would be printed, with the discussion upon it, to go to all the North of England members; but, in addition to that, either the paper might be considered of such general interest that it might be decided at once that it was worth sending out to all the other Institutions, or any Institution which desired could have copies of that paper, and the discussion might be supplied with it on certain terms to be arranged. Then he thought it would be very well if their Transactions could be sold by a publisher or by the Secretary of the General Institution at a certain price fixed by the Council or the author of the paper. Nothing had been said yet about discussions. Now, he thought it would a great pity if they were to confine themselves to the publication of the papers without the discussions, or at least abbreviated discussions. Sometimes very erroneous ideas were promulgated in papers, and if such a paper went forth with the *imprimatur* of the Institute, it might seem as if they were to some extent giving their sanction to ideas which were perhaps erroneous, premature, or badly digested. Now, would they just take an illustration which had been alluded to at that meeting? They had been having papers at the Midland Mining Institute on new flameless explosives. The discussions had been, he thought, very much more valuable than the original papers,—at any rate, quite equally so. Different gentlemen have brought before the Midland Institute the result of their individual trials and their individual experiences under very wide and different circumstances, and the net result had been a very valuable amount of information brought together, which certainly would not appear in the original papers.

Then he thought that the discussions, so far as those discussions appeared to be pertinent (and the General Secretary might strike his pen through anything that was said that did not appear to be pertinent) should be published in smaller type with the papers. As to Mr. Daglish's remarks about their not knowing what papers were being read in different Institutes, he presumed that a general circular would be sent out each month saying what papers were going to be read before each of the affiliated Institutes; and the question of time was rather important. There must not be too much delay before they got their papers. In the Midland Institute they made it a rule to always have papers month by month. Thus they were received before the meeting, and members went to the meeting prepared to discuss the papers. If any great amount of delay were to take place before members got their papers, then the result would be that a very great deal of interest in the subject would be lost, and perhaps time wasted which it was desirable to save. Then it was proposed to call that Institute the Imperial Institute. If it was really meant to be an Imperial Institute, was there any idea of extending its operations to the Colonies? Some of them was becoming more and more interested in mining in other parts of the world, and if it was to be an Imperial Institute it certainly seemed that if its scope could be extended to Australia and Canada, it might be a great help to those of them who had any dealings with coal-fields in those parts of the world. For instance, the coal-fields in New South Wales appeared to be very different in geological structure and conditions from those in this country, and the publications which had come under his own notice with regard to these matters were very scanty, so that if their Proceedings could be opened to reliable reports by engineers on the spot as to the structure of the Colonial coal-fields and so on, it seemed to him that it might be exceedingly valuable to so extend the scope of the Institution. It would be rather a mistake to call it "Imperial," unless they intended to embrace in their scope something more than the four corners of the United Kingdom.

The CHAIRMAN said he believed Mr. Mitchell, the Secretary of the Midland Institute, had just come into the meeting, and they would be very glad to hear if he had any opinion to offer upon the subject.

Mr. MITCHELL said he quite agreed with Mr. Cochrane, and he could only endorse his remarks so far as he was personally concerned.

Mr. SMITH, with regard to the question of the Colonies, asked if they were not rather taking it that the whole of the Transactions would have to be published, and the expense borne by the membership fees of the Central Institution? That was not the idea of the paper; because, in addition to the fund derived from the membership fees, he took it that each federated Institution would have to contribute to the Central Institution for the publication of the Transactions, and they could well do that if they were spared publishing their own, because it would be a material saving to them.

Mr. MILLS said he would like to ask whether they could not do something that day to bring the matter to an issue? They must have funds. It had been proposed by the late Mr. Bunning that the fees, in the first instance, should be a call upon the funds of the Institution. He did not think that would be at all an advisable step. He thought that if they could raise a little money in the first instance to have a secretary for their Committee, they might afterwards ascertain the extent of support the new Institution was likely to receive. He had looked into the question of subscription and finance to some extent, and he would propose that there be several classes of members. In the first instance there might be the first class, called "fellows," or anything of that description—that was to say four guineas; then members at two guineas; non-resident members, such as Mr. Walker proposed, in the Colonies and different places, one guinea; also associates, who would be the under-viewers and people who could not attend, under-viewers in the Colonies, and students as well. He thought if they could come to some idea as to the subscription the Institution would require, and then ask each one of their own members of their own Institutions whether they would join the Institution at the subscription, it would be a great thing.

Mr. CHAMBERS saw very great difficulty in the suggestion which Mr. Mills had made. He thought they wanted a scheme first, and thought the proper thing to do would be to appoint a small Committee to draw up one to meet the views of all the gentlemen; then they could have some idea of what the conditions would be, and what amount of subscription would be required. He thought that, taking the moderate scheme which Mr. Daglish had sketched out as a basis, it was quite possible to draw up one which would unite them to a certain extent, though not to the full extent which the late Mr. Bunning desired, nor to the full extent which some of them hoped it might develop into by and by; but still he thought a scheme might be drawn up which would only necessitate a moderate expense, and which would be the first stage in uniting the Mining Institutes of the country.

The CHAIRMAN said Mr. Daglish had placed in his hands a motion, which he would leave him to bring forward afterwards. He (the Chairman) reminded them that in the first place they must prepare a scheme; but not until they received a preliminary and conditional assent of every one concerned. They had had the matter now before them, he ventured to say, in a fairly complete form; they had heard the opinions, with many of which he entirely agreed,—some he was perhaps not inclined to adopt without modifications. Their observations would all be printed in a condensed form, and they would have an opportunity of considering the general views of those who had taken a part in that morning's proceedings, which might, as far as possible, be embodied in some scheme that ought to be drawn out; and it ought to be drawn out, as had been suggested by two or three of the last speakers, by a small Committee. He thought a copy or copies of that scheme might be sent to the different Institutes, to ascertain how far they would feel inclined to join in the federation.

Mr. DAGLISH said the only feeling in his mind about appointing a Committee was, that it would only result in a scheme emanating from those gentlemen, individually, who formed the Committee. because they had not really any scheme as yet before them, although they were met there that day as a Committee appointed by the several Mining Institutes favourable to federation for the purpose of presenting some tangible scheme to their respective Institutes. They had a suggested scheme for a central confederated body for all purposes, with a special subscription, and another suggested scheme purely confined to publishing, and these were two entirely different things.

The CHAIRMAN said he left it to the discretion of the Committee to propound a scheme either on the lines spoken of in the one case or on the other as they thought fit.

Mr. DAGLISH said, if they would allow him, the only object he had in suggesting that a definite resolution should be proposed was, that it was easier to speak for or against and to a resolution, rather than to deal broadly with various facts. He was going to suggest a resolution for discussion which, if agreed to, the delegates present that day could submit to their Councils as something definite.

The CHAIRMAN: They would see that if that resolution was carried they rather confined the Committee to draw the scheme upon the lines of that resolution, whereas he wished to leave them entirely free.

Mr. DAGLISH said he did not propose to limit it to that, but simply to commence with that as something definite.

The CHAIRMAN: Supposing the first step were to annul it?

Mr. DAGLISH: They could suggest something else.

The CHAIRMAN: It would be rather an awkward thing for a Committee which had been appointed by a particular resolution if their first step would be to cancel that resolution.

Mr. SMITH said he thought the Chairman was quite right. They were not as delegates in a position to pledge their Institutes to any course. As the Chairman has said there was no actual scheme before them.

The CHAIRMAN: No; but he gathered that on the whole they were all favourable to federation.

Mr. SMITH: To the principle.

Mr. COCHRANE said he would propose something if they would allow him which, he dared say, would commend itself to the meeting. It was—" That this meeting recommends that each society appoint its secretary to form a Committee, and that each society bear the expense of its own secretary, in order to formulate a scheme to submit to a future meeting of this Committee." The object in doing that would be—each secretary would go back again and have an opportunity of consulting in his own way his own Council at the mimimum of expense and at the maximum of convenience to the members, and also with the power of learning in the best way what his particular Institution wished. Those secretaries ought to meet, and they were best capable to formulate the matter. Let each society bear its own expenses. He proposed that the meeting appoint each secretary as a member of a Committee to formulate a scheme, and then, having got their report, they might go back again to their societies simply with that as a recommendation; and as to the meeting place, he forgot to suggest that.

Mr. MARLEY: Say, the presidents and the secretaries.

Mr. COCHRANE: No; he should only ask for the secretaries—it was practically their duty to formulate the scheme. They could collect the best opinion from their own Councils; and he proposed that the meeting place be Derby.

Some conversation as to the place of meeting then took place.

Mr. COCHRANE: Let that be a subsequent matter. If the idea be that the secretaries should do that, they being the responsible people for formulating a scheme, and each responsible to his own Council, it met the proposition that far.

Mr. JACKSON said he would be very glad to second the motion, because he thought it was a very practical way of bringing the matter to a start.

MR. FORSTER BROWN said although he quite agreed with Mr. Cochrane that the secretaries of the different Institutions were the gentlemen who certainly should be members of the Committee to formulate the scheme, he did not think that it should be limited to the secretaries of the Institutions, because they might have certain views, whilst the members of their particular Associations might have different opinions, and he thought the Committee ought to be very much wider, because, after all, it all hinged upon the report of that Committee as to whether the matter was to be carried out, and whether it was to be carried out on sound lines; therefore, he thought, the first step was to have a representative Committee of all the Institutions who were favourable and who would promulgate their scheme for the consideration of the Committee, and he begged to propose that.

Mr. CHAMBERS said he quite agreed with Mr. Forster Brown that the Committee consisting of the secretaries only was not wide enough. He was going to suggest a comparatively limited Committee himself, and was afraid that it would not be as some gentlemen would possibly desire it, but he thought it would be sufficient for the purpose— namely, two members from the North of England Institute, and one from each other Institute, in addition to the secretaries. That would be a small workable Committee. If they got a very large Committee the members were liable to leave the work to others; then, if they all attended, there would be a very great difficulty in getting through the business.

Mr. COCHRANE said the expense was an important item. At the present moment nobody was subscribing any money. He did not want to interfere with the conditions.

Mr. MARLEY moved that a Committee of three or six be appointed by each respective society to consider the necessary details for carrying the matter out, and then that a general meeting of the whole be held to formulate the result of what they in their opinion thought was best to promote a federation. It was, practically, Mr. Forster Brown's suggestion.

The CHAIRMAN said that would make about ten more members than the whole meeting then present. He asked if Mr. Marley did not think that he was suggesting somewhat too largely?

Mr. MARLEY thought he might say one word more, just by way of explanation. In proposing from three to six in each society, he proposed that each society should then send a deputation out of that six to meet together; therefore, practically, it was, say two out of the number.

Mr. DAGLISH said that each society should see that two of its members did attend the Committee meeting.

Mr. MARLEY: Yes.

Mr. FORSTER BROWN said he had no objection to that.

Mr. MARLEY said each society would appoint its own member.

Mr. SMITH said practically, if Mr. Cochrane's resolution was accepted in the amended form, appointing other representatives besides the secretary, the societies themselves would do that. He thought they should leave it to the Institutions themselves. As Mr. Marley suggested, they would meet and instruct their representatives how to convey their ideas to the Committee. He did not think that from that Committee they ought to stipulate that the several Institutions should appoint Committees.

Mr. COCHRANE, addressing the President, said the secretaries were the people to formulate—not in any way to determine. After that they wanted some tangible resolutions.

Mr. DAGLISH: They must have a meeting.

Mr. MARLEY: They must have a meeting.

Mr. COCHRANE: Of the secretaries?

Mr. DAGLISH said they thought there should be some other members present as well as the secretaries.

Mr. COCHRANE: Yes, at each Council. What he proposed was—that that meeting should recommend each society to appoint its secretary, then those secretaries would be instructed by the Council of each society fully. Then they would meet, they would formulate a scheme in writing—resolved this, that, and all the rest of it. He might say it was peculiarly their province to do that.

Mr. DAGLISH: The only further condition was, that in addition to the secretaries there should be one or two members of the Council.

Mr. COCHRANE said he did not object to it except on the ground that they were putting each society to very much more expense than what was necessary. In the one case they simply sent the secretaries to that meeting, and in the other case also they would not get the attendance of those members at the meetings, whereas the secretary's official position was such that he could always go to those meetings, and the members they might appoint might not be capable of doing so.

Mr. DAGLISH said he thought Mr. Marley and the Chairman had put that right. It was proposed that each society should see that two of their members attended by arrangement.

Mr. MARLEY said he thought they were all pretty well agreed.

Mr. HAINES: What was proposed had been really done, and they had it before them. The representation had already been done. It was not only the representatives, but what they might or might not do.

The CHAIRMAN asked if he meant in point of numbers?

Mr. HAINES: Yes; he thought they had provisionally three names from each Institute before them.

The CHAIRMAN said it did not follow that those gentlemen would continue their attendance. They could see that what they wanted was to have the sanction of that meeting.

Mr. HAINES said he was merely speaking of their Institute; their representation was prospective rather, he should say.

Mr. WALKER said it would be necessary for each Council to have their secretary, he thought, so that on each point that might arise they could discuss such questions amongst themselves as there might be a little doubt about.

Mr. FORSTER BROWN: Simply two representatives.

Mr. MARLEY: Neither name nor office.

The CHAIRMAN asked Mr. Cochrane if his motion was that the secretaries be appointed?

Mr. COCHRANE: Yes; and the secretaries should be empowered to prepare the draft.

The CHAIRMAN said the motion before the meeting was that by Mr. Cochrane, namely, that the secretaries be appointed in order to form a Committee to draw up a scheme. To that an amendment had been moved that each Institute send two of its members, neither of the two being of necessity the secretary, to attend the meeting in order to draw up a scheme; and, as usual upon such occasions, he would put the amendment first.

The amendment was then put to the meeting, when 10 voted for it and 3 against it, and the Chairman declared the amendment carried.

The CHAIRMAN said that seemed to him to complete the business; and he thought, before they left the room, they ought to allow him to convey the thanks of the meeting to the Institution of Civil Engineers for the use of their room.

After further conversation as to the proposed place of meeting,

Professor BENTON begged to move, as a representative of South Staffordshire, that Sheffield be the place of meeting.

Mr. MARLEY seconded that.

The CHAIRMAN said the motion was that Sheffield be the place of meeting.

The resolution was put to the meeting, and carried unanimously.

Mr. DAGLISH: As Sir Lowthian Bell had been kind enough to act as Chairman on that occasion, should they ask him to be so good as to convene the meeting?

The CHAIRMAN said he should be very glad to do that, if they wished it.

Mr. DAGLISH said he would propose that.

Mr. FORSTER BROWN said he would second it.

The resolution was put to the meeting by Mr. Daglish, and carried unanimously.

Mr. FORSTER BROWN begged to propose a vote of thanks to the Chairman, for fulfilling the duties of the chair.

The resolution was put to the meeting and carried unanimously.

The CHAIRMAN said he had had very much pleasure in occupying the post that had been assigned to him.

Mr. DAGLISH said he thought it should be known that the Chairman had come up specially that day from his place in Yorkshire to attend that meeting.

JOINT COMMITTEE OF THE NORTH OF ENGLAND INSTITUTE OF MINING AND MECHANICAL ENGINEERS, MIDLAND INSTITUTE OF MINING, CIVIL, AND MECHANICAL ENGINEERS, AND THE SOUTH WALES INSTITUTE OF ENGINEERS "ON MECHANICAL VENTILATORS, 1888."

MEETING HELD IN THE COUNCIL CHAMBER OF THE INSTITUTION OF CIVIL ENGINEERS, 25, GREAT GEORGE STREET, WESTMINSTER, WEDNESDAY, 6TH JUNE, 1888.

Mr. A. L. STEAVENSON proposed that Mr. Daglish take the chair.

Mr. W. COCHRANE seconded the proposition.

The resolution was put and carried.

The CHAIRMAN said they all knew why they had met, and he did not think he need make any preliminary remarks. Mr. Walton Brown had been in communication with the secretaries of the Midland and South Wales Institutes, and he would commence the proceedings by asking Mr. Brown to state exactly the position in which they stood just now.

Mr. M. WALTON BROWN said he had corresponded, on behalf of the North of England Institute, with the secretaries of the Midland and South Wales Institutes, and the preliminary details were all thoroughly understood and arranged on behalf of the three co-operating Institutes. As to the expense of carrying on the experiments, each of the three co-operating Institutes had agreed to subscribe not more than £100, and it was agreed that the Report, when completed, should be the joint property of the three Institutes. Three engineers would be appointed, one by each Institute.

Mr. HORT HUXHAM: That is right.

The CHAIRMAN: How many engineers is that then?

Mr. M. WALTON BROWN said three experimenting engineers would be required. Each of the three Institutes would appoint one, and each pay the expenses of their respective engineer; that would be the simplest way of proportioning the cost.

Mr. THOS. EVANS asked if it was understood that each engineer would be paid for his time, as well as his hotel and travelling expenses, or were they supposed to give their services?

The CHAIRMAN said so far as the North of England Institute was concerned up to the present time they had never paid anyone. They had always been able to obtain the services of suitable gentlemen who were good enough and able enough to undertake the duties; but it was possible they might not be able to do so, and he thought that it was left to each Institute to do as it liked; each paying their own engineer.

Mr. HORT HUXHAM: Not out of the £100 or £300?

The CHAIRMAN: No; I think not.

Mr. HORT HUXHAM: That is just the point. We are a little doubtful about it.

Mr. CHAMBERS said he understood the Institutes joining in these experiments were three—the North of England, the Midland, and the South Wales, and no others.

Mr. M. WALTON BROWN: Yes; no others have been asked.

The CHAIRMAN asked if there was any scheme formulated?

Mr. M. WALTON BROWN said he had drawn up a programme of observations to be made, and instructions to the engineers, which had been sent to the committees of the co-operating Institutes.

The CHAIRMAN asked Mr. Brown if he proposed that they should go through the programme *seriatim?*

Mr. M. WALTON BROWN said he would suggest that it should be gone through *seriatim*, and ascertain whether they approved of it or not. He would read it through.

The CHAIRMAN asked if it was the general wish of the gentlemen present that they went through the programme *seriatim?*

Mr. G. B. WALKER asked if the general principles were agreed to as to what should guide the investigation? These were instructions, as he took it, to the engineers. The Midland Institute had been under the impression that they intended, in these experiments, to confine themselves very much to two things—first, to places where there were two ventilators of different descriptions working on the same mine, so that their results could be very accurately compared; and, second, to fans, which had not hitherto been dealt with. They felt that the experiments which had previously been published (for instance, those which the North of England Institute published some four or five years ago) were not quite final, and that it was very desirable to take into account all the ventilators which were now in successful operation, and that a sufficient number of each type should be experimented upon, in order to arrive at some reliable data as to their general characteristics, effects, and advantages. The Midland Institute thought that the conditions of the mines where these ventilators were at work should be clearly stated, as there were many sources of error, which have probably crept into former experiments through sufficient information not having been given when the results were published, in order to enable any one to decide what were the conditions under which the fans were worked, and under which the results were obtained. The idea of the Midland Institute was, therefore, to rather extend the scope of the investigation beyond what he understood was the idea of the North of England Institute.

The CHAIRMAN said would not the three great fans be the basis of the tests—the Schiele, the Waddel, and the Guibal? Did not that cover nearly the whole ground of previous experiments? They would not require to test another Schiele, and another Waddel, or another Guibal in any way, because they would have tried these in the first sets of the experiments they made.

Mr. WALKER: Take for instance the Waddel at Celynen. He understood (he said it with all due reserve) that it was by far the best Waddel that had ever been erected. The results of that particular Waddel were the only ones which were given in the report of the North of England experiments.

The CHAIRMAN: Yes; but you see the Waddel will now be tested as against another fan, under exactly the same circumstances.

Mr. WALKER said there was a second set of considerations—that was durability. A fan which, after five or six years wear, was considerably shaken, was not so valuable a fan as one which had run for a very much longer period without any perceptible deterioration.

The CHAIRMAN asked if Mr. Walker proposed at present to move a resolution to extend this, or would he bring it forward afterwards as the work went on?

Mr. WALKER said he simply made those remarks because he thought they were going on to instructions to experimenters before they finally decided what the scope of the investigation was to be.

The CHAIRMAN said Mr. Walker had mentioned that the scope of the enquiry included two things: that was to say, the fans of different kinds upon the same mine, and the new fans that had never been tested. Did he propose to add a third to that, namely, to go over some of the other fans?

Mr. WALKER: That was the idea of the Council of the Midland Institute.

The CHAIRMAN: Are we not then going into a very large question, and possibly a question that will give rise to some degree of dispute and squabbling if we are going to test individual fans again?

Mr. WALKER said he simply mentioned it because they had already appointed a Committee to experiment with a certain number of fans in the South Yorkshire district, and they suspended those experiments in consequence of the invitation received from the North of England Institute to co-operate with them, but, at the time they agreed to co-operate, it was very clearly mentioned by his Council that the enquiry ought not to be of too restricted a character. They thought that there were not before the world as yet any reliable statistics respecting the principal types of fans.

The CHAIRMAN asked if he (Mr. Walker) did not think that they, or rather most of them, had come to the conclusion that they could not place much confidence in those isolated experiments on account of the very fact of the circumstances differing so much; and unless they could get different fans under exactly the same circumstances, that really these experiments were of no value? He only mentioned that; he did not wish to put a stop to the investigation any further than it might be their wish. Perhaps Mr. Walker would test it by moving a resolution at once.

Mr. WALKER said he would rather do so, if there was time, after a little more expression of opinion. If they liked he would move the general resolution, " That the object of the investigation be to ascertain as thoroughly as possible the relative efficiency and value of the various kinds of fans now in operation." That was a somewhat comprehensive resolution.

The CHAIRMAN: Yes; that would carry it, certainly. Would any gentlemen second that resolution?

Mr. ARMSTRONG, Jun., seconded it.

The CHAIRMAN: It being understood that at present the investigation is limited to two different fans on the same pit, and to fans that have not hitherto been experimented upon. It is now proposed to extend this investigation further. Those who are in favour of that, please signify the same by holding up their hands.

Mr. HORT HUXHAM : Before putting that resolution, he should like to ask what the words " experimented upon " refer to?

The CHAIRMAN: Published, I suppose.

Mr. HORT HUXHAM : Published, you mean?

The CHAIRMAN : Yes.

Mr. HORT HUXHAM : Published in the Transactions of any particular Institute, or not?

The CHAIRMAN: The resolution covers everything. It is as wide as possible. There is no limit to it.

Mr. HORT HUXHAM said what was passing in his mind was simply this, he apprehended every fan had been more or less experimented upon.

The CHAIRMAN : No ; every *system* of fan, not every fan.

Mr. HORT HUXHAM : Every system of fan; and those experiments had been more or less published.

Mr. ARMSTRONG, Jun.: " Recorded " would be perhaps a better word.

Mr. HORT HUXHAM : Recorded in some particular Proceedings or Transactions?

Mr. ARMSTRONG, Jun.: Yes; in the Transactions. That is better than " published."

Mr. WALKER : The inference is, that the three Institutes combining to make these experiments would probably publish the results.

Mr. FORSTER BROWN: They would be the joint property of the three.

Mr. ARMSTRONG : Recorded in the Transactions of the three Institutes.

Mr. HORT HUXHAM : Quite so.

Mr. FORSTER BROWN said he was going to suggest this: Would they not obtain all the practical objects they sought by testing different fans, where there are duplicates on particular pits? By that they would get definite results as regarded those particular fans. But inasmuch as those particular fans probably comprised the principal fans which had otherwise been experimented upon, they would obtain all the objects required without going into an unlimited enquiry.

The CHAIRMAN said probably, like every other gentleman present, he had made a number of experiments, and he found that the conditions were so utterly different that they could not compare two fans on different pits. They had engines underground with steam only; they had engines underground with a boiler. They could not tell how much of the ventilation was due to these actions: therefore, to take an experiment with a fan upon a pit was no indication of its relative value as compared with another fan on another pit.

Mr. A. L. STEAVENSON said his impression of the origin of this Committee was that it was merely to test the fans where there were two fans of different descriptions on the same shafts, to satisfy the want that had been felt by mining engineers, and to prove whether two fans, worked under exactly the same conditions, gave different results. For his part, by a mere calculation alone, he thought they had satisfactorily solved the question; but then it would be much more satisfactory if different kinds of fans were tested on the same pit. He rather thought that they should first give attention to that point. If they began to test different kinds of fans they would get into a very extended range of examination, for there were a large number now of different kinds; but that should be considered before they started. He should like to suggest before they went to any very great extension of their work, they should decide as to how the cost was to be divided.

The CHAIRMAN: That was arranged not to exceed £100 for each Institute.

Mr. CHAMBERS said he could not help thinking that Mr. Walker was leading them a little further than even the Midland Institute intended to go by his comprehensive resolution; and he was bound to say, after hearing what other gentlemen had said, that he thought it would be almost better for the Committee at present to confine itself to the scheme which the North of England Institute proposed. He should like also to suggest that they should make some experiments with the original fan, now forty or fifty years old—that was the Biram fan; the first fan, he believed, put up in the country, and which had been running from the day it was put up to the present day. It would be very interesting to know what that fan was doing. He thought they could get permission to have it tested.

Mr. COCHRANE: They would find the whole of the experiments upon the Elsecar fan in the Transactions of the North of England Mining Institute, made by the late Mr J. J. Atkinson and himself, before they adopted the Guibal type of fan.

Mr. CHAMBERS said he was not aware of it; therefore he withdrew his suggestion.

The CHAIRMAN asked Mr. Walker if it would meet with his views to let this matter rest for the present? If the Committee had energy left, after they had finished the objects for which they were started—

Mr. COCHRANE: And money.

The CHAIRMAN: And money; or can get more. He quite agreed with him it would be very advisable not to let it drop. It would be a pity; but he thought they should limit themselves to the very large undertaking they had in front of them at present, otherwise they would never get to the report stage.

Mr. WALKER said he should like to add some limited proposal to the effect that the Committee might experiment with fans which present certain novel features in their adaptation. Mr. Garforth, of the West Riding Colliery, had a Schiele fan at the top of a very small shaft, whose friction would be entirely abnormal, and the results of that fan should be very instructive.

The CHAIRMAN: Yes.

Mr. WALKER: And if any fan was, in the opinion of the Committee, so placed that it presented new features, he thought it would be a pity to neglect to get the particulars of the working of such a fan.

The CHAIRMAN: You will move no resolution then?

Mr. WALKER: No.

Mr. GARFORTH said as the money that was voted was very limited, he thought it would be better to go step by step, and take it in two stages—first, the fans in duplicate at each pit, and make that the scope of the Committee's investigation at first; then, if the money ran to it, they might go into the other. The same Committee would continue the experiments.

The CHAIRMAN asked if Mr. Garforth would kindly move that resolution, and state, as the expression of the opinion of the meeting, that the Committee should report as soon as they had completed the experiments of the duplicate fans?

Mr. GARFORTH said he should be very happy to do so.

Mr. COCHRANE: With the present money?

Mr. GARFORTH said he should be very happy to move "That the operations of the Committee be confined to those cases where two fans of different constructions were erected on the same mine, but the Joint Committee at the same time express their hope that they will be able to extend their operations to newly invented fans after this series of experiments are completed."

The resolution was put from the chair and carried.

Mr. M. WALTON BROWN then put in a schedule of observations to be made, and instructions to the engineers, which, after discussion and amendment, were adopted by the meeting. (See Appendix p. 310.)

The CHAIRMAN said in their case they had appointed Mr. M. Walton Brown, and the Joint Committee had asked Mr. Brown to act throughout as General Secretary also. Who would he communicate with on behalf of the other Institutes?

Mr. M. WALTON BROWN said he had communicated with Mr. Mitchell and Mr. Huxham.

The CHAIRMAN: Quite right; so long as that was understood.

Mr. HORT HUXHAM asked if it was understood that the Secretaries should accompany the experimental engineers?

The CHAIRMAN: Not unless they like; but it was expected that some of the Committee would be present always at these experiments.

Mr. THOS. EVANS: I think so.

The CHAIRMAN hoped that they would be present, both to assist and to see that the thing was carried out properly.

Mr. M. WALTON BROWN: Three engineers were to be appointed, one from each Institute.

The CHAIRMAN asked if it was the pleasure of that meeting that each Institute appoint one engineer?

Mr. GARFORTH: Yes.

The CHAIRMAN said the next thing was where were the experiments to be made?

Mr. HORT HUXHAM: Is it left to the engineers to decide where they commence their experiments first?

Mr. M. WALTON BROWN asked if that could not be done by correspondence, so as to avoid any further meetings of the Joint Committee until the experiments were completed?

Mr. COCHRANE: The Secretaries could prepare lists of fans proposed to be tried in each district for approval by each Committee.

The CHAIRMAN : Yes ; there was no need to call a meeting for that purpose.

Mr. COCHRANE : No ; It would be entirely done by correspondence.

The CHAIRMAN : Yes ; so that each Institute might agree.

Mr. GARFORTH moved that the best thanks of the meeting be given to Mr. Daglish for his kindness in presiding there that day.

Mr. THOS. EVANS had much pleasure in seconding that.

The resolution having been put and carried,

The CHAIRMAN said he was much obliged to them. He thought they had done a very good day's work.

Mr. STEAVENSON moved a vote of thanks to the Institution of Civil Engineers for granting them the privilege of meeting in their rooms.

The resolution was unanimously carried and the meeting separated.

APPENDIX.

OBSERVATIONS TO BE MADE, AND INSTRUCTIONS TO THE ENGINEERS.

Six separate experiments shall be made upon each ventilator, in which the friction of the mine is varied, as follows:—

(a) The return to the ventilator closed.
(b) The return to the ventilator closed, with the exception of an opening of 3 square feet.
(c) The opening doubled in area.
(d) The mine under ordinary working conditions, with all machinery at rest (hauling, winding engines, etc.).
(e) The entrance of air facilitated by opening some doors.
(f) Air admitted as freely as possible from the atmosphere.

In each trial the six experiments shall be made in the above-named order, and as nearly as possible at the normal speed of periphery, subject of course to the ability of the engines to drive the fans at the required speed when passing large volumes of air.

Two more experiments shall also be made with the mine under ordinary conditions, and the fan running at higher and lower speeds.

The normal speed of periphery shall be taken at 6,000 feet per minute.

In each experiment observations shall be made of—

(a) The number of revolutions per minute of the fan and engines.
(b) The volume of air.
(c) The water gauge.
(d) The indicated horse-power.
(e) The height of barometer.
(f) The temperature.

NOTES.

(a) *The revolutions* of the fan and engines shall be counted by an ordinary engine counter, and, if possible, two independent observers shall undertake this duty.

(b) *The Volume of Air.*—A Casella air meter or Biram's anemometer shall be employed, provided with some simple form of stopping and starting gear, say, started by the tension of a string and stopped by the reaction of a spring; that is to say, the revolutions would be recorded so long as the string was pulled tight.

The measurements shall be made at the same point in (1) the return air-way and in proximity to the inlet of the fan, and also at (2) the inlet (or inlets) and in (3) the shaft.

If possible a length of arching shall be taken, and the place of measurement must be of some regular geometrical form.

If all parts are not accessible to the observer, the place of measurement must be reduced in size by a rectangular wood frame or doorway.

The area of the place of measurement must be divided into 16 equal areas, and a reading of the anemometer taken in each at its centre of gravity. The division shall be made by means of horizontal and vertical strings, thus—

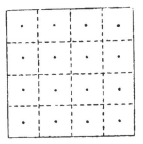 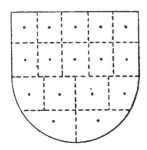

The anemometer shall be held for 30 seconds in each position, without intervals between the readings. Two observers shall attend to this, one to handle the anemometer (standing at one side) and the other to observe the seconds' watch and book the results.

When the resistance of the mine is varied, the position of the fan shutter, or other appliance for modifying the useful effect of the fan, shall be tested, if practicable, to ascertain the position which yields the highest water gauge at the normal speed.

The anemometers shall be tested at intervals with the same efficient machine.

(c) *The water gauge* readings shall be made at the centre of the drift (where the air is measured) with the end of the tube pointing to the fan, and at the same time as the anemometer readings.

Simultaneous readings must also be taken at the centre of the inlet to the fan.

The end of the tube shall be protected by a flannel cap from the effects of velocity.

The readings shall be made every 30 seconds.

The water gauge used in the experiments shall be of the ordinary form. Distilled water shall be used in the water gauge.

Flexible rubber tubing will be required to connect the instruments with the points of observation.

(d) *The indicated horse-power* shall be obtained by means of a Richard's indicator, made by Negretti & Zambra, and costing about £7 10s.

Both ends of the cylinders shall be connected thus, with a three-way

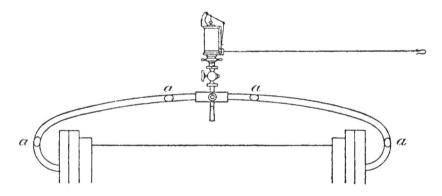

cock at the point of union, and above which the indicator shall be placed.

If there are a pair of cylinders, two indicators shall be simultaneously employed. By this means both cylinders and both ends of each cylinder will be indicated almost simultaneously.

Three sets of diagrams shall be taken during each experiment, at the beginning, middle, and end.

Experiments shall also be made to determine the friction of the engine without any air passing through the fan, or by detaching the fan from the engine where possible.

The indicators shall be tested by weights in the ordinary manner.

(*e*) and (*f*) *The readings of the barometer and thermometer* in the open air, and that of the thermometer alone in the drift, shall be registered. The hygrometric conditions of the inner and outer airs shall also be recorded.

Generally, all the experiments shall be made under similar conditions either when pits are idle or otherwise. All time observations shall be made with a centre-second watch, costing about 13s. 6d.

Additional information shall be obtained as under :—

(1) Depth and diameter of downcast and upcast shaft.
(2) Obstructions (if any) in shafts, with sketches.
(3) Distance apart of the shafts, with working sketch of seams.
(4) Difference in surface level of shafts.
(5) Temperature at tops of upcast and, downcast shafts.
Temperature at middle of do. do.
Temperature at bottom of do. do.
If boilers, etc., are in use underground, the temperatures should also be observed (where possible) at the point where the smoke is delivered into upcast, with sketch and dimensions of the smoke drift and volume of air passing through it.
(6) Dimensions of fan, distance from pit, and dimensions of fan drift (with plans).
(7) Dimensions of engines.
(8) A record of the steam pressure at the time of taking the indicator diagrams.
(9) A record of the water gauge at the bottom of the pit, where possible.
(10) The date of erection of the fan and engines.
(11) The original or estimated cost (and date) of fan, engine, boilers, building, etc.
(12) The cost of maintenance, being the actual cost of stores and repairs of fans and engines.

(13) Particulars of all accidents, and duration of stoppages of fan since erection.

Instruments required :—

2 water gauges.
100 feet india-rubber tubing with wire core.
2 flannel caps for tubing.
2 thermometers, wet and dry bulb.
3 anemometers.
2 Richard's indicators.
1 set of reducing gear.
2 Bourdon steam gauges, 60 and 150 lbs.
2 centre split-second watches.
1 aneroid barometer.
2 Harding's counters.
2 three-way cocks

Tool chest, ratchet brace and 4 drills, screw spanner, pipe tongs, pincers, pipe cutter, callipers, stock, dies, taps and key, oil tin, short lengths of steam pipe of various diameters.

Memoir.

The late Mr. JOHN BROWN, who died on the 24th August, 1888, was a member of the Institution of Civil Engineers, a member of this Institute, and a Fellow of the Geological Society, London.

Mr. BROWN was born at Stafford in 1823, and after completing his education was articled to the late Mr. J. T. Woodhouse, and was one of that eminent engineer's first pupils, and afterwards held the appointment of his principal assistant for some years. In 1853 he was engaged with Mr. Homel in mining matters in connection with Lord Granville's Collieries in Staffordshire, and afterwards practised in Barnsley and Sheffield in co-partnership with the late Mr. T. W. Jeffcock, who was killed at the Oaks Explosion in 1866, and when they dissolved partnership he went to Cannock Chase Collieries, and ultimately carried on his profession in the neighbourhood of Birmingham. He was one of the original members of the South Yorkshire Viewers' Association, which Association was the pioneer of the Midland Institute.

Mr. BROWN rendered valuable assistance at the terrible explosion at the Oaks Colliery in 1866, in conjunction with Mr. T. W. Embleton, Mr. Parkin Jeffcock, and others; he was also engaged upon many important arbitrations.

vi

OF MINING, CIVIL, AND
L ENGINEERS.

MEETING.

,D, ON FRIDAY, DECEMBER 21ST, 1888.

resident, in the Chair.

ere read and confirmed.
ted me   .e Institute,
n   ,, Sefton Park,

Jeffcock," *read*

, F.G.S.

s been successfully applied
ges and for a shorter period
ite recently that substantial
t as a motive power.
as yet in the empirical stage; the
nknown, and we can only utilise its
phenonema of its action which have
found to be constant.

ERRATUM.

On page 315, Part 99, line 10, *for* "Mr. T. W. Jeffcock," *read* "Mr. Parkin Jeffcock."

MIDLAND INSTITUTE OF MINING, CIVIL, AND
MECHANICAL ENGINEERS.

GENERAL MEETING.

HELD AT THE VICTORIA HOTEL, SHEFFIELD, ON FRIDAY, DECEMBER 21ST, 1888.

C. E. RHODES, Esq., President, in the Chair.

The minutes of the last meeting were read and confirmed.

The following gentlemen were elected members of the Institute, having been previously nominated :—

Mr. ROBERT E. BANCROFT, Mining Engineer, Tivoli, Sefton Park, Liverpool.

Mr. H. J. DURNFORD, Colliery Engineer, Swaithe Main Colliery, Barnsley.

Mr. EDWARD BROWN, Mechanical Engineer, St. John's Colliery, Normanton.

Mr. ROBERT WORDSWORTH, Assistant Colliery Manager, St. John's Colliery, Normanton.

Mr. G. BLAKE WALKER read the following paper :—

ELECTRICITY AS A MOTIVE POWER,
WITH SPECIAL REFERENCE TO ITS APPLICATION TO HAULAGE IN MINES.

BY GEORGE BLAKE WALKER, F.G.S.

ELECTRICITY has now for some fifty years been successfully applied to the transmission of telegraphic messages and for a shorter period for lighting purposes, but it is only quite recently that substantial progress has been made in utilising it as a motive power.

The science of electrology is as yet in the empirical stage; the actual nature of electricity is unknown, and we can only utilise its force by the application of the phenomena of its action which have been observed, and which are found to be constant.

I cannot enter upon the present paper without acknowledging in the most explicit manner my want of adequate knowledge to deal with it properly from a technical point of view; but the subject is one of such importance, and of such general interest, that I have no doubt its introduction and discussion will be welcome to the members of the Institute.

What we term electricity is evidently a condition inherent in a greater or less degree between the particles of matter—the atoms or molecules of which all bodies consist. It pervades the globe we inhabit, and is probably co-extensive with the universe. The earth itself is a great magnet, and to the action of electricity may be referred many features of its structure. It must be remembered that the conditions of matter which we term solids, liquids, and gases, are relative terms only, and depend on the preponderance of the forces of attraction and repulsion which exist between the molecules of every body, whatever its nature. These relations are largely dependent on temperature; thus ice is a solid, water a liquid, steam a gas, but the constituent molecules are the same in all the three. The difference is that in ice the molecular forces of attraction preponderate, in steam those of repulsion, while in water equilibrium exists.

The molecules, then, of which any body consists, no matter how solid, may be regarded as being distinct and separate from each other, and as having a form proper to them.

The more physicists learn of the great forces of nature the more their simplicity and uniformity become evident. We have learnt that matter is indestructible, and that we can at most only change its form; more recently it has been found that energy in like manner is indestructible. The evaporation of the tropics furnishes the burden of the rain clouds of temperate latitudes, and the course of prevailing winds and currents are dependent on the same simple causes. The conservation of energy naturally suggests a law of the *conservation of electricity*, and although the means of demonstrating such a law are not yet at our disposal, its truth is believed by many most competent authorities.

We have already mentioned the fact that the earth itself is a magnet, and we know it to be permeated by magnetic currents. What function these perform, or how they are stimulated, we are as yet ignorant of; but the fact that the earth is whirling round on its axis, and that motion is impossible without the expenditure of energy, affords one plausible clue to the solution of the problem. At any rate the fact of magnetic attraction is familiar enough to everyone, and is made use of in the mariner's compass and in the miner's dial.

When a piece of iron is magnetised it is found to possess a north and a south pole, and if two pieces of magnetised iron are brought into proximity to each other it is found that the poles of like name tend to repel each other, while those of unlike name attract each other. If then a bar of magnetised iron be bent into a U form, one end will be a north pole and the other a south pole. When attraction or repulsion takes place, force is evidently exerted beyond the limits of the magnet itself. This may be demonstrated as follows. If a sheet of paper be laid over the ends of a horse-shoe magnet, and a thin layer of iron filings sprinkled upon it, the filings will be found to arrange themselves in magnetic curves indicating the passage of an electric or magnetic current from one pole of the magnet to the other (Fig. 1). These curved lines shew the diverted flow of electricity from pole to pole of the magnet, which but for the proximity of the poles would pass off into space. According to Professor Hughes, what takes place visibly with the iron filings on the sheet of paper actually takes place invisibly in any electrical conductor through which a current of electricity flows. The molecules of the metallic or other body constituting the conductor are so many minute magnets, each possessing a north and a south pole. When magnetised they tend to place themselves in polarity, forming a chain of molecular magnets in which the north and south points are in proximity, and along which the current of electricity flows. In proportion as the molecules are able to adapt themselves to the polar position is the goodness of the conductor. Thus copper and soft iron are good conductors, it being believed that the molecules readily assume the required position. Hard steel is less good as a conductor, as it is supposed molecular polarity is less easily effected.

There is then an unlimited reserve of electricity present everywhere, and the question is how it is to be captured and converted into work. It may be useful, in order to obtain a concrete conception of the object in view, to institute a comparison between electricity and water as agents by means of which work is to be performed. Neither water nor electricity can perform work while quiescent; but when water is raised to a height its weight, in its tendency to seek a lower level, or to restore equilibrium, may easily be utilised. A steam engine working a turbine by means of a belt may raise water to a cistern; the same water may be conducted through a turbine and reproduce an equivalent quantity of work, friction and other mechanical loss being deducted. In the same way electricity may be utilised. The normal condition of equilibrium may be disturbed say by the action of a dynamo driven by a steam engine; the electricity may be stored in accumulators, as water in a cistern, and from these it may be drawn off at pleasure and converted into work by being passed through a motor,—that is, a dynamo reversed. We have, then, in the dynamo and motor the means of accumulating electricity and expending it at pleasure in the performance of useful work.

Let me endeavour, then, as simply as possible to describe how this is accomplished. A magnet, the two poles of which are in proximity, is the first essential. As already stated, there will flow between the poles of such a magnet streams of magnetic force, from the pole N in one direction, from the S pole in the other. To intercept this stream there is introduced between the poles a member called the *armature*. To illustrate the action of this contrivance, I will borrow from Mr. Kapp's "Electric Transmission of Energy," the theoretic diagram shewn in Fig. 2. Between the poles of a magnet, of which N represents the north pole and S the south, lines of force are thickly scattered, being denser in the middle, and getting less dense towards the outside. If a frame of wire, as shewn in the diagram, is placed in a horizontal position, no lines of force will pass through the frame. If now the frame is moved into a vertical position, the lines of force will pass through it in greatest quantity. If twisted suddenly from the vertical to the horizontal position, a current of electricity will be produced, which will travel round the wire and be absorbed in heating it. Again, if the wire be twisted from the horizontal to the vertical position, exactly the same effect will be produced, only the current will travel round in the reverse direction, being derived from the opposite pole. A continuance of the operation would therefore produce an alterating current constantly neutralised by one in the opposite direction. If, therefore, the current is to be drawn off and utilised for any purpose, it is necessary to have what is called a *commutator* or *collector* to collect the current. The commutator in this case consists of two metallic discs, placed side by side but apart from each other. Pressing against each disc there is a metallic spring, and from these springs the current would be taken. With a commutator of this kind the current produced on rapidly revolving the frame in a magnetic field would be alternating, *i.e.*, when the frame was moving from top to bottom the current would be generated in a reverse direction to when it was moving from bottom to top. This reversal of the current may be obviated by the adoption of a commutator such as that shewn in Fig. 3. It consists of a metallic tube sawn in two parts, and separated from each other by an insulating material. The two ends of the frame are connected to the two semicircular plates as shewn. If the two springs are made to press against this commutator, it will be seen that in revolving the frame the springs will press on the plates alternately, and the current which before was alternating will now become direct and constant. Although the current in the frame will be the same, it is, on being collected by the springs, made to travel in the same direction as the plates change their position, at the same time that the current changes its direction.

If the frame instead of having only one turn has two, and is revolved as before, the current will be twice as great; if three turns, three times as great; if four, four times; and so on, the strength of the current being directly proportionate to the number of turns. This being so, the number of turns is in practice made as great as possible, and becomes in fact a *coil* (Fig. 4). If instead of increasing the turns on the coil we increase the intensity of the magnetic field, we increase the current in the same proportion. And if we revolve the coil at a faster speed, the current is again increased in the same proportion as the speed. From this we learn that the current given from any dynamo machine depends on three things:—

1. The number of turns on the armature.
2. The speed or rotation of the armature.
3. The strength of the magnetic field in which the armature rotates.

For simplicity it will be best to divide armatures into two classes, as they are almost invariably of one or the other for motor work.

These are "drum armatures" and "ring armatures." They are sometimes known as *Siemens* armatures and *Gramme* armatures, from the inventors of the earliest types.

The drum armatures consist of a large number of thin discs of soft iron separated by a non-metallic disc, such as paper, and keyed on to a central shaft; the core (as these finished discs are called) is then well covered with varnished tape or other good insulating material, when it is ready for winding the wire over it. The winding of armatures is rather difficult to describe; it can best be understood from an inspection of the armature itself. (Figs. 5 and 5A.)

The commutator is made of a brass cylinder sawn up into a number of strips. These are held firmly together, and a thin strip of insulating material, such as mica, is placed between the sections. At equal distances round the armature the wire is bared, and a strip of metal connected to one section of the commutator. In practice the wire is wound round in a number of sections, to facilitate connecting up to the commutator, and not in one continuous length.

If such an armature is now revolved in a magnetic field, and springs or "brushes," as they are called, are pressed upon the commutator (Fig. 6), a current of electricity will be generated which will be carried to the brushes, and can be used for any purpose required.

The ring or Gramme armature is almost identical with the drum, with two exceptions: the core of the drum armature is solid, the ring armature has a cylindrical core; the winding is entirely on the outside in drum armatures, partly outside and partly inside in ring armatures (Fig. 7). The winding is commenced the same as before, but instead of passing right round the outside of the core it is wound inside the cylinder.

Both these armatures have special advantages of their own, but taken as a whole there is little to choose between them. The Siemens is usually used for high speed, and the Gramme for low speed work.

The use of the iron core is to concentrate the magnetism, and cause the lines of force to pass through the armature.

On rotating the armature of a dynamo, a pressure of electricity is produced, but so long as the brushes are not connected up to any circuit no current will flow from the armature, but directly they are connected, the current passing in the armature tends to resist the mechanical power which is producing it, the armature becomes strongly magnetised, the south pole of the armature is forced in the direction of the south pole of the magnet, and the north pole is forced against the north. As like poles repel each other, power is absorbed in forcing them against each other's repulsion.

The action of the motor is the reverse. If instead of generating a current from the machine, a current is passed through it from another source, the action is simply reversed. The south pole is repelled by the south and attracted by the north, and *vice versâ*, and mechanical power is given off by the armature.

So much for the broad principles of the action of dynamos. There are other various actions of great importance in practice, but it is unnecessary to describe them here.

Motors may be actuated either direct from a dynamo or by means of primary or secondary batteries. Of the latter (usually called accumulators) there are many kinds, but the best is what is known as the lead accumulator. In this cell the positive plates are composed of a lead grid filled with peroxide of lead. The negative plates have the grid filled with spongy lead. The accumulator is placed in a wooden box lined with lead and filled with a mixture of sulphuric acid and water (gravity 1·200). When an electric current is passed through the cell the water is partly decomposed into its constituents, oxygen and hydrogen. The negative plates absorb the hydrogen, and the positives the oxygen, until they are fully charged. When fully charged the gas is given off, the plates being unable to absorb more than a certain amount. On discharging the accumulator the oxygen and hydrogen re-combine to form water, and in doing so yield on an average about 70 per cent. of the energy with which they were charged. This is the main cycle of operations which take place in an accumulator, but there are other chemical actions taking place at the same time.*

---

* Electricians have had to have recourse to a new set of units, several of which are always made use of in speaking of the quantity of work effected by electrical means. These may perhaps be most easily comprehended by means of analogy to familiar examples in hydraulics.

Ampere = Strength of current, and is equivalent to the *quantity* of water flowing through a tube.

We must now consider the advantages and disadvantages which are likely to attend the application of electricity to the purpose of actuating machines underground in collieries.

For certain purposes there are undoubted advantages to be derived from its use. In the first place electricity can be conducted with a lower percentage of loss, and with greater certainty, to a distant motor than either steam or compressed air can, and it is free from any objection from heat, and costs much less for mains. Apart from the application of electricity to locomotives for haulage, its employment for other purposes must, when some initial difficulties have been overcome, be arranged with very great advantages. For example, as the workings of a mine are pushed further and further from the shaft, the difficulties of ventilation increase, and sooner or later auxiliary fans will be more and more employed for the purpose of providing for the more efficient ventilation of the outlying districts. Now to supply compressed air to actuate an auxiliary fan at a distance of a mile or more from the pit bottom, must necessarily be a very expensive and unsatisfactory arrangement. The amount of compressed air required would be large, and its supply would be constant and imperative. A suitable electric motor seems to offer special advantages for such a purpose as this. No attendance would be required, and the speed of the fan might be regulated with accuracy from the primary motor. Again, the transmission of electricity for the purpose of working pumps at a distance from the shaft is equally advantageous. Indeed, this has already been made a practical success at the collieries of Messrs. Locke & Co., Normanton. With regard to fixed haulage machines, there is probably somewhat greater difficulty. The work in this case is more variable

---

Volt = Electro-motive force, and is equivalent to the *pressure* of water in a tube. 1 Volt = the pressure produced by one standard Daniell cell (copper in sulphate of copper, and zinc in sulphuric acid).

Ohm = Unit of resistance, and is equivalent to the obstruction which water would meet with in flowing through a tube (for example, the resistance to flow in a large tube choked with corrosion or deposit might be the same as that of a very small tube unobstructed).

Coulomb = One ampère flowing for one second is the same expression as one gallon per minute might be in hydraulics.

$$\text{Current} = \frac{\text{E M F}}{\text{Resistance}} \quad i.e., \text{ 1 ampère} = \frac{1 \text{ volt}}{1 \text{ ohm}}.$$

and rough, and an electric motor would probably suffer more from shocks, such as are almost inseparable from the working of underground haulages, than other classes of machinery. There would also be more difficulty in protecting the machinery in this case from dust and dirt, than in either of the preceding uses referred to. But besides these applications, the possibility of conveying power enclosed in accumulators to various points must lead to a revolution in the way in which many difficulties now press upon the colliery manager may hereafter be surmounted. A colliery equipped with powerful primary motors, and having mains from these to charging stations in various parts of the mine, can, as it were, tap a moveable power, and apply it to any purpose for which it may be needed. Every colliery manager knows how troublesome small feeders or accumulations of water are, when situated in out of the way corners of the mine, and where water lading in by tubs is a most costly and inconvenient proceeding. If he were provided with a travelling carriage on which are placed a number of accumulators, these could be charged and taken to any part of the mine, and there used to actuate a motor for pumping or for many other purposes. Many other applications will suggest themselves to every practical man. Indeed, so obvious is the convenience of portable power in this form, that it only needs that a few minor difficulties should be overcome to make electricity the leading power for use underground in collieries. The ingenuity of electricians has enabled them within the last few years to bring this section of the subject out of the regions of the imagination into those of practical possibility; and there can be no doubt that the same ingenuity will not be wanting to overcome any of the smaller difficulties which yet remain.

The preceding part of this paper covers ground so familiar to many members of the Institute that I feel I owe them an apology for trespassing on their patience at such length, but I have had in view those of our members to whom electricity is somewhat of a mystery, and the modes in which an electric current is converted into work, quite unknown; for their sakes I trust the digression will be pardoned.

Coming now to the more immediate subject matter of this paper, we may ask—What is necessary in the construction of electric motors for haulage and other kindred purposes? For stationary electric motors any dynamo can be used, but where a portable motor is required, as in the case of locomotives, tramcars, or steam launches, lightness of construction combined with the requisite efficiency is a primary consideration. The massive iron frames of ordinary dynamos must therefore be dispensed with. A *Siemens* electro-motor of the D2 type developes 7 horse-power on the belt pulley if a current produced by an expenditure of 9 horse-power is led to it, and its weight is 650 lbs. The useful effect is, therefore, 78 per cent., and the duty per pound weight 330 foot-pounds per minute, or 92 lbs. per horse-power. A *Reckenzaun* motor weighing 125 lbs. will yield at 1,550 revolutions per minute 1·37 horse-power, whence its electric work represents 67·5 volts and 31 ampères. Its duty per pound weight is, therefore, 350 foot-pounds per minute, or 91 lbs. per horse-power. An *Immisch* motor at the Inventions Exhibition was tested by the jury and gave the following results:—

Speed.	Volts.	Amperes.	Brake HP.	Commercial Efficiency.
1,240	40	19·0	.794	·778
1,500	49	19·5	·960	·750
1,800	58	19·5	1·150	·761
2,100	68·5	27·0	1·340	·732
2,100	69·9	20·0	1·340	·719
2,220	84	27·0	2·308	·759
2,320	90	26·0	2·413	·771
2,400	93·5	27·0	2·496	·733

The weight of the motor tested was 156 lbs., being at the rate of 62 lbs. per horse-power for full load. This motor, then, is the most efficient of the three in proportion to its weight.

The electro-motor, as we have already said, is a dynamo reversed: it reproduces work communicated to it either directly from a dynamo or from storage batteries. In the case of the application of an electric motor for haulage it will be placed either stationary in connection with drums, pulleys, or gearing for actuating ropes; or will be the motive part of a locomotive. In the latter case a stationary dynamo is required to produce the initial current, and a motor is carried on a wheeled vehicle, with the wheels of which it is connected by suitable gearing. Such a vehicle constitutes an electric locomotive.

The various types of electric locomotives hitherto constructed receive the motive current of electricity in one or other of the following ways:—

1. *The Three Rail system*, in which between the rails on which the wheels of the locomotive and its train of wagons run, there is an insulated central rail, which acts as the positive conductor, while the rails which carry the load serve to conduct back the return current and constitute the negative conductor. The locomotive takes the current from the central rail by means of a rolling or rubbing contact. This system is open to the objection that by means of crossings, &c., wagons in passing over may create a short circuit, and so cut off the flow of electricity to the locomotive; besides which, men and horses by treading on the rails are liable to receive electric shocks.

2. *The Two Rail system*.—In this case one of the bearing rails constitutes the positive and the other the negative conductor, to accomplish which one rail must be perfectly insulated,—a thing very difficult of attainment and maintenance. Edison and Daft in America have sought to improve the conductivity of the rails by introducing copper conductors into a recess prepared for the purpose.

3. *The Channel Conductor system*, in which the conductor between the rails is laid in a channel sunk beneath the surface of the ground.

4. *The Overhead system*.—The firm of Siemens & Halske have adopted a system in which the conductor is a split copper tube, carefully insulated, which is hung in the air above the locomotive. In the tube there is a contact piece on a small traveller, and the current is communicated to the locomotive by means of wires.

5. *The Accumulator system*.—The last system is that in which the power is stored in secondary batteries or accumulators carried on the locomotive, which supply the necessary current until they are exhausted, when they need re-charging,

These being the various methods in which the electric current can be supplied to a locomotive, it must be evident at once that serious difficulties are inseparable from the application of any of them to underground haulage with the exception of the last. To carry a naked conductor between the rails of an underground haulage road, along which a powerful electric current is constantly passing, would constitute a permanent source of danger. The construction of the depressed channel in system No. 3 would be difficult in most mines, especially where there is a hard floor. The overhead conductor possesses marked advantages over the other types, but an elaborate arrangement of timbering is required to support it, and it is constantly liable to be thrown out of order by falls of roof. Where the motive power is contained within the locomotive itself, the advantages are obvious. In the first place no special road is required. The locomotive can be used on any of the roads in the mine, providing that the permanent way is well laid, the curves not too sharp, the width and height sufficient, and the gradients suitable. These simple conditions being satisfied, the locomotive can run to any siding in the pit, no matter whether the amount of coal to be sent out from that siding is much or little, the expense is no greater if one train per day only is required than if there were thirty. The locomotive can be run first to one siding and then to another, and it is immaterial how many branches there may be. This applies to no other system of haulage except horse traction and compressed air locomotives. In every other case a certain outlay must be incurred for every branch. If rope haulage be adopted, there must be a rope, return wheel, side sheaves and rollers; with any of the primary systems of electric haulage the conductors have to be laid in, with the secondary battery system all this is unnecessary. This constitutes the great feature in favour of haulage by locomotives—*the expenditure of power is directly proportional to the work to be done*. In almost every other system of haulage there is more or less dead weight and wear and tear over and above the paying load, and often bearing a quite abnormal proportion to it. This is probably greatest in the case of tail-rope haulage, or single-way endless rope, and least in that form of double-way endless rope haulage where the rope is carried over the top of the corves. In cases where branch ropes are in use, but where the traffic is insufficient to keep up a constant supply, the waste of power is very important, and here the economy of the locomotive becomes very manifest. The great obstacle to the employment of locomotives in mines is, of course, the generally unfavourable gradients, and it may be admitted at once that where the gradients exceed 1 in 50 a locomotive cannot be economically employed. In order to utilise its tractive force a locomotive must possess considerable weight to secure adhesion; the power absorbed in raising the weight of the locomotive itself on a steep gradient leaves but little power available for dragging the load. A second objection is the necessity for periodical re-charging which applies to locomotives of all kinds. With those in which compressed air is the motive power, pipes capable of sustaining a pressure of perhaps 500 lbs. per square inch have to be led into the recesses of the mine; where electricity is employed, suitable cables carefully protected and insulated must, in like manner, be carried inbye to the points at which the locomotives have to be re-charged. A third objection to the use of electricity, as at present applied, is the generation of sparks when the machine is at work, an evil which must necessarily be greatly aggravated where the roads of a mine are dusty. From this it follows that it would not be wise to use an electric locomotive (or indeed any electric machine) in any part of a mine where gas is likely to be met with. I do not go so far as to say where safety lamps are used, because in many mines these are used in the main intake airways where gas is never seen, and in such places there would, of course, be no objection to the use of electricity.

The foregoing seem to me to be the conditions which chiefly determine the suitability of electricity to locomotive haulage in mines. In the case of the Lidgett Colliery, the conditions seemed unusually favourable to its successful application. The strata at Lidgett are exceptionally flat,—the effect partly of certain small local faults, besides which as these faults run lengthwise throughout the royalty the field is divided into several long narrow strips practically on the level line. These conditions, it will at once be seen involve great complication in the adoption of any system of rope haulage, and an excessive amount of dead weight and wear and tear. They are, however, well suited for the application of locomotive power, and the seam worked being shallow, damp, and free from gas, there was no objection to the employment of electricity from the point of view of safety. With the view of putting the application of electricity to a practical test, I made arrangements with the well-known firm of Immisch & Co., London, to supply a motor and accumulators for a locomotive which I designed to meet the circumstances of the case.

The locomotive is shewn in Plates III. and IV., and the principal details in Plates V., VI., and VII. It consists of a strong oak platform supported on a rigid wrought iron frame and four 16 in. coupled wheels. The motor is situated beneath the platform and between the wheels, actuating them by means of an endless chain between the armature and one of the axles. In order to secure a maximum degree of steadiness, four sets of accumulators are suspended in front and rear of the wheels from the platform, and this device has been remarkably successful. Nearly all pitching is prevented, and the acid in the accumulators is very slightly agitated. The principal groups of accumulators are placed upon the oak platform, as will be seen from the illustrations. They are arranged in boxes, each containing three accumulators, with slide contacts, the total number of cells being 44. The weight of the locomotive in running order is about $2\frac{1}{2}$ tons; of this the cells weigh about 2,000 lbs. The engine is carried on springs of the ordinary type, the bearings being outside the wheels. The gauge is 1 ft. 9 in.

The electricity was generated by one of Immisch's 7 horse-power dynamos (Fig. 12, Plate III.), driven by a small Willans engine having 3 cylinders 5 in. diameter, and making about 350 revolutions per minute. The plant was not, therefore, a very costly one.

The process of charging accumulators is a slow one—it takes as long to charge as to discharge them, but with this difference, that while the charging is of course continuous, the discharging will only take place intermittently and in proportion to the work to be done. For a period of say 8 to 9 hours work, a period of say 6 hours is required for charging at the rate of 35 amperes.[*] The locomotive is capable of working up to 50 ampères, so that of course at that rate if kept continuously at work it would be exhausted in say 3 hours, but such a condition of things could hardly arise in practice.

Let me here say incidentally that, in the case of a pit where several locomotives were employed, they can all be simultaneously charged if put in circuit; and that, if desired, duplicate sets of accumulators would enable the locomotives to be kept constantly at work, the only delay being the time required to change the accumu-

---

[*] The dynamo employed by the writer gave out in practice 120 volts × 35 ampères = $\dfrac{120 \times 35}{746}$ = 5·63 HP. given out by dynamo *electrically*. Charge put into locomotive = 200 ampère hours, *i.e.*, 5·7 hours at 35 ampères ∴ 5·7 × 6·62 = 37·73 HP. hours put into locomotive *mechanically*—that is by steam.

lators. This could be done by an experienced man in 15 minutes.

In order to actuate and control the machine a switch of simple construction is placed at one end. This switch passes over a series of contacts in connection with a set of resistance coils, whose function it is to retard the flow of electricity, and so regulate the speed as may be requisite. It is also provided with a powerful brake.

The other features of this machine will be best understood from the illustrations accompanying this paper; and I will, therefore, at once proceed to an analysis of the results of numerous experiments.

At the Wharncliffe Silkstone Colliery there is on the surface an incline formerly used for running coal down from the No. 2 Pit. Its gradient varies as follows:—

    About 200 yards ... ... 1 in 70
      ,,   150 ,, ... 1 in 40
      ,,   250 ,, ... 1 in 25
      ,,   200 ,, ... 1 in 40

At the top is a short piece of level where formerly the siding was situated.

The locomotive would move a train of 20 corves loaded with coal, and each corf weighing about 11 cwt. gross, that is, 11 tons for the train on the gradient of 1 in 70 with an expenditure of 55 ampères. This is rather too heavy a load. With 15 corves or $8\frac{1}{2}$ tons a speed of 3 miles an hour was obtained with an expenditure of 45 ampères. This could have been maintained *continuously* for about $3\frac{1}{2}$ hours, when the cells would have run down. In working such a gradient in practice, allowing for descending alternately with ascending and stoppages at termini, the duration would be about 8 hours. Taking $8\frac{1}{2}$ tons as the weight of the train, and adding $2\frac{1}{2}$ tons (the weight of the locomotive), we have 11 tons 7·7 miles up 1 in 70 in $3\frac{1}{2}$ hours. In 7·7 miles the 11 tons is raised 581 feet, and 2 HP. is required to raise the weight alone.

Taking the efficiency of the motor itself at 80 per cent., the power given by it in propelling the locomotive is—

$$\left[\frac{45 \text{ ampères} \times 87 \text{ volts}}{746*}\right] \times \frac{80}{100} = 4\cdot 2 \text{ HP.}$$

---

* W (746) is the number of watts required to make 1 electrical horse-power. 1 volt × 1 ampère = 1 watt.

This power is given out for $3\frac{1}{2}$ hours: consequently HP. hours given out by locomotive $= 4\cdot 2 \times 3\cdot 5 = 14\cdot 7$.

$\therefore$ HP. hours put into locomotive from dynamo $= 33\cdot 1$
,, ,, given out ... ... $= 14\cdot 7$
Efficiency, 42 per cent.

We have said it absorbs 2 HP. to raise 11 tons 581 feet, and the power given out by the motor is $4\cdot 2$ HP.; $\therefore$ $4\cdot 2 - 2\cdot 0 = 2\cdot 2$ HP. remaining to propel the 11 tons, and overcome friction.

On the level we found that about 1 ampère was absorbed for each corf (or 11 cwt.), and 15 ampères for the locomotive: *i.e.*, the locomotive could draw about 30 corves on the level at 45 ampères, or 35 corves on the level at 50 ampères.

That is, 30 corves ... ... $= 16\frac{1}{2}$ tons.
Locomotive ... $= 2\frac{1}{2}$,,
$\underline{19}$,,

On the gradient of 1 in 40 we found 8 corves a maximum load; on the 1 in 25, 6 corves. The speed with 6 corves on 1 in 25 was but little over 2 miles an hour.

Without regarding these results as constituting a successful solution of the problem, they may be fairly considered as being a long step in that direction.

The chief fault in the experimental locomotive is the slow speed attained in the experiments, of which the results have been given. With regard to its weight, which has been thought by some of those who have seen it to be excessive, I have already pointed out that a considerable weight is necessary for adhesion, and that the weight is not too great in the case of the experimental locomotive is proved by the fact that when the wheels were not coupled the load could not be drawn for slipping. The slow speed is the consequence of an insufficient current, and this can only be remedied by substituting accumulators of greater power. The trial accumulators gave most excellent results, and kept up their charge for several weeks with scarcely any loss while the locomotive was standing, but the amount of space occupied by them can, it is thought, be considerably economised, and Messrs. Immisch & Co. are now remodelling them with a view to obtaining a current of considerably greater intensity in the same space.

ARMATURES.

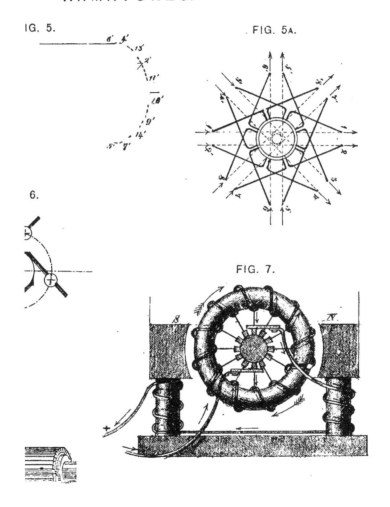

ARMATURES.

FIG. 1.
FIG. 2.
FIG. 3.
FIG. 4.
FIG. 5.
FIG. 5A.
FIG. 6.
FIG. 7.
FIG. 8.

MIDLAN *Plate III.*

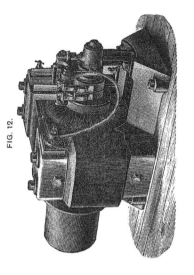

FIG. 12.

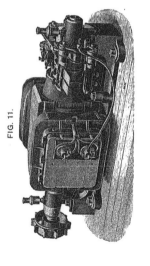

FIG. 11.

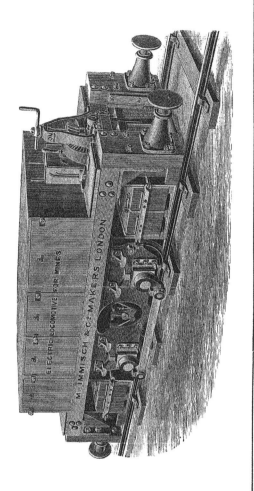

Midland Instit. *Plate I.V*

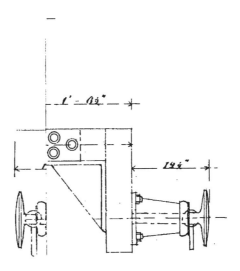

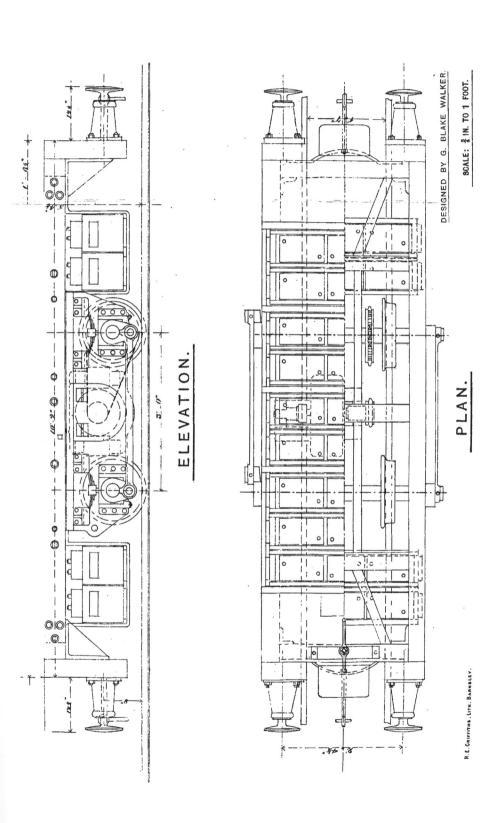

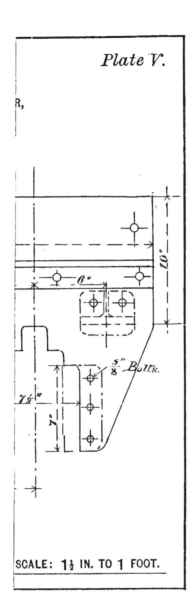

Plate V.

SCALE: 1½ IN. TO 1 FOOT.

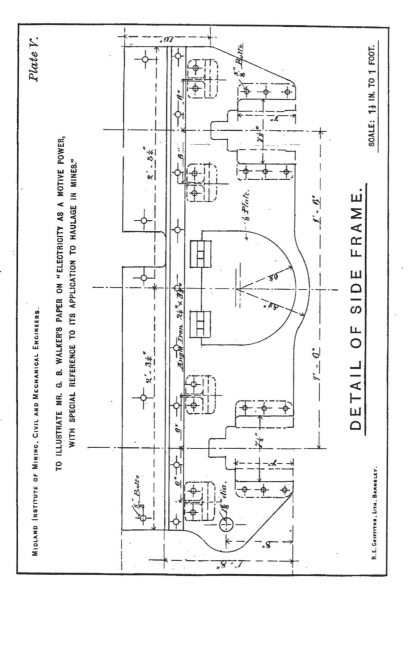

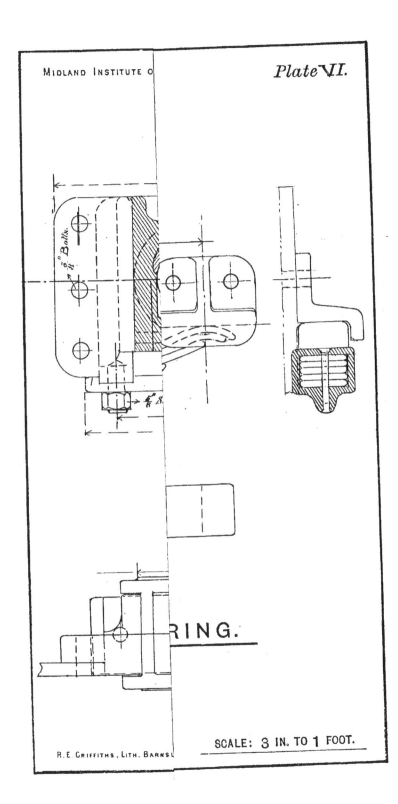

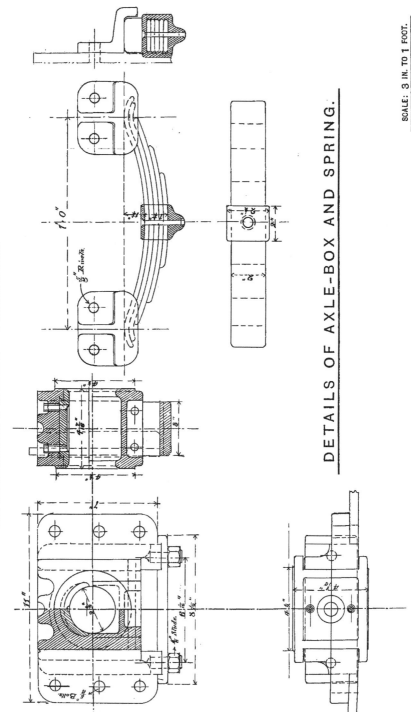

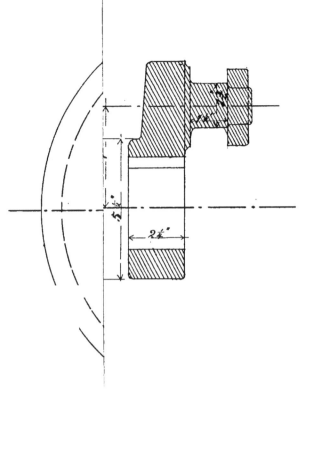

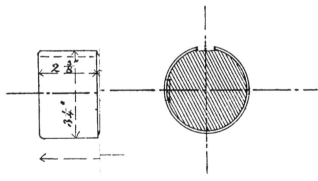

R.E. Griffiths, Lit

SCALE: 3 IN. TO 1 FOOT.

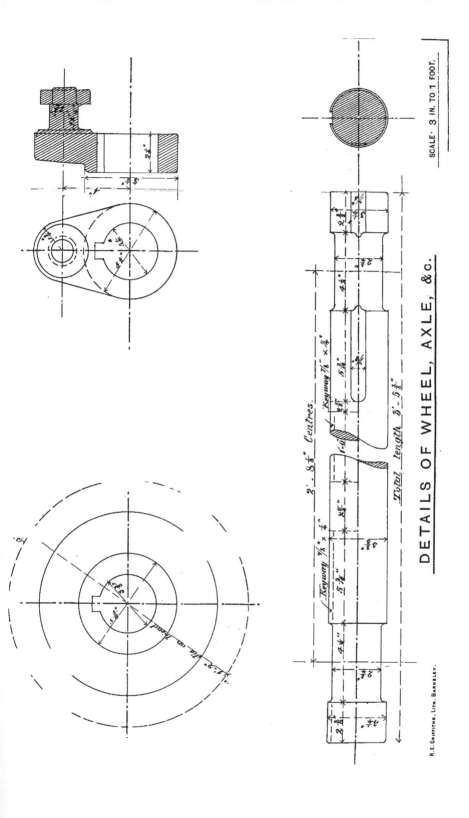

Plate VIII.

S."

r

E: ¼ FULL SIZE.

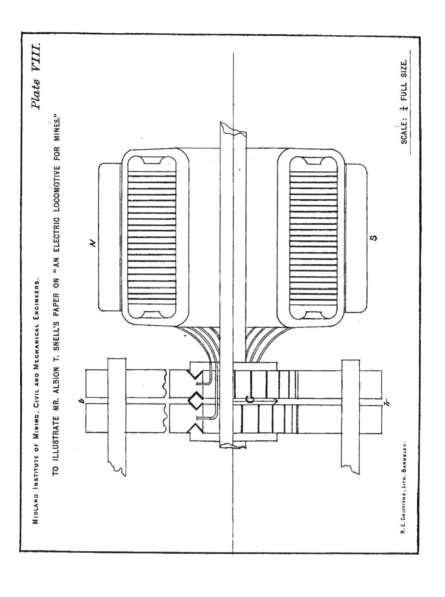

Plate IX.

MINES."

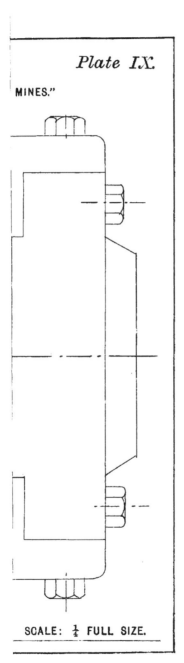

SCALE: ¼ FULL SIZE.

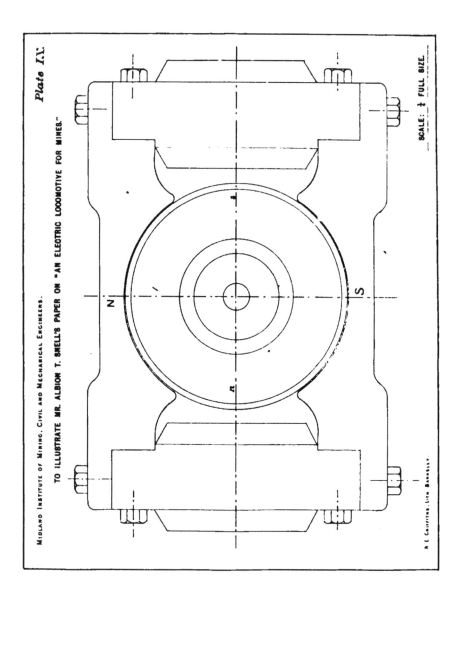

M

ORS,

.55"

.24"

FULMINATE OF MERCURY MIXTURE.

.9   1

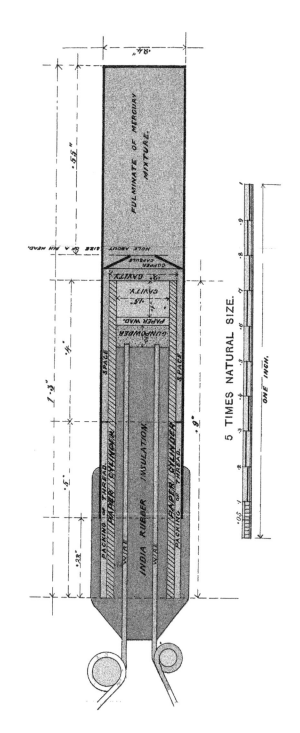

All who have seen the locomotive at work have been struck by the extreme facility with which it is manipulated, the action of the switch being simplicity itself.

The experiments which I have made are, I may say, only preliminary to others which will I have no doubt be of far greater interest. They have been chiefly useful in affording a practical test for the experimental locomotive, and for finding out its weak points. These are now being remedied, as far as they are known, by Messrs. Immisch, and when completed the locomotive is to be subjected to much more important and practical tests by our President, who with the enterprise and shrewdness for which he (in common with Lord Charles Beresford) is remarkable, has promised to put the machine to work under practical conditions underground, and so prove conclusively whether it does or does not contain the germ of a new system of haulage for coal mines.

Mr. A. M. CHAMBERS: I beg to move that the thanks of the Institute be given to Mr. Walker for his paper, and that it be printed and discussed at next meeting.

Mr. J. T. JEFFCOCK: I have great pleasure in seconding that.

The resolution was carried.

The following paper by Mr. ALBION T. SNELL was next read:—

AN ELECTRIC LOCOMOTIVE FOR MINES.

BY ALBION T. SNELL.

THE problem of electric haulage in mines by a self-contained locomotive presents several difficulties of no mean order. Messrs. Immisch & Co., in attempting this class of work, were fully aware of the obstacles to be met, and the result obtained at Wharncliffe Silkstone with the "Precursor" fully confirmed their expectations.

It was necessary that the locomotive should not only be of small dimensions and light weight, but the duration of the charge was necessarily required to be far greater than that usual with electric tramcars run on the ordinary street lines. The character of the rails, too, is not favourable to traction, owing to the general want of stability of most underground planes. The dirt further increases the resistance, and the grades common in pits are severe for locomotives. The weight of the locomotive was fixed between 2 and $2\frac{1}{2}$ tons, and it had to run in a drift 4 ft. high by 4 ft. 6 in. wide. To meet these several difficulties it was necessary to depart from all previous experience gained in tramcar work. Mr. Walker, of the Wharncliffe Silkstone Colliery Co., kindly placed his extended knowledge of pit work at the disposal of the firm, and designed the frame and general dimensions of the trial locomotive. The space being thus given, it only remained to fit as powerful a motor as convenient, and to arrange the accumulators to the best advantage. This seems simple enough, but really the difficulty was considerable. There were, in fact, three distinct problems put before the engineer. The *first* was to design a light motor of small dimensions which should run at a relatively high speed, while the locomotive itself was not required to much exceed five miles per hour. The *second* was to connect the high speed motor with the slow running axles, the space being so confined as to exclude a countershaft and the rolling wheels so small that a high transmission by means of ordinary spur gearing was impossible. The *third* lay in the accumulators, which had not only to give energy for a shift of 8 hours, but were limited as to size and weight by the conditions of the problem.

The motor difficulty was easily met, Messrs. Immisch & Co. having had so large an experience of this branch of electrical engineering. The general shape is shewn in the diagrams, and will be readily understood. Mining engineers not versed in electrical matters will note that there are two distinct axes of magnetism, the one vertically through the field poles from N to S; and the other lying approximately in a horizontal position across the armature, along the line $n\ s$. The first of these is formed by the electric current flowing through the field magnet windings; the second by the current in the armature. It is the attraction of these two axes which gives rotation to the armature and propels the locomotive. The axis $n\ s$ is apparently always trying to coincide with a line drawn through N S, but it never can, owing to the action of the commutators and brushes, marked b and b. By this arrangement the axis $n\ s$ is allowed to move towards N S by a few degrees of angular motion, and is then brought back again to the point of starting. This continually recurs, and causes a practically constant torque in the motor shaft.

The particular motor used weighs about 4 cwt., and gives about 4 HP. at 800 revolutions per minute.

The direction of motion is reversed by changing the polarity of the field so that the axis N S is inverted. This is easily accomplished by the switch shewn at the front of the locomotive behind the brake lever. A suitable resistance of iron wire is connected to the switch, so that the speed of the locomotive can be varied at pleasure, and the load can be started in either direction without shock or undue strain. The gearing was perhaps the most troublesome part of the mechanical details. Owing to the limited space none of the ordinary forms of simple spur gear were applicable. It was necessary to design something which, while sufficiently strong to bear the rough usage of a coal pit, should be both efficient and give as large a ratio of speed between the driving and rolling axles as possible. The form of nest gear finally decided on has fully answered the several duties expected of it. On the motor axle is keyed a phosphor bronze pinion: this gears into four steel pinions placed in the same plane and 90° distant from each other. These pinions are bushed with gun-metal and run on steel pins carried on a cast iron disc. The disc revolves on a journal turned outside of the end of the motor bearing. Outside of but in the same plane as these pinions is fixed an annular casting of gun-metal, with teeth cut on the inside. The steel pinions gear into the ring, which forms a fulcrum on which they revolve when the motor spindle turns. The power is transmitted from the cast iron disc by a sprocket pinion keyed to it on the inside next the motor. A steel chain connects this sprocket pinion to a suitable wheel mounted on the rolling axle. The four rolling wheels are coupled by outside rods. The whole arrangement is very compact, is entirely cased in, makes little noise, and the wear seems to be very little. The accumulators were, of course, purely an electrical problem. The only matter for doubt was how far the plates would stand the expected rough usage. The cells were especially designed for the job, and were fitted in sections of three in strong wood boxes. The electrical contacts were made by brass runners at the bottoms. The results obtained fully answered the expectations. The contacts gave no trouble whatever during the three months' run. But few boxes were short-circuited, and these were easily repaired; indeed, they stood the severe work remarkably well. Good though these cells were, further progress has now been made in the storage cells, and better results can be given with the new plates. Indeed, the locomotive as a whole has given better results than was anticipated. It must be recollected that this was the first self-contained electric locomotive put to run under such trying conditions of space, power, and weight. Complete success was not, in the nature of things, to be expected, but the three months' trial run of the experimental locomotive argues well for the future of this form of electric haulage.

Messrs. Immisch & Co. are giving the matter their constant attention, and look forward to a large extension of electric haulage in the near future.

Mr. A. M. CHAMBERS: I move that the thanks of the Institute be given to Mr. Snell, that the paper be printed, and that it come up for discussion with Mr. Walker's paper.

Mr. J. GERRARD: I second the motion.

The resolution was carried.

Mr. J. GERRARD: May I ask if the locomotive which Mr. Snell describes is the same which Mr. Walkes describes?

Mr. G. B. WALKER: Yes.

The PRESIDENT: I may say that the locomotive sent to Mr. Walker has been sent away with the view of improving it. They say they hope to send it to me in the course of three weeks for trial, so that I may be able to supplement the remarks made by actual trials with it in its improved form.

Mr. J. GERRARD: By next meeting?

The PRESIDENT: Yes.

DEFECTIVE DETONATORS.

The PRESIDENT: We have concluded the business on the agenda. There was a matter brought forward at last meeting with reference to detonators, and I said I had taken the matter up to see if I could explain an occurrence that took place in some blasting operations of ours where a detonator had exploded several minutes after the electric current had been disconnected, and after the men had returned to the working place. Since then I have had one of the detonators supplied by the Roburite Company taken to pieces and an enlarged drawing made of it. I have made several experiments in connection with the subject which I think explain the reason of the detonator not firing the explosive mixture at once. At the time when we discussed it I said I thought it was due to the fulminate of mercury not being in direct contact with the electric spark, but with some substance which acted as a slow match. I am confirmed in that opinion by the sketch and some experiments I have made, and which have been further confirmed by a communication from the suppliers of the detonators. From the sketch you will see that the detonator differs very much from the detonators supplied by Nobel's. There is a copper capsule in direct contact with the fulminate of mercury, and the spark from the powder mixture which the electric current lights has to blow through a hole the size of a pin's head. The paper wad shewn on sketch has blown up against the hole, but was sufficiently lighted to heat the copper so that it could fire the fulminate of mercury, that accounting for the length of time that elapsed.

Mr. J. T. JEFFCOCK: How long did it delay firing?

The PRESIDENT: Three or four minutes. The men had time to disconnect the wires, walk from a refuge hole 30 or 40 yards away to the place where the shot was, and the charge exploded, severely injuring one man. If it had been a point blank charge it would have killed him at once. I thought the matter so important that I took it up at once, so as to bring it before others using the detonators. I wrote to the Roburite people who came over to see me about it, and referred me to the makers of these detonators. I find the cap part, containing the fulminate of mercury, is imported from Germany, and they are put together at Cinderford. The following correspondence has passed on the subject:—

" Cinderford, Gloucestershire,
" 15th Dec., 1888.

" Dear Sir,—Mr. Bigg Wither has sent us your sketch of the
" electric fuze, and detailed the circumstances attendant upon the
" serious accident which occurred to one of your workmen. We have
" been manufacturing electric fuzes for fifteen years, and never before

"in our experience did such a thing occur. We have given the
"matter our most careful attention and consideration, and have
"satisfied ourselves as to the cause.

"The detonating cap used is a powerful one supplied to the
"Roburite Company from Germany. It is a speciality,—a patent,
"we believe,—and this consists of a copper capsule fitting inside the
"detonator and enclosing the fulminate charge, only a small circular
"aperature about the size of a pin head being left in it, through
"which, in the properly constructed detonator, the spark from the
"electric fuze reaches the fulminate. It is obvious, therefore, the sur-
"face of fulminate exposed to this spark is not more than one-fourth
"that exposed in the ordinary make of detonator. Now we have
"found occasionally that through some mishap in the manufacture
"the small circular opening in the copper capsule is in some instances
"wholly and in others partially covered with metal, so that in such
"instances the fulminate would be quite covered or only a very
"small portion of it exposed. When noticed by us these defective
"ones have been rejected, and we have by us quite a number of such.
"Now we believe the ill-timed explosions have occurred with detona-
"tors such as these, which have escaped our notice. The electric
"fuse has exploded, but the circular opening in the centre of the
"patent capsule being, as before explained, wholly (or nearly so)
"covered over, the small explosion caused by the electric fuze failed
"to reach the fulminate; the paper wad which confines the explosive
"mixture in the fuse head continued to smoulder, having been
"ignited by the explosion; the heat from this ultimately came into
"contact with and exploded the fulminate, with the result described.

"We are now using the ordinary detonators, without any inside
"copper capsule to intervene between the explosive mixture in the
"fuze and the fulminate. With this we think such a mishap would
"be impossible. We are sending Mr. Bigg Wither a copy of this
"letter, and shall strongly advise him never again to use the patent
"capsule detonator for electric fuzes. We shall be pleased,—having
"given our views,—if you will favour us with an expression of your
"opinion. Even the bare possibility of such a catastrophe must be
"prevented, and we venture to think the abandonment of this
"special make of detonators will do it.

"Faithfully yours,
"The Electric Blasting Apparatus Co.,
"F. Brain, Managing Partner.

"To C. E. Rhodes, Esq.,
"Aldwarke Main Colliery."

"Aldwarke Main and Car House Collieries,
"Rotherham, 20th December, 1888.
" F. Brain, Esq.

" Dear Sir,—I am much obliged for your favour of the 15th "inst., with reference to the detonators in use by the Roburite "Company, and more particularly with respect to an accident we "had with one and which I brought before Mr. Bigg Wither.

" The opinion I expressed to Mr. Bigg Wither, and which was "arrived at after careful experiments had been made by my assistant "with various detonators, is exactly in accordance with that set out "in your letter, and is the only logical explanation of the occurrence "in question, which so far as our experience goes is by no means "singular, that is, in a shot firing after the electric current has been "taken off.

" Before expressing an opinion as to the abandonment of this "type of detonator altogether, can you inform me the object of the "copper capsule ? Has it been introduced to secure greater safety in "the handling and the carriage of the detonators, or for what other "reason ?

" So long as the liability of these mis-fires or 'hang-fires' is "recognised, little or no danger need be incurred by using them, "because it can be avoided by simply allowing a little time to elapse "between the disconnection of the wires and the entering of the "working place.   "Yours faithfully,
"*Signed*, C. E. Rhodes."

"Cinderford, Gloucestershire,
"22nd December, 1888.

" Dear Sir,—We are much obliged for your esteemed favour of "the 20th. The copper capsule, we take it, enables the makers of "the detonators to press and retain in the cap the fulminate charge, "—there is no risk of any dropping out, as sometimes occurs with "the ordinary detonator. It is a safeguard to those inserting the "fuze, as there is no possibility of its coming into contact with the "fulminate. An employé of ours lost part of his hand some time ago "while engaged inserting the electric fuze into the ordinary make of "detonator. It exploded, and no doubt this was caused by the fuze "coming into contact with some loose fulminate,—an impossibility "with the capsule detonator.

"For the Electric Apparatus Co.,
" To C. E. Rhodes, Esq.,   "Yours faithfully,
" Aldwarke Main Colliery."   "F. Brain.

"Aldwarke Main and Car House Collieries,
"Rotherham, 31st Dec., 1888.
" F. Brain, Esq.

" Dear Sir,—Referring to our correspondence *re* detonators.

" As this subject is looked upon with much interest by a number "of gentlemen in this district, would you have any objection to our "correspondence being published in the Proceedings of the Midland "Institute of Mining Engineers, of which I am the President.

" Believe me yours faithfully,
"*Signed*, C. E. RHODES."

"Trafalgar House,
"Drybrook, Gloucestershire,
"2nd January, 1889.

" Dear Sir,—I have no objection to our correspondence being "published. I should esteem it a favour if you could arrange for "two or three copies of the Proceedings containing same to be sent "me.

"Truly yours,
" C. E. Rhodes, Esq., "FRANK BRAIN.
"Aldwarke Main Colliery."

Mr. G. B. WALKER: I think it is desirable that this communication should appear in the Transactions.

Mr. GERRARD: With the diagram?

Mr. WALKER: Yes.

Mr. GERRARD: The paper wad you refer to, is there not a paper in so to speak between the gunpowder and the copper?

The PRESIDENT: No; the paper cartridge is a paper wad practically. The drawing shows it, but we had to pull the cartridge in pieces to get out this, you know. You will see the wad is really the packing.

The meeting then ended.

MIDLAND INSTITUTE OF MINING, CIVIL, AND MECHANICAL ENGINEERS.

GENERAL MEETING.

HELD AT THE INSTITUTE ROOM, BARNSLEY, ON WEDNESDAY, JANUARY 16TH, 1889.

C. E. RHODES, Esq., President, in the Chair.

The minutes of the last meeting were read and confirmed.

The following gentlemen were elected members of the Institute, having been previously nominated:—

Mr. JAMES ANGUS, Mining Engineer, Rockingham Collieries, Barnsley.

Mr. WM. JAMES BELK, Mining Student, Thorncliffe Collieries, near Sheffield.

DISCUSSION ON MR G. BLAKE WALKER'S PAPER ON "ELECTRICITY AS A MOTIVE POWER, WITH SPECIAL REFERENCE TO HAULAGE IN MINES," ALONG WITH MR. ALBION T. SNELL'S PAPER ON "AN ELECTRIC LOCOMOTIVE FOR MINES."

The PRESIDENT: The next business is the discussion on Mr. G. Blake Walker's paper on "Electricity as a Motive Power, with special reference to Haulage in Mines," and also on Mr. Albion T. Snell's paper on "An Electric Locomotive for Mines." Everybody who has read Mr. Walker's paper will appreciate the vast amount of trouble he has been at in working out this question, and placing it before us in such a form as it now is. Electricity as a motive power, especially as applied to colliery operations, is such a very new subject that I am very much afraid that few of us will be able to speak with anything like authority upon it. I may say that Mr. Snell, who was good enough to read us this paper, is here, and will give any explanation that is wanted; and as Mr. Snell represents the firm of Immisch & Co., who have done more of this class of work in this country than anyone else, anything that he may say will carry great weight and give information to us unobtainable otherwise, and I am sure we shall therefore appreciate the more the trouble he has been at in coming here. I shall be glad if anyone will ask any question they wish, or give us their views on the subject.

Mr. JOSEPH MITCHELL: I had the pleasure of visiting the works, and Mr. Snell was good enough to explain all the working parts of an electrical machine. So far as I could see there was every appearance of a motive power that in the near future would supply our wants much more cheaply than power obtained from compressed air. What I saw was very simple, but there was one serious drawback—the liability to sparks from the shutting off and putting on of the electricity, and also the question of liability to sparks arising from the working of the motor. I pointed out these difficulties to Mr. Snell, and suggested that if he could convey the cold air from the main intakes to and from the motor in pipes, any question of liability from using the motor would be avoided. He said that could be done, and also that he would make a valve by which he could adjust the throwing on of the electricity to the motor the same as admitting steam to a steam engine through a throttle valve. I think if he can overcome these difficulties, the motive power we have been looking forward to, that we can apply for working loads along the gateways, is near to us, and we shall soon be able to have the benefit of this machinery in place of using horse power. I shall be pleased to know if Mr. Snell has given the subject any further consideration, and what has been the result.

Mr. J. GERRARD: Mr. George Blake Walker has been indefatigable in conveying information to the members of our Institute. I am sure that he offers every encouragement to other members to do something in the same direction. I have read this paper with very great pleasure. Electricity is, as our worthy President puts it, comparatively in its infancy, and therefore, though we may not be able to criticise it, we are all anxious to know more of it. Mr. Walker goes to the very beginning of it; he puts it so simply and so tersely that anyone who reads his paper cannot but be wiser. Already in mining we have derived advantages from electricity. With regard to signalling, I am sure no one would care to go back—in fact it would be impossible to get through the work with the old form of signalling on our modern engine-planes. The ease and distinctness with which signals are conveyed is remarkable, compared to the old form of conveying signals. To electricity we are indebted for that, and with regard to lighting, I am sure there are some in this room who can speak upon that point. The flaming lamps and open torches which lit the bottoms of our shafts, and threw that lurid, flickering light only a very short distance away for the hanger-on and his helpers, has now passed away, and we have the electric light lighting up many of our pit bottoms; and we have also upon the pit tops beautiful installations of electric light. Having in connection with our mines seen what electricity can do with regard to signalling and lighting, we must be disposed to give every help and countenance to this further development of electricity—the using it as a motive power. I am quite sure that the question of sparks must be recognised in the beginning. It may not be an insuperable difficulty. It may be overcome, and we must not allow it to slide. We must be plainly spoken with regard to it. We are taking out open lights, we have made vast improvements with regard to safely lighting our mines, we are taking out furnaces, going in for mechanical ventilators, and so on, and we must not introduce electricity as a motive power with a danger and liability to sparks, if there is a danger attaches to it, without that being brought to the mind of these inventors that they may grapple with it and overcome it. I am sure that so far, from my investigation, there is some danger in that direction. It is all very well, perhaps, when the machines are running smoothly, but we know that brushes are burnt up. How are they burnt up? There is development of heat. We know that if a current is shortened that sparks more intense are given off. I am not sure whether any tests, any experiments have been made to show whether these sparks will light an explosive mixture or not, but in the absence of such conclusive test, there is a grave question of danger attached to it. If that development of heat from something going wrong with the machine will destroy the brush; if by shortening the circuit there is an intenser series of sparks thrown off, there must be danger involved in it. I am not prepared to accept one or two lines in this paper with regard to the possible uses to which electricity may be applied as a motive power. Leaving out any question of the advantage of auxiliary fans in bye, in the workings, the 3rd General Rule says: "Where a mechanical contrivance for ventilation is introduced into any mine after the commencement of this Act, it shall be in such position and placed under such conditions as will tend to ensure its being uninjured by an explosion." It may be said that this rule prevents the introduction of auxiliary fans into the workings. As to haulage, that offers itself as a very good field, because the haulage engines are often placed in the main intakes, and not far from the downcast shafts. With regard to pumps, there is another fair field for the application of electricity. Compressed air is a very expensive power. It is very comfortable and cool, but it is very expensive; and if electricity can be applied with regard to raising water in bye, I think there is a very fair field for it there. But we want this question of danger from sparks settling. I do not know whether any experiments have been made to prove what intensity of spark, or what amount of sparks will explode an explosive mixture. I heard the other day of an experiment being made with electric signalling—trying to bring two naked wires together in an explosive mixture, but they failed to explode it; they tried over and over again, a number of times, and utterly failed. It was on a circuit of a thousand yards of wires. Whether if they had shortened the circuit the result would have been different I am not prepared to say; but before we adopt or encourage the application of electricity as a motive power in our workings, we ought to have some conclusive test with regard to the sparks. I do not wish to discourage the application of electricity at all; on the contrary, to encourage it as far as possible; but let it be open-eyed, let the people who are moving to make these improvements know what they have to overcome, and I think they may remove the danger.

The PRESIDENT: With regard to the question of sparks. The first mechanical application I ever saw of electricity to underground haulage was at Newcastle Exhibition. I had the pleasure of meeting Mr. Snell there, and that was the objection I then raised to him. As to whether the sparks will light an inflammable mixture, I think there is no doubt they will, because I have seen repeatedly our man at the collieries light gas,—putting his fingers on the edge of our fan belts, and light the gas with a foot rule, the electric spark passing through his body. The sparks I saw from the electric motor would, I cannot help thinking, light an inflammable mixture. Then you have the common contrivance for lighting the gases, fusees, &c., and in these cases there is no doubt the sparks fire an inflammable mixture. Mr. Snell had the idea in Newcastle that it would be practicable to enclose the machine in a box of such strength as to resist and keep inside an explosion. That is obviously impossible, because if there was a crack small enough to admit into the box an explosive mixture, it would let the mixture fired out again to communicate with the surrounding atmosphere. Another idea was, whether it would be possible to encase it in safety lamp gauze, because Mr. Snell pointed out you could not enclose the motor in an airtight case, air being necessary to keep it cool. It was suggested whether by enclosing it in two or three thicknesses of safety lamp gauze, it would be possible to obviate the danger. In my opinion no casing of safety lamp gauze would stand the explosion which would take place in anything like the area a motor would occupy. You are then driven to keeping the engines at the pit bottom, assuming the pit is so fiery you cannot do with them out, and in that case you are a long way from making electricity the success it should be. If steam or compressed air can be used direct, it is obvious that it is more economical than passing that steam or air through the engine and using electricity. There is nothing so economical as the first use of power, whether steam, water, or anything else. My idea of electricity is that its great advantage lies in its extreme portability. The fact of being able to take an electric cable down steep gradients, and being able to use it in any direction, is its great charm to me. The only objection I can see is the spark question, and it seems to me at present a very difficult one to get over, because the locomotive at Wharncliffe Silkstone which I saw, and intend to try shortly, gave off a spark,—not just a flash, but a thing apparently you could have lit a cigar at,—when the current was turned on and off. Unless something can be produced to obviate this sparking, it seems to me there would be very great risk in having a motor far in bye in a fiery mine. After the great expense that has been gone to in many collieries in providing safety lamps of the latest possible pattern, in doing away entirely with the use of explosives, and in dispensing with everything in the form of an open light, it would appear to me, that after making all these efforts in the direction of safety, it would be a backward step to introduce a machine into the mine giving off sparks capable of igniting gas. The question that I should like to see made quite clear is whether these sparks are innocuous or not, and I certainly think we should direct our attention towards clearing this matter up, and I should be glad to hear what Mr. Snell has to say on this important subject. I may say, although I never like to differ from my friend Mr. Gerrard, who has good ground generally for any statement that he makes, that I do not agree with his remarks about fans. We all know there are many instances where you do want a little auxiliary fan,—say in going up a fault side where it is almost impossible to get your air up by bratticing whilst the fault is being proved, a small hand fan may then be used to keep the place clear till the slit is got through; or in headings in very steep work where it is almost impossible to keep the place clear, and where the men cannot work, nine times out of ten, without something of the kind, it does not seem clear to me why, if you can have electricity, and get over the spark business, you should not have a small auxiliary fan in such cases as these.

Mr. GERRARD: When I spoke of the sparks, it was not of the greater sparks, which it is an absolute conviction to my mind involve danger. I spoke of the small sparks given off in the ordinary working from the ordinary dynamo, and it was not proved to my mind that these smaller sparks given off in the ordinary working——

Mr. G. B. WALKER: From the brushes?

Mr. GERRARD: Yes. It was not clearly proved to my mind that they involved that danger which has been so clearly and so amply depicted by our worthy President, and which not for a moment shall I differ from. With regard to auxiliary fans, I do not know that I said anything on the question of the advisability of putting a fan into your workings; I was only putting it as a sort of question whether the 3rd General Rule under the new Act would not prevent the erection of fans in bye, in our workings, from the fact that such appliance must be placed under such conditions as would tend to ensure its being uninjured by an explosion. That is only a minor question after all; but I am not going to do anything to encourage the development of ventilation by small hand fans, nor by small electrical fans. I have seen a little of such work; whilst having highest respect for Mr. Rhodes, we will differ upon that point if he pleases.

The PRESIDENT: Your point is in the abuse.

Mr. SNELL: Sparking is of two kinds—there is high tension and low tension sparking, and all the cases you have spoken of have been high tension. I have not seen the experiment of taking electricity through a man, but I can understand it; that is statical electricity, which we have not to do with. Statical electricity is a thing we have not to deal with in dynamos or motors, except when we have a large new belt, and then only for a few hours till the leather gets softened. I fully admit what you say, and those little machines for lighting gases are all based on that principle—the electricity is at a very high tension and very small in quantity. Then again, the point at which gas will fire depends largely on the temperature of the points at which the spark is emitted. If we took two pieces of carbon, where the carbon is a bad conductor we should have no difficulty in lighting the gas—the spark would have considerable heat itself. But when we take a dynamo or motor, and find sparks on the commutator, the sparks are in contact with such a large body of metal that the temperature is low. I can assure you it is possible to put your fingers to such sparks and not feel the slightest effect as regards heat. I do not mean to say it would not ignite gas under certain conditions, but I mean to say this—that although you may see a few sparks it does not follow there is danger. It rather bears out what was said about signal wires. The reason the sparks on the wires would not light the gas was because they were in contact with an amount of metal sufficiently large in bulk to conduct the heat that was produced instantly away, and so the spark was not up to the flashing point of the gas. When we come to put a motor or dynamo in a gaseous place, we have really two ways of getting over the difficulty. I think the President has rather misunderstood my views on the subject. I may have told him that we could build a box strong enough to resist an explosion, but if so, I have certainly grown wiser since then. At any rate that is not the plan we intend at all. We intend building boxes, not necessarily air-tight, but nearly so. The box will have an opening at the bottom, in which fresh air will be brought, probably in an inch or two inch pipe. On the top there will be an orifice slightly larger through which the air, after passing round the motor, will be emitted; and I think you can see then that it is not at all necessary to make it perfectly air-tight. Suppose you had a leakage of one per cent.—the air coming into the bottom hole; I think there would be no difficulty then whatever, because you could only dilute the fresh air to the extent of one per cent. with the atmosphere outside, and there is no chance of the ventilation being reversed, because the motor gets heated and with the heated air inside you would draw the draught from the bottom to the top. I think we should get over the difficulty in that way entirely. Another idea has been suggested by Mr. Brogden, mining engineer, South Wales, which is worthy of consideration, and that is to surround the whole motor by three skins of gauze, with a distance between each of half an inch, so that if any explosion did take place, it would only take place to the extent of the space between the casings; it might damage the outer casing, but probably would not damage the inner casing at all. This method, if carefully examined, seems to be by no means equal to the first proposal based on the experience we have at present. We have three plants at work now. We have them placed in the intakes; we have built walls round them, leaving a door open to the upcast, and brought a six inch pipe from the intake side, so that the air from the six inch pipe passes over the motor. By this means we have not had the slightest difficulty, although at times there has been sparking, and I would not attempt to tell you the contrary. You must get sparking sometimes, which does not exist in a normal state. The men who handle the machines are, of course, of an uneducated type; as a rule you may trust them when you see them, and no further. I have been down at uncertain times to see how they were going on, and found the man asleep and the motor doing as it liked; but the motors are well under control by reason of the general design, and nothing happens from this neglect. Indeed I consider an electrical plant would be more likely to run with success without constant supervision than any other class of machinery, assuming the man at the top maintains the steam pressure.

Mr. J. NEVIN: There is no doubt in my mind that one of the great wants of our profession, a greatly increasing want as pits get deeper and of greater extent, has been a ready means of transmitting power from one part of the pit to another. Of course, in coal mines it is generally pretty well out of the question to use water power, and therefore the primary power for us is steam, and when we want power some distance from the pit bottom we cannot take steam in pipes for more than about a thousand yards. The next mode which we have had has been that of transmuting steam power into compressed air. That plan is very expensive, as the pipes are costly and the loss is great, in fact when you have done all you can you do not get above 30 per cent. of the power. Another useful way of transmitting power is by means of endless ropes, but you can only transmit power by those ropes when the pit is at work. If electricity can be brought into use for transmitting power it will probably be the greatest benefit that we have had for many long years. The difficulty to my mind is that which has been mentioned by both the President and Mr. Gerrard. I do not see how sparking is to be got rid of. So long as we are obliged to have that sparking with the electrical machinery, after all the pains we have taken in improving safety lamps, and taking away everything that can possibly cause an explosion, its adoption would be a step in the wrong direction. If its adoption can be carried out and sparking done away with, it will be the greatest improvement in mining there has been for years.

Mr. GERRARD: I had intended to ask a question. In several places Mr. Walker says that "some initial difficulties have been overcome," and speaks of "some minor difficulties." If he could give some idea of what he recognises as difficulties, I should be glad.

Mr. WM. HY. CHAMBERS: We have had no difficulty at Denaby Main with the electric light. Of electric haulage I cannot speak with authority. I was in Germany a short time ago, and had intended going over the Zaukeroda mine to look at the electric haulage in use there, but was prevented carrying out my plan. From inquiries I made in the neighbourhood I found the method adopted there was not altogether satisfactory, and that it could be very much improved. It is the overhead wire system and made, I think, by Siemens, of Berlin. I was looking the matter up since I came home, and I found in one of the Transactions of the North of England Institute, —Abstracts of Foreign Papers, No. 134, Part 1, p. 5,—an account of some experiments made by Professor W. Schutby, which is very interesting indeed, and which I will quote:

Underground Haulage by Locomotives.

	Steam Engines.			Compressed Air Engines.		Electric Engines.
	With Fire.		Without Fire.			
	Domau Mine.	Cessous Mine.	Honigman System.	Petan System	Mekarski System	Zaukeroda Coal Mine.
Cost of plant, in £	1,600	1,600	550	650	1,000	800
Interest and depreciation per mile ton, in pence	·42	·14	·50	·79	·81	·44
Cost of haulage per do., in pence	·60	·51	·82	1·18	1·35	1·10
Total cost per do., in pence ..	1 02	·65	1·32	1.97	2·16	1·54
Daily performance, in mile tons	303	882	88	66	99	146
Length of run, in yards	2,530	5,030	676	676	676	676
Time per run, in minutes ..	2·3	3·3	1·5	1·5	1·5	2·6
Weight of the machine, in cwts	86·3	156·3	47 2	53·1	45·2	31·4
	Ft. In.	Ft. In.			Ft. In.	Ft. In.
Height of do.	6 3½	6 10¼	(?)	(?)	5 1	4 11
Breadth of do.	4 3	5 3	(?)	(?)	3 7¼	2 7½

If Mr. Rhodes and Mr. Walker would be good enough to give us a similar comparison of cost and performance with Messrs. Immisch's locomotive with other systems of haulage it would be of great value. There is another paper, read by Mr. Frank Brain before the Society of Arts, mentions the same system, which is worked in some of the gold and other mines of America.

Mr. McMurtrie: On page 328 Mr. Walker states, "where the gradients exceed 1 in 50 a locomotive cannot be usefully employed." This will greatly reduce the benefit of this system of haulage, because if a new system of haulage is needed it is needed in seams with a steep inclination. In the Rotherham district the inclination is often 15 to 18 inches to the yard, and the system of rope haulage adopted there is subject to carelessness on the part of the engineman and consequent smashes. If Mr. Walker's statement is correct the electric locomotive could not be adopted there, and in the case of any but a slight gradient would be of no use.

The President: That only applies to locomotives; not to stationary engines. It struck me one of the places where a locomotive would be of great service would be in cases where you want to take a main rope say 3,000 yards from the pit bottom, and where to have a tail rope means 6,000 yards. If you could,— where you have an undulating gradient,—take out the empty train by one of these, it would be a great advantage. There are places where we may have a road through a fault, or where you have a bad roof, and where the cost of making a double road would be very great, when such a power as this would be of inestimable advantage. I do not think they will do away with horse power for tramming purposes, because the difficulty of keeping the roads in and out of working places would be too great. I do not see how a machine can be a financial success that has to take the place of a pony working for one or two sets of men, when the great expense of the machine is taken into consideration, with the extra cost in wages necessary to work it. Where they could be applied to main roads where it is level, and on undulating roads to trail out the rope, I can see great economy and saving in them.

Mr. G. BLAKE WALKER: I will make very few remarks, because Mr. Snell has a good deal to say. First then, with regard to this sparking scare, if I may call it so. I think perhaps we are going to an extreme in the question of making safety lamps universal throughout some pits. I do not refer especially to safety lamps themselves, but to the argument that because safety lamps are used for certain reasons, therefore open lights are, *per se*, dangerous, especially with regard to sparks which only exist whilst the machine is at work in the main intake air roads, where gas is never seen from one year end to another and where it never would be seen except under abnormal conditions. I think we may dismiss without very much anxiety this question of sparks. If we are practically working in air as fresh as at the surface,—if you have, for instance, an intake airway where 30,000 feet of air per minute is passing, even if it be a mile or more from the pit bottom,—I think myself that to raise a great outcry against the making of sparks in such conditions as that is excessive caution. With regard to working auxiliary fans, I infer from what Mr. Gerrard said that he supposed the whole thing, motor and fan, would be in the return. That was not my idea at all. I assumed it would probably be possible to so place the fan that the motor could be at one side of the stopping in connection with the intake current, and the shaft by which the fan was actuated carried through the stopping into the return, the pressure, of course, being always from the intake to the return. If there was any small amount of leakage, it would not be from the place where gas might exist, but the place where the air was pure. That auxiliary fans might be very serviceable in outlying districts where the principal fan was unable to give a sufficiently high water gauge I have always maintained, and believe that they will be more and more used. With regard to the 3rd General Rule, I think that the meaning of the Act is confined to the principal ventilating machine, by which the ventilation of the pit as a whole is produced. As to using the locomotive which has been spoken of in this paper on gradients exceeding 1 in 50 in steepness, it would certainly be a bad application of power. The idea is only to use the locomotive under conditions where the adoption of a locomotive would be advantageous, and certainly it would not be advantageous in drawing up anything like the steep gradients of our dip planes. But there may be long levels, or roads approaching to levels, on which they might be very usefully made use of. Then with regard to the application of electricity to steeper gradients than these, we should then call into operation a machine in which the motor would be established at a fixed point, and would actuate a drum by means of gearing, and the rope would be worked in the ordinary way. As to taking in tail ropes, that is a point we nearly put to a practical test at Wharncliffe Silkstone before the locomotive was sent to London to be altered. Mr. Snell went down and saw the place where we thought of applying it, and we had only this difficulty, that we had some 700 yards of cable to provide, and Mr. Snell was not able to obtain a second-hand cable at a cheap price, and it did not seem wise for the sake of an experiment to go to the expense of a new cable, and so the proposal fell through. There are lots of cases where the corves would not take the rope where a locomotive could be used admirably and be most handy. As to its economy with regard to horses, it is a pretty well understood thing that with short distances no system of haulage is so economical as horses; but as soon as a certain maximum distance is exceeded, where longer runs and heavier loads are required, the economy of the horse rapidly declines. Mr. Gerrard has asked a question—what I mean by overcoming certain initial difficulties. In the first place there is the difficulty in the construction of the accumulators. The accumulators have been very much improved within the last two or three years, but they are by no means perfect as yet, and they involve a great weight in proportion to their efficiency. Therefore electricians are now devoting themselves as much as possible to increase the efficiency of the accumulators in proportion to the amount of weight. This is an initial difficulty which I have no doubt will be gradually overcome to some extent,— there is a certain amount of finality about the possibility of decreasing the weight below a certain minimum. Another thing is the best form of conductor, if some other systems of haulage are adopted where the locomotive obtains its supply of electricity from a conductor. This has not been thoroughly worked out in a perfectly satisfactory manner, and there are many improvements to be made in the form of the conductors to be adopted. But the thing I have set myself to do in this paper more than anything else was to try to make clear the action of the electric motor. Mr. Snell tells me I have rather failed in making it clear, that I have made it rather obscure than otherwise,—(No, no,)—and he has promised that he will try to remedy the obscurity by means of a piece of chalk and the black board, and, therefore, I would like to leave the rest of the explanation to him.

The PRESIDENT: One remark Mr. Walker made in the beginning of his address I personally rather take exception to, and that is that this question of sparks should be called a scare. To my mind anything that affects the safety of your underground operations is very far indeed from being a scare. If we can demonstrate and be perfectly assured that the sparks can be rendered innocuous, then if we discuss the thing for twelve months the time will have been well spent; and if we prove that the sparks are dangerous, even then it has been well spent time. I do not think it matters much where these engines are placed, if they are placed far away in bye, whether in the intake or the return, because I have seen it myself where, against more than 30,000 feet of air, the gas has backed into the main intake for a very considerable distance, fouled 30,000 feet of fresh air within 650 yards of the pit bottom. You cannot get over facts of that sort. No doubt these were abnormal cases, which perhaps only take place once in a generation, but these big accidents have taken place from abnormal causes. Nobody has ever worked a colliery in a generally normal condition of being ready for an explosion; I cannot believe that any man could ever be found to do that; it is the abnormal case we have to try to guard against. To shut our eyes to these sparks as a Mining Institute of Engineers would be placing ourselves in a false position. I think the subject should be thrashed out; if the sparks are dangerous try to remove them; if we cannot remove them it is a strong point against electric haulage.

Mr. SNELL: I do not intend to bring before you any point of practical detail, because I hope to have the pleasure of bringing before the Institute at an early date a practical account of the work which my firm has been doing for the past twelve months. We have now on hand some four installations in mines. One is a hauling job from fixed engines; one a pumping job at Locke & Co.'s, which is probably well known to you. This has been running for the last twelve months without the slightest hitch in the electrical work. We have had one or two break-downs in connection with the engine, which would not have occurred if we had had an engine sufficiently powerful for the long continuous work,—we run through 22 hours out of the 24, so that it is stiff work. There is another pumping plant in South Wales of about 20 horse-power, which embodies the new type of machines. These machines have been improved on those supplied to Locke & Co., and sparking has been reduced to such a small amount that it is difficult to see the sparks even in the dark. I do not mean to say that they would not spark if the brushes were not put on properly, but I take that to be no fault in the matter, because in using machinery of any particular type one must expect to have men to understand them. Motors are not difficult things to handle by any means, and an ordinary man with common intelligence could be taught in a few weeks to handle a machine with ease, and there would be no excuse for any sparks being found there. Again we are arranging signals by means of bells between the motor and the engine-house. Put in circuit with the machinery is an instrument which measures the quantity of electricity flowing at any time, and the engine driver has only to keep his eye on the index and can tell exactly what is going on below. If, for example, an installation was arranged to run with 66 ampères there would be a red mark upon the dial at this point, and the needle would stand at or about the red mark. If the pumps by any means had more work put on them, say a bearing was heated, and more work had to be done, more electricity would be required to drive them, and the needle would move to 70 or 80. The driver would see that, and signal below "Look at your pump." The pump would be looked at, and if necessary to stop, the signal would be given to the top to stop the engine. That gets over the difficulty as to sparking on breaking circuit; that was a difficulty Mr. Mitchell raised in connection with the switch of the locomotive. At Locke and Co.'s I do not think they stop the machinery from the bottom once in three weeks, and even then it is only done in case of necessity, such as the engine driver being away from his post for any reason. The bell has been up for the last eight months, and a regular code of signals arranged, by which it is perfectly simple and easy to stop the engine at any moment, and the whole machinery goes slowly to rest. Other signals are arranged to start the engine, and the machinery can be slowly started again. We have another plant in Scotland, which will be interesting inasmuch as we transmit power over 4,000 yards of cable. That involves the question of efficiency in a much greater degree than an installation over shorter distances, and gives electricity a greater chance of comparing favourably with compressed air or hydraulic power. A stationary haulage plant we are putting down, also at Locke & Co.'s, will be a 50 horsepower plant. There will be no difficulty about sparking there, since the machine is to be near the pit bottom in the main intake. It is walled all round with the exception of two slits through which the cables pass in and out, so that the difficulty of sparks does not enter into the case. But we are hearing this point in mind; in fact, I may say I feel this question as deeply as any of you do, and although I am inclined to take Mr. Walker's view, that at present we may regard it as a scare, for noboby would think of putting an electrical plant on a gaseous face at present. Yet electric coal getters are being put in in some mines, and as far as I am concerned, I should have no hesitation, in some mines, in putting machines of that type. There are other places where I would not recommend or advise it.

I think with these few remarks I will give you an idea how the armature works. I have often been asked how it is the armature goes round, and it does seem astounding that the armature should go round and do mechanical work when there is no evidence of anything happening beyond a certain amount of external magnetism. It is really done by means of two magnets. [By means of diagrams on the black board, Mr. Snell then explained the principle of the motor in its simplest form. By reference to a drawing of two bar magnets, one fixed and one free to rotate about an axis such that the two magnets lay in the same plane, he shewed the principle of attraction and demonstrated the method of commutating the axis of polarity so that a continuous motion was produced; and then describing an enlarged copy of Plate VIII. in the December number of the Transactions, said]: Instead of having a couple of bar magnets I have two electro-magnets, one called the field magnet being fixed, and the other known as the armature free to rotate. By means of the current flowing through the copper windings, on both field and armature, two axes of magnetism are produced at right angles to each other, N S and $n\ s$. This arrangement is really only a more complicated case of the simple drawing on the black board. The line of polarity on the armature $n\ s$ is determined by the position of the brushes by which the current enters and leaves the windings—a north pole is formed at one brush and a south pole at the other. When a current is caused to pass through the motor, the armature starts into motion by reason of the attraction between the two magnets. The axis $n\ s$ advances towards the axis N S by the width of one coil, and then the segments of the commutator in connection with this coil cease to touch the brushes and the axis $n\ s$ retreats the width of the coil. This is continually recurring, and so gives an even and regular motion. This rough illustration of the principles of the electric motor will perhaps help to familiarise you with a class of machine likely to come into general use at no distant date.

Mining engineers frequently ask me why we measure electrical quantities in volts, ampères, and watts rather than in foot-pounds. Well, the main reason is that the practical units, the *volt, ampère,* and *ohm* are so easy to measure and so simply connected by the equation

$$C = \frac{E}{R},$$

in which C = the ampère, the unit of current;
E = the volt, ,, ,, pressure;
R = the ohm, ,, ,, resistance.

Now, the product of one *ampère* and one *volt* = one *watt*, and 746 *watts* = one horse-power.

In the mechanical units 33,000 foot-pounds = one horse-power per minute; and if we are doing electrical work at the rate of 746 watts per minute, we are doing 33,000 foot-pounds per minute. The electrical unit of work is then related to the mechanical unit by the ratio $\frac{746}{33,000}$, or one *watt* is equal to 4·4 foot-pounds.

We could thus measure all electrical quantities in foot-pounds if it were desirable, but it is far more convenient to measure the volts and ampères and then estimate the horse-power. If anyone, however, wishes to express electrical quantities in foot-pounds, he will now be able to do so; but bear in mind that the electrical horse-power is equal to the mechanical brake horse-power.

The PRESIDENT: I am sure that we have had a most interesting discussion and one that we have all benefitted by. We have had an opportunity of exchanging our crude ideas, and also had the further opportunity of hearing a clear and explicit illustration of the nucleus from which we hope to get so much advantage either above or below. I have very great pleasure in proposing a vote of thanks to Mr. Snell for coming here, at no doubt considerable personal inconvenience, and giving us the benefit of the research of not only himself but of the very enterprising and able firm with which he is connected.

Mr. T. W. H. MITCHELL: As a junior member of this Institute I have enjoyed this lecture more than any we have had since I have joined. We have had a very instructive meeting. I am glad to see the friends who have come to-day, and I hope they will bring other friends in future and help on the work of the Institute. I have great pleasure in seconding the motion.

The resolution was carried.

Mr. SNELL: I have very much pleasure in thanking you for the kind manner in which you have received my name and the very poor attempt I have made to explain what is very difficult indeed. You must bear in mind that the most eminent scientists of the day are not satisfied as to what electricity is. Sir William Thompson has told us that he has thought over the subject 45 years, and he tried to tell us the other night, before the Institution of Electrical Engineers, what electricity was, but I do not think that he succeeded. He may know himself, but nobody outside himself does. The advantage for us practical men is that we don't want to know what it is but what it will do, and I hope at some future time to bring before you some results of what electricity can do.

The PRESIDENT: I think in view of this subject being so important it would be a pity to strike it off the agenda altogether. I think if we adjourn the discussion until we have had a further opportunity of seeing some of these machines at work; if we could arrange an excursion to some place where they have the pumps at work, and arrange for the Institution to go it might be of great advantage. I think we might arrange an excursion before next meeting to Messrs. Locke and Warrington's. If we could have it in the morning of the day of our next meeting at Leeds, I think it would give us food for discussion and thought, and lead to us having as interesting a meeting at Leeds as we have had to-day.

Mr. COOPER said it would be a great advantage if the papers could be gone through and the discussions taken on the days on which they were announced. It would be a great advantage to those who, like himself, could not get to every meeting.

The PRESIDENT said that was followed as far as practicable; discussions had not been deferred save when the writer of the paper was absent or some other unavoidable cause.

The meeting then ended.

MIDLAND INSTITUTE OF MINING, CIVIL, AND MECHANICAL ENGINEERS.

GENERAL MEETING.

HELD AT THE QUEEN'S HOTEL, LEEDS, ON TUESDAY, FEBRUARY 19TH, 1889.

C. E. RHODES, Esq., President, in the Chair.

VISIT TO ST. JOHN'S COLLIERY, NORMANTON.

Previous to the meeting a large number of members—by kind permission of Messrs. Locke & Co.—had the privilege of inspecting the Electrical Pumping Plant in operation at St. John's Colliery, Normanton. Mr. Snell, of the firm of Messrs. Immisch & Co., Electricians, London, and Mr. Brown, Mechanical Engineer at the Colliery, accompanied the members, and explained in detail the working of the machinery.

The minutes of the last meeting were read and confirmed.

Mr. JOHN McADOO, Colliery Manager, Ravensthorpe, was elected a member of the Institute, having been previously nominated.

Upwards of 50 members attended the meeting.

ADJOURNED DISCUSSION ON MR. G. BLAKE WALKER'S PAPER ON "ELECTRICITY AS A MOTIVE POWER, WITH SPECIAL REFERENCE TO ITS APPLICATION TO HAULAGE IN MINES," ALONG WITH MR. ALBION T. SNELL'S PAPER ON "AN ELECTRIC LOCOMOTIVE FOR MINES."

The PRESIDENT: The next business is the adjourned discussion on Mr. George Blake Walker's paper on "Electricity as a Motive Power." I am sorry to say Mr. Walker is not here. He has been called away to London, consequently we shall have to dispense with his attendance. Through the kind courtesy of Messrs. Locke and Warrington, a large number of our members have had an opportunity of seeing in operation to-day the pumps at St. John's Colliery, and I have no doubt they will be able to give some very interesting information with respect to them. We shall be glad to have the opinion of anybody who has seen the plant.

Mr. T. W. EMBLETON: I have had very little time to obtain particulars of this apparatus, but it seems to me from what I learnt there that it is far cheaper than using compressed air. The machinery itself will go 22 hours without any stoppage, and the only regulation required is that the quantity of electricity shall only be to a certain extent.

The PRESIDENT: Can some gentleman favour us with figures as to the approximate cost, the initial cost, and the cost afterwards of producing and maintaining the amount of power required to drive these pumps. Can you, Mr. Snell, give us any figures as to the cost of such a pump as we have seen to-day at St. John's Colliery, or of any pump to raise a gallon of water any height you like?

Mr. SNELL: Mr. Brown can give you the cost of the plant we have at Normanton, and it will come from him with greater authority than me, because I am to a certain extent a prejudiced person. He can speak from an *ex parte* point of view, and anything he may say will carry more weight than what I say.

Mr. BROWN: With regard to the electrical machinery at St. John's Colliery, it is twelve months this week since we started the plant, and during that time we have turned up the commutators upon the machine four times. The question of wages in turning up the commutators amounts to about 5s. per commutator. Then the next thing is the wear of the brushes. I find we have worn out five sets of brushes, which cost 25s. the set, which covers the cost of the brushes for the two machines. With regard to the life of the commutators, as far as I can see they will run another twelve months, making two years. Mr. Snell says he can put a new commutator in for £15, and that it will not take more than 24 hours to do it. The motor commutator will run $2\frac{1}{2}$ to 3 years, so that the cost for commutators will run to £5 a year, and the brushes about the same. With regard to attention, I do not think it takes more attention than a compressed air plant to do the same work.

The PRESIDENT: What sort of engine does it require to drive the plant?

Mr. BROWN: The present engine is 30 horse Robey, indicating 76 or 77 horse now.

The PRESIDENT: How much water are you pumping with it?

Mr. BROWN: There are from 117 to 120 gallons per minute, lifted about 900 feet high.

The President: What had you previously?

Mr. Brown: We had an air pump pumping about 40 gallons a minute, but when the feeder increased it became a question of extending the compressed air plant, or putting another plant down, and we decided to try electricity.

The President: What size was the pump?

Mr. Brown: The one we had down was a four inch pump, and delivered the water 172 yards out of the Haigh Moor seam to the Stanley Main.

The President: What was the size of the cylinder?

Mr. Brown: A 16 inch cylinder.

The President: How many yards did you lift 120 gallons?

Mr. Brown: The distance is 172 yards between the seams; we lifted it about 180 with the sump as well.

The President: And with the compressed air you were pumping the same height?

Mr. Brown: No, it was pumping 180, but the electrical is pumping 300 yards. We did that on account of the water being so salt. When it got to the Stanley Main it mixed with the water for the boilers, and we had to get it out of the mine.

The President: So you are pumping three times the quantity twice the height.

Mr. Brown: Not quite twice the height,—110 yards higher.

The President: You never calculated what size pump and what size cylinders it would have taken to have pumped it out by means of air?

Mr. Brown: I roughly calculated it last night. To pump the quantity we are doing now, we should require a 20 in. air cylinder in the bottom, and I do not think the compressing cylinder at the top would do less than 20 in. to keep the pump going in the bottom.

Mr. Snell: That is with a 3 ft. stroke?

The President: Have you made any calculation as to what amount of horse-power you require to drive this?

Mr. Brown: I have not.

The President: You are using practically an 80 I. HP. engine to get the work out of your electrical plant?

Mr. Brown: Yes.

The President: Would it have taken more or less horse-power than that to have got it by air or steam?

Mr. BROWN: I think it would have taken nearly twice that with air.

The PRESIDENT: There would be so much loss?

Mr. BROWN: Yes.

The PRESIDENT: You would not get more than 40 per cent. out of your air cylinder?

Mr. BROWN: We could not get 40 per cent. Taking the quantity of water delivered, we could not get 40 per cent. of the indicated power in the steam cylinder. With compressed air I do not think it would average more than 20 per cent. With the electrical plant we get more than 40 per cent.

The PRESIDENT: So the difference is in the useful effect you get?

Mr. BROWN: The useful effect, and beyond that I should think to put down a compressed air plant to do the same work we do would cost more money than the electrical plant has cost.

The PRESIDENT: I suppose you may have some objection to answer this question, if you have do not answer it—What was the cost? I have generally found that the estimates of the patentees and their engineers is a trifle under the mark.

Mr. BROWN: They gave us a tender to put the plant down, so it does not matter to us whether it paid them or not.

Mr. SNELL: We would like a good many at the same price.

Mr. GERRARD: I believe your compressing plant was much behind modern ideas, and not at all in good order, which only gave 20 per cent. of useful effect?

Mr. BROWN: It was in very bad order, but did not give 20 per cent. I estimated if it was put in good order it would not exceed 20 per cent.,—that is taking the indicated power of the steam cylinder in the compressors and the actual quantity of water delivered at the top, which, of course, covers all losses by friction, &c. We did not get 20 per cent, but it was not a fair test in consequence of the bad order of the compressors.

A MEMBER: But what was it when you made the test: was it under 10 per cent.?

Mr. BROWN: We got about 10 per cent., and with another compressor we got about 13 per cent.

Mr. PEARCE: And how much did you say with the electrical machine?

Mr. Brown: About 40 per cent.

The President: Have you, Mr. Brown, been able to compare hydraulic pumps with the electrical plant?

Mr. Brown: Roughly we have, but we have not tried any exhaustive experiments. There are so many corrections and allowances to make that it would be difficult to get accurate experiments. So far as I could tell, hydraulic power is better than compressed air, but not so good as the electrical plant.

The President: If your water was as bad as you have told us, it would militate against the success of the hydraulic plant, would it not?

Mr. Brown: Where we had the hydraulic plant was in another pit with better water.

The President: But in this case it would have been difficult to have applied the hydraulic plant with this very bad water?

Mr. Brown: Yes, it would.

Mr. Embleton: Could you have applied either hydraulic or air power where you have this electric pump?

Mr. Brown: We could.

Mr. Embleton: With great difficulty?

Mr. Brown: Yes, with great difficulty. We should require very large cylinders for the height that we have; and if we had got them down the pit we should have had the trouble of freezing.

Mr. Nevin: I have been much pleased with what I have seen at Messrs. Locke & Co.'s. I have not worked out any figures, but so far as I can see the thing seems a great success. As I have said before, the only objection I see to the use of electricity as a means of transmitting power is the spark. I do not think one would be safe in taking it far into a pit, especially one in which there were outbursts of gas, or where gas was given off. But certainly it seems to me from the evidence we have had to-day, that so far it is in advance of anything we have had for transmitting power long distances. I think there can be no doubt that it gives better results than compressed air; we cannot take steam long distances; and I cannot speak from my own knowledge of the power of taking power by ropes a great distance. The whole thing seems to work very nicely.

The PRESIDENT: At our last meeting, when this excursion was arranged, we very fully discussed the question of sparking, and one or two gentlemen present were of opinion that that was likely to be one of the main difficulties with regard to the application of electrical power in fiery mines far away in bye. Mr. Snell expressed an opinion that the sparks given off from the motor were innocuous practically, and doubted whether they would fire an explosive mixture. It was arranged that experiments should be made with a view to demonstrate this, and if anybody can give us further ideas on the subject we should be glad. It seems to me to be a very grave difficulty at present. When we think of the pains, trouble, and expense many people have had to do away with naked lights and explosives of all sorts, to find the very best kind of safety lamp, to lay dust, and do everything in the way of safety, it does seem to me an anomaly to introduce a machine, far away in bye in a fiery mine, that throws off a spark. I think we ought to be quite clear as to whether those sparks are innocuous or not. If any gentlemen can give us any information on that, it will be of great value.

Mr. HEDLEY: Has Mr. Snell made the experiments?

Mr. SNELL: I had expected the principal question to-day would be that of sparking, and had arranged with Mr. Rhodes to make experiments at Aldwarke Main, but before we had finished the electrical arrangements for making the test, I decided to make trials at my own place. I may tell you there is some difficulty in properly estimating the danger of sparking, for we all know that a spark will ignite gas under certain conditions; but myself, and the better part of the electrical profession are also of opinion that a great many sparks are innocuous, are inert in gas,—their temperature is so low they will not ignite gas. My opinion is borne out by what Mr. Gerrard said at the last meeting. He mentioned a case where they endeavoured to explode gas by means of sparks from signal wires, and they could not do it. Undoubtedly I could do it if I wished. You all know the little arrangement for lighting gas,—there is a very small spark, but it ignites gas with ease. But the mixture of gas there is not what we get in a coal pit,—it is far richer in marsh gas, which is the explosive element of coal pit gas,—and so, with a view of making these experiments accurate and having them based on scientific reasoning, I have rigged up a certain arrangement at our place. I have built a box about six feet long and ten inches square, inside measurement; I fixed a glass door to it at one side, and a wooden slide at one end to control the admission of gas. I have fitted a small motor in the middle of it, which I can arrange to spark to any degree. I can get such sparks as you saw at St. John's Colliery, or sparks such as are used in the little arrangement for igniting gas. In the one case I get an induction or extra spark of high tension, and on the other a spark of low tension. I have not finished these experiments, but so far they are hopeful. I had the other day an explosive mixture regulated by one of Ansell's gauges, and I had the gauge varying from a 5 per cent. to a 20 per cent. mixture, and had the motor running in it for half an hour, but could not explode the gas. I put in a candle and exploded it in a few minutes. That showed that under certain conditions the electric spark was safe. I am not prepared to state at present that it is safe. I should be loth to gloss over the difficulty and to ask a gentlemen to put an electric machine in bye, where there is danger of explosion happening. I feel so certain that electricity will come into use, that there is no occasion for hurrying the matter. It will come. It is cheaper, it is more efficient. Mr. Brown bears me out in that point too. I feel the only difficulty we have is that of sparking, and we mean to get over that. Indeed, at present there are in existence motors which run without sparking, because they have no brushes—the armature is in connection with the field only through the journals and bearings; and there is no current of electricity between the two; but these machines, unfortunately at present, are not so efficient as machines of the type you have seen to-day. At present we must put all electrical plant into the intake, where I think you will agree it is perfectly safe, and if we can give you a plant at 25 per cent. less cost than that obtaining with compressed air, and with an efficiency of 40 per cent. as against 20 per cent., you will see strong *prima facie* reasons for using electric plant. I am going now to the question of cost. If my memory serves me, the cost of the plant you saw at Normanton was £1,600. I have run out the cost of the compressed air plant for several cases to make a comparative test with electrical plant, and I find in nearly all cases it comes out something like 25 per cent. higher. In relation to this particular case, the compressed air plant would cost £2,100, and this had cost

£1,600. If they had an efficiency of 40 per cent. against 20 per cent., that means we shall use only half the coal. I know coal is not a serious item, but I think in the nineteenth century we ought to look at the question of fuel, and if you can get a certain result with 100 tons of coal, instead of 200 tons, I think most men would be inclined to use the plant which consumed least fuel.

Mr. PEARCE: It seems a very satisfactory result to attain 40 per cent. of useful work. Will you kindly explain how you arrive at it?

Mr. SNELL: We measure the water delivered at the surface, and the indicated power at the engines.

Mr. PEARCE: Did you take the weight of water lifted, and multiply it by the height?

Mr. SNELL: Yes.

Mr. PEARCE: Then the actual percentage would really be in excess of what you have given, considering the water would be driven some distance in pipes underground. There would be the power required to drive the water through the length of pipes and also to overcome the inertia of the water in the pipes.

The PRESIDENT: You have not made any allowance for friction?

Mr. SNELL: I have not.

Mr. PEARCE: Would you kindly give us some information through the Transactions as to the general arrangement of the pumps for getting the water to the surface.

Mr. SNELL: I have in course of preparation a practical paper which will give details, and if Messrs. Locke & Co. will allow me to give drawings and data of the installation at St. John's Colliery, I shall have pleasure in bringing it before the Institute. I may say we have half a dozen other plants of similar character, but involving various modications. I shall also bring them before the Institute.

Mr. PEARCE: I suppose from what has been said power is transmitted from engines on the surface, the electricity passes down the shaft and into the workings. Have you any difficulty underground in keeping the pumps at work?

Mr. BROWN: We have had no difficulty at all up to now. We have had one illustration how the thing could go wrong, that was when the hanger-on by some mis-management sent up the empty cage in such a way that it broke the cable, and a short circuit was made which fused the covering.

Mr. PEARCE: I suppose you would not have anyone to attend to the pumps in the night?

Mr. BROWN: We always have somebody to attend to them.

Mr. SNELL: In this case it is such a big plant that it pays to keep somebody there, but with a small plant it is not necessary. We have put a plant in Scotland where we are transmitting power through 4,000 yards of cable, and a five days' trial run there has come off perfectly successfully. The machines there are considerably in advance of those at St. John's Colliery. We have paid great attention to the question of sparking, and almost entirely done away with it. My assistant writes me, "There is absolutely no sparking," but I know there is a little because we cannot avoid it; but there is so little that practically it is none. Anyone seeing the new hauling plant, which we are about to put down at St. John's Colliery, after it has been running a short time, will see we have made considerable improvement in the matter of sparking. The machines tested in the factory gave 90 per cent. efficiency; they ran for twelve hours without their temperature being raised more than 40 degrees above the surrounding air, and the sparking was not half what we have at St. John's now, which is a considerable advance in the matter.

Mr. GERRARD: The attendance at this meeting shows the interest that is taken in this question of electricity as applicable to mining motive power, and I am sure those who have, by the kindness of Mr. Warrington, had the advantage of visiting the colliery and seeing the plant at work, will be very grateful to Mr. Warrington, to the managers, and to Mr. Snell, for the attention they have received. In discussing this application of electricity we have had it compared with compressed air. Whilst thinking that the statements made as to compressed air may not be quite fair, inasmuch as it is admitted the air compressing machinery was out of order and not of the most approved form, so as to ultilise only from 10 per cent. of useful effect; and passing the thought that steam might possibly have been used with even greater economy, being close to the shaft bottom, the important point to my mind is that Mr. Warrington has applied electricity under conditions not involving any danger, so that we are enabled to judge how far it is applicable under other conditions. The inventors have an opportunity of perfecting by improvements,

—devising checks or safe-guards,—which may enable us to extend the application to places where steam could not well be used. Again, we cannot help but discuss this question of sparks in connection with the application of electricity. At the last meeting I mentioned some experiments made with electric signalling, and these so far as they had then gone had not succeeded in firing an explosive mixture. But they have since, and something rather interesting in connection with it. There are two different forms of batteries—one was the Leclanché, a battery of twelve cells and a wire about one sixteenth of an inch, and a short circuit of about five yards. With this they got a spark a twelfth of an inch long, which did not fire an explosive mixture. They tried another battery, a Bunsen, with a wire of one thirtieth of an inch in diameter, and the same length of five yards. With this they got a flame of one-eighth of an inch in length which did fire gas. The wires were fixed one-eighth of an inch apart, so that you got fixed conditions which you would not get ordinarily, but it was done purposely to see if an explosion could be brought about. That leads me to what Mr. Snell has said—ordinarily with the machinery working as it ought to do you may have only a low temperature spark; but if something goes wrong you have a spark increased in intensity, and which might at any moment be a source of danger. We must not shut our eyes to the fact that the plant may get out of order, and that you may have irregularity in working which in mining operations must be taken into account. On ordinary occasions the spark may be harmless, but if something goes wrong, if there is an irregularity, you may have a dangerous flame. This should be thoroughly elucidated by experiments, or provided for and made impossible by further appliances. With regard to the £1,600 the plant is said to have cost, I should like to know more in detail with regard to that. Did the whole plant cost £1,600?

Mr. SNELL: It did.

Mr. GERRARD: Cables, motors, Robey engine, the whole thing?

Mr. SNELL: Yes.

The PRESIDENT: The Robey engine would make a great hole in that?

Mr. SNELL: Yes; £665.

Mr. BROWN: Yes, everything is included, pipes, cables, motors, Robey engine, &c.

Mr. ROUTLEDGE: Does it include the pumps?

Mr. BROWN: No; simply what was required for the electrical plant.

Mr. GERRARD: The Robey engine, the dynamo at the surface, the cables, the motor at the bottom and the pipes from the pump?

Mr. BROWN: The pipes were not taken into consideration, because they would be required for compressed air or any other system. Everything is in that would be required for an electrical plant, but things required for every plant are not put in.

The PRESIDENT: You include everything necessary to set the pumps going, but not such things as pipes?

Mr. BROWN: No; they would be required in any method of pumping.

The PRESIDENT: Of course, if you had compressed air you would want a steam engine to drive your machinery,

Mr. SNELL: But it would have to be a larger one.

The PRESIDENT: In any case you must have an engine whether you use compressed air or whatever it is.

Mr. SNELL: If the efficiency were double in the electrical plant, as we presume it to be, you would have to have an engine to indicate double the power in the case of compressed air.

The PRESIDENT: Yes; my point is in favour of your contention of economy—in any case you must have an engine, and you would have to have two engines for the compressed air, so that it is all in favour of the economy of this power.

Mr. SNELL: That is where the other £500 comes in; the cost of taking the compressed air or hydraulic power down would be greater than the cost of the cables.

The PRESIDENT: Five times I should think.

Mr. SNELL: This cable cost £165 per mile, and it was run down the shaft in six hours by three or four men.

Mr. PEARCE: What amount of power would the cable transmit?

Mr. SNELL: It is designed to carry 100 horse.

Mr. PEARCE: Suppose you wanted to apply underground haulage, you could raise the power for that purpose if you thought proper.

Mr. SNELL: Certainly.

Mr. PEARCE: Would the cost of cable for increasing the power underground increase very fast?

Mr. SNELL: No; cable to carry 200 horse-power would cost £200 per mile.

Mr. PEARCE: Have you to increase the cable to carry a further distance?

Mr. SNELL: If you wanted to carry the same power double the distance you would have to have double the cross section of cable, so that it would cost three times as much, or you would have to be content with twice the present loss.

The PRESIDENT: Is there a gauge by which the loss of power can be calculated—suppose we wanted to carry power three miles could you give us an estimate of the loss?

Mr. SNELL: The bigger the cable the less we lose per mile per horse power. If we take the current as there, 64 ampères, we lose $2\frac{1}{2}$ horse power in 1,000 yards.

The PRESIDENT: Then it is a question of losing $2\frac{1}{2}$ horse power down the shaft or spend £100 at first?

Mr. NASH: Does it not follow that if you have the cables cut you will have a flash in any case?

Mr. SNELL: Yes, a flash certainly.

Mr. GERRARD: There have been fires where electric lights are used by the severance of cables near wood.

Mr. J. E. CHAMBERS: We have had some experience in lighting in the erection of an installation at the pit top. By some means the two cables came in contact and there was sufficient spark from them to light the covering of the two cables. I understand there is some difference between high and low tension, and that you would not get sufficient heat from a low tension machine to light the gas.

Mr. SNELL: You might to fire the covering.

Mr. J. E. CHAMBERS: Ours were low tension machines but gave off sufficient spark to light the covering, and even to fuse the cable itself.

Mr. SNELL: I may say something relative to short circuits. The case this gentleman has mentioned is one of short circuiting. I must ask you to bear in mind a little equation I put on the black board at last meeting, that the current is equal to the electro-motive force divided by the resistance. That resistance is made up of the long line of cable and the motor as it runs. At normal we have 550 volts taken up to drive the motor. Imagine an accident causing a short circuit, in the shaft say, so that the entire load of the motor and the resistance due to the underground cables were reduced to something very small. Suppose that the resistance has fallen to one-hundredth of what it was, you get a current one hundred times as strong, and that will set fire to the covering. That means that our current of 64 ampères would rise to 6,400 ampères, or would try to do so. We have tried to meet that. We have arranged on the top a cut-out, which consists of two pieces of copper wire, No. 16, about four inches long. These have to carry the whole current. They will carry 64 ampères without heating, but if you had 200 or 300 ampères they would melt and cut off the circuit. They break the circuit, and the supply of electricity to the cable ceases; the man in the engine-house would see the speed of the engine rise because all work would be taken off, and danger would be at an end. The question of short circuits is a difficult one, and that is why we use lead-covered cable. If you broke a lead-covered cable you would get a bad spark, but if nothing happened then in the way of explosion the cables would fall apart, and directly they were apart all danger would cease because you cannot set fire to lead. If you have the cables in wooden casing you might have fire, and we have taken care nothing of that sort shall happen.

Mr. J. E. CHAMBERS: In this case we had no covering.

Mr. SNELL: Except the ordinary material in which the cable is enveloped, and which is inflammable.

Mr. J. E. CHAMBERS: Would there be any difference in the heat of the current coming in contact near the machine?

Mr. SNELL: The current would not be greater, but the distance would be less and the resistance would be less.

Mr. BROWN: We short-circuited the machine in trying to change an accumulator. As soon as the switch was turned on it pulled the engine up, but did not melt the fusible plug. I think it would be better if we had the fusible plug a little less, because we pulled the engine up; the engine could not drive the load.

Mr. SNELL: The reason your fusible plug did not go was that you had your connections on the dynamo side of your fusible plug.

Mr. BROWN: It was.

The PRESIDENT: If your cables were cut off by a fall of roof, would not a spark be given off?

Mr. SNELL: One spark as you broke the circuit.

The PRESIDENT: It would be so instantaneous that there would be little more danger than in the sparks given off when the motor is running?

Mr. SNELL: It would be simply an instantaneous spark, like a flash of lightening, and it would be all over.

The PRESIDENT: That would be very different to the wires coming in contact.

Mr. SNELL: Very different. One is an open circuit and the other a short circuit.

The PRESIDENT: With the open circuit would there be a spark of such virulence as in a short circuit?

Mr. SNELL: It would be worse. In either case it would be an awkward affair undoubtedly.

The PRESIDENT: The difficulty of the wires coming in contact with one another seems to me very easily got over, either down the shaft or when carrying cables in bye, because you could have them on either side of the shaft or road.

Mr. SNELL: A short circuit is a thing which never ought to happen in a well designed installation, but you cannot guard against the roof coming down, and I may say, by the same rule, you cannot guard against a fall of coal coming down and smashing the miner's lamp.

Mr. NASH: The only way to guard against it would be to have the cables buried in the floor instead of suspended from the roof.

The PRESIDENT: Then you would have rot.

Mr. GERRARD: You can say how your cables are, Mr. Blackburn— you are carrying cables some distance underground?

Mr. BLACKBURN: Our cable is on the ground covered over so that no harm can happen. We have had no misfortune yet. We have run them eighteen months and have had nothing wrong with them. So far as sparking on the dynamo is concerned we have an air-tight cover, and you can cover them so that you can find no sparks.

The PRESIDENT: I thought we understood you to say that it was impossible to have the motor in an air-tight cover?

Mr. SNELL: I should say it would be impossible to case a motor the size we have at Normanton in an air-tight case. It would be undesirable, but as I pointed out we could case in the commutator without any trouble. Perhaps that is what this gentleman refers to?

Mr. BLACKBURN: That is what I do refer to. We cover the commutator for this reason—the commutator is in bye and the dynamo on the surface.

Mr. ATKINSON (submitting photographs of coal cutting machine driven by electricity): The portraits will show how the commutator and armature are covered in. The armature revolves in a safety lamp. There is one covering of thick glass protected by wire, or simply a double thickness of the wire gauze used on ordinary safety lamps, so that should there be an explosive mixture this affords the same protection that a safety lamp does. In the case of an explosion inside the thing there is no fear of the flame coming outside.

The PRESIDENT: I doubt that very much.

Mr. ATKINSON: In some of them instead of a glass covering they have a wire gauze as on a safety lamp, and in that case the air can percolate through, but there is no difficulty experienced with these machines from the fact of their being closed in air-tight; there is no undue heating whatever. There is almost as much cooling surface on the metal covers as is sufficient to carry off any heat generated in the electric work of the machine, so that they run practically as cool as if they were uncovered.

Mr. SNELL: There is an apparent inconsistency in what Mr. Atkinson and I have stated, but we are both correct from our points of view. I have been speaking of a very large machine, but Mr. Atkinson is not speaking of a machine by any means so large. What is its size?

Mr. ATKINSON: Ten and a half horse power.

Mr. SNELL: We are comparing $10\frac{1}{2}$ with 50 horse power—this machine would probably weigh altogether a ton and a half.

Mr. ATKINSON: Twenty-five cwt.

Mr. SNELL: The machines I am speaking of would weigh five tons, and it would be impossible to protect one of such a size as you would a smaller one, but we can protect the commutator in the same way if necessary. If one of our big machines had to be put in bye we should have to use some arrangement, but I mentioned in a former meeting a better way, that was to put the motor in a practically air-tight case, and supply fresh air from the bottom, so that there would be always a current.

Mr. BLACKBURN: You would not expect sparks anywhere except at the switch box and the commutator?

Mr. SNELL: Certainly not.

Mr. GARFORTH: We have noticed that much better results have been got on a cold day or than when the engine-house windows were closed.

Mr. SNELL: I cannot explain that, and I cannot imagine there should be any difficulty. I am inclined to attribute it to the question of steam or the piston running slower on a hot day.

Mr. GARFORTH: We have had the idea of taking compressed air to the commutator.

Mr. SNELL: If a current of cold air was carried to the dynamo the whole machine would be cooler and you would get a higher voltage at the terminals. With a cold blast, thus keeping the temperature down, we should have greater efficiency on a hot day.

Mr. GARFORTH: Yes, ventilate your dynamo or motor in the same way as a working place?

Mr. SNELL: That is why I suggested compressed air for the air-tight case if possible.

Mr. GARFORTH: I have had the same idea.

The PRESIDENT: Mr. Holliday might tell us something of their machinery.

Mr. HOLLIDAY: I think I cannot say more than has been said. The machine is working satisfactorily and great credit is due to Immish & Co. I have not had the slightest trouble in the last twelve months. We hope by the end of next month to be working the engine plane by it and hauling by electricity. Of course the machine now is under different conditions to what it will be. We are now ventilating by a six-inch pipe, but when we get the new machine to work the place will be ventilated with fresh air, and it will be kept quite cool.

The PRESIDENT: I suppose if it was a long way in bye and absolutely necessary to enclose it in an air-tight case, that it would be possible to do it so as to avoid the possibility of sparking?

Mr. HOLLIDAY: I should think so.

The PRESIDENT: I heard Mr. Snell say that it is impossible to get them air-tight; I do not know whether that is right or not, I have **not sufficient experience to say.**

Mr. HOLLIDAY: You might get them air-tight, but the cooler they are kept the better without enclosing them.

Mr. BLACKBURN: Mr. Snell does not draw a strong objection against an air-tight covering. I think you said after looking at the photograph there would be no difficulty.

Mr. SNELL: Not at all, because it is so small.

The PRESIDENT: We are not dealing with one individual case but are discussing these machines as a whole. We are not dealing with the question of electricity in these machines, but whether it is possible to protect large machines in bye and make them innocuous, supposing there is a dangerous spark.

Mr. T. W. H. MITCHELL: The work Mr. Blackburn is speaking about is it the same work?

Mr. BLACKBURN: Proportionately.

Mr. T. W. H. MITCHELL: Because the machine we saw was working at the high temperature of 125 degrees, and Mr. Blackburn says his is not high even when quite enclosed.

Mr. ATKINSON: The temperature of a machine does not depend on the size of the machine, but on the amount of copper put on. If it is thought desirable to make a large machine to go in bye, and thought desirable to keep it covered, there would not be any difficulty in making a machine of that size to run absolutely cool. It is a mere question of expense.

The PRESIDENT: But if you have a large casing you would have to have the casing hermetically sealed or have a pipe.

Mr. ATKINSON: All that is necessary to prevent an explosion from the flame penetrating from the inside of the gauze to the outside of the gauze is that the bearing surface of the joint is of sufficient length to cool the flame before it penetrates. In a number of safety lamps the flame is prevented penetrating from the inside to the outside by having a gauze which takes the heat away before the flame comes through. If you hold a piece of gauze in a gas flame until it is hot the gas will ignite on the outside of it, but as long as that is cool enough to carry the heat away no flame can pass through. I fancy it was in the Stephenson lamp that he used a small piece of brass tube so that the gases were cool before passing outside. In this case it is only necessary to have the jointing surfaces fitted together so as to carry off the heat before passing outside.

The PRESIDENT: I do not agree with your deductions at all as to that.

Mr. HEDLEY: Can Mr. Blackburn give us the cubical contents of the case inside the hermetically sealed portion?

Mr. BLACKBURN: The air-tight box inside will be about 18 by 12.

Mr. HEDLEY: My experience is when you get a certain volume of gas inside a chamber and explode it, the force of the explosion forces itself through the gauze.

Mr. BLACKBURN: Our idea is to keep the gas out.

Mr. HEDLEY: You said you had gauze on.

Mr. BLACKBURN: Mr. Atkinson has been speaking of gauze; I am not of the same opinion.

The PRESIDENT: If you once got a case 12 ft. by 6 ft., and an explosive mixture was allowed to explode inside, there is no case that could stand it; it would be blown to pieces.

Mr. T. W. H. MITCHELL: Where does the heat go that is generated inside the casing? There must be some.

Mr. BLACKBURN: We have found no difficulty with the heat generated, where it goes to I am not able to say, only I know it keeps about one regular heat. It may pass off in the wire; it may be carried off by the current, I do not know.

Mr. ATKINSON: The casing is practically like a surface condenser, and if you only make the outside surface of the casing sufficient it will carry off any heat generated inside.

Mr. EMBLETON: I do not know much of electricity, but I am delighted with the discussion. I think from what Mr. Snell has said there will be no difficulty to get rid of such a spark as will ignite gas underground. There is only this thing strikes me—the incident mentioned by Mr. Brown at St. John's Colliery, that the cage broke one of the cables, and that a considerable quantity of lead covering was melted in consequence.

Mr. BROWN: Four inches.

Mr. EMBLETON: That heat would be sufficient to ignite gas. The same accident might take place underground where there is gas, so that there will have to be some provision made for the protection of the cables. Suppose Mr. Snell is able to carry out the question of sparks safely, what is to be done when the cable is broken? The heat in such a case would be sufficient to ignite the gas, and the cables should be protected in some way, either by burying them underground or in some other way, so that neither of the cables will be affected by any fall of roof or any fall of coal. That seems to be the only danger. No doubt Mr. Snell will be able to carry out the other part successfully, and this is the only danger to my mind that can occur to the cable.

The PRESIDENT: I think if cables carried on a long way in bye were encased in wrought-iron pipes it would protect them and prevent any danger of being cut by falls of roof. It would not, in some mines, be safe to have them buried in the floor, as you would, with a lifting floor, have the liability to have them broken.

Mr. BLACKBURN: What heat could be got from the current in the cable?

Mr. SNELL: From the direct current none at all.

Mr. THIRKELL: How long does Mr. Snell consider a cable would last in the shaft, because we had an electric cable in the shaft and it did not last long? It was covered with lead.

Mr. EMBLETON: Would not that depend on the quality of the water passing down the shaft dissolving the lead?

Mr. SNELL: I think it would if there were any acids present.

Mr. THIRKELL: The water coming down was coming from the Melton Field seam. The cable in question was laid down the upcast pit, and the gases coming up there might have some slight effect upon it. In this case the cables lasted only some 18 months, and we had to take them out altogether and put them down the downcast pit. If Mr. Snell could give us an idea how long he estimates a cable would last in a wet shaft it would be of great use to several?

Mr. SNELL: We have had little experience of the time that a cable will last, but from what I have seen there is no reason why they should not last ten, twenty or thirty years. If you have got any chemical fumes which will form any acids which will decay the lead I will not say how long they would last, but you can paint them or cover them with tar or pitch, and I believe they will wear very well then. There are a good many ways of getting over the difficulty; it is only a question of putting it in the hands of men who have had experience in the matter.

Mr. NASH: In the case of putting them in pipes, would it not be better to put them in wood pipes, so that in case of any leakage, the wood, not being a ready conductor of electricity, would not conduct so much of the power away as if it was an iron pipe?

The PRESIDENT: It might set it on fire.

Mr. SNELL: If the cable was properly put in it would be safe whether in iron or wood.

Mr. PEARCE: In the application you have made of electricity for underground pumping, have you applied it to electric lighting at the same time?

Mr. SNELL: Yes; at St. John's they are coupled in parallel with the motor.

The PRESIDENT: I think we have had a most interesting discussion, and one which I am sure we have all enjoyed. I only hope that when the electric hauling engines are started Mr. Warrington will add another to the favours he has already conferred upon us, and allow the members to see them at work. This question of electricity is one of such tremendous importance that I think it is our duty as it will be our pleasure thoroughly to thrash it out. I am pleased to see the photograph of the electric coal-getter. The application of electricity to getting coal is a new departure, and one of great interest to us. I think the large number of people here to-day ought to be at any rate gratifying to Mr. Snell and those gentlemen who have explained to us so fully the power which is now under our notice. I feel sure if we have meetings like this we shall benefit not only from observation, but from interchange of views. We are all here to learn, for so far as electricity goes, I think the mining engineering profession as a rule are somewhat ignorant, practically wholly ignorant of it except as applied to lighting. I should like, although I have not visited the colliery to propose the best thanks of this Institute be accorded to Mr. Warrington and his staff for their kindness in allowing the members to go there, and in coming here to give us the valuable information we have had; and I should also propose this discussion be adjourned again so that we may have another opportunity of going into the question.

Mr. EMBLETON: I have great pleasure in seconding your resolution. I think an adjournment is quite necessary. I am glad to find that Mr. Warrington allows mining engineers to go and see what he has done that they may profit thereby.

The resolution was carried.

Mr. WARRINGTON, junr.: I must thank you for the kind way in which you have mentioned the few words in our favour. We were glad to see you at the colliery, and shall be glad to show you the new hauling plant when it is put down. Electricity has given us every satisfaction and I hope others will follow our example and use it.

Mr. EMBLETON: I did not know much about electricity, but I am much pleased, and have learnt a great deal that I did not know before I entered this room. I think we are altogether indebted to Mr. Snell for the information he has given us, and I have pleasure in proposing that the best thanks of the meeting be given to him.

Mr. PEARCE: I am very glad to second that.

The resolution was carried.

Mr. SNELL: I thank you for the kind way in which you have received my name. It is a pleasure to me to give information, because I have a *bonâ fide* love for the profession.

Mr. GREAVES proposed a vote of thanks to the President.

Mr. GARFORTH seconded the motion.

The motion was carried, and the President acknowledging the vote the meeting closed.

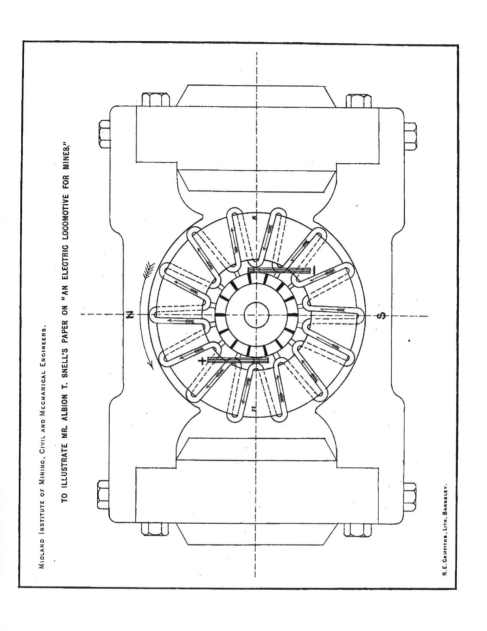

MIDLAND INSTITU'

SKE

T

N SEA

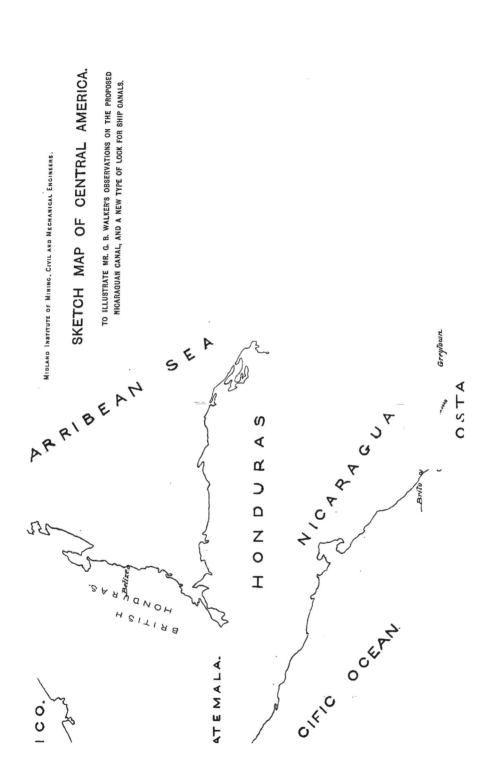

Plate II.

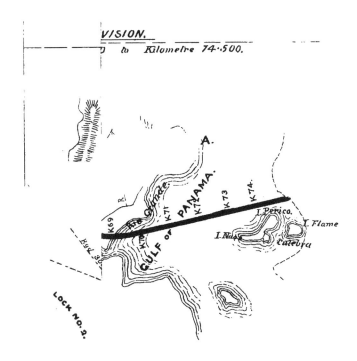

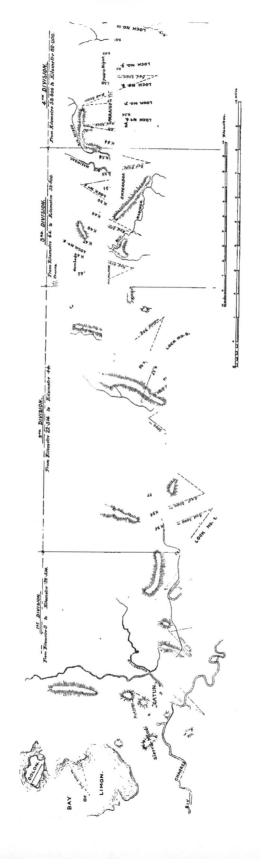

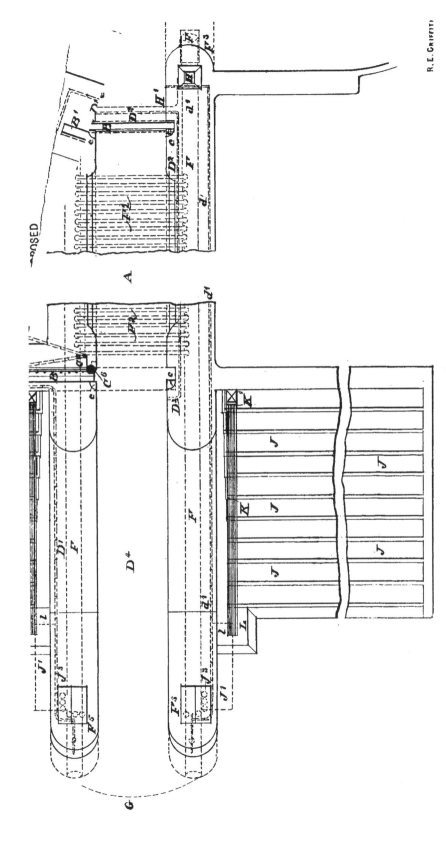

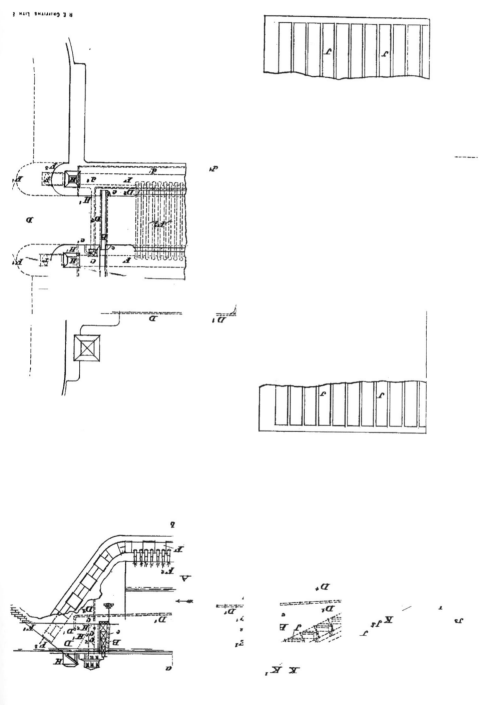

TO ILLUSTRATE MR. G. B. WALKER'S OBSERVATIONS ON THE PROPOSED NICARAGUAN CANAL, AND A NEW TYPE OF LOCK FOR SHIP CANALS

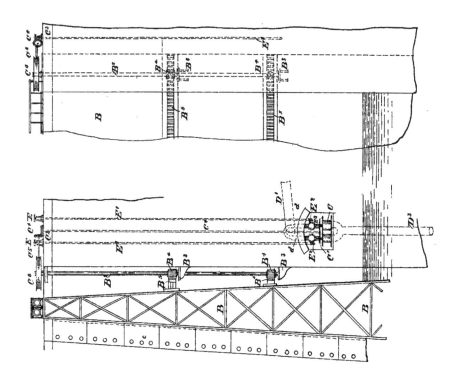

Plate V.

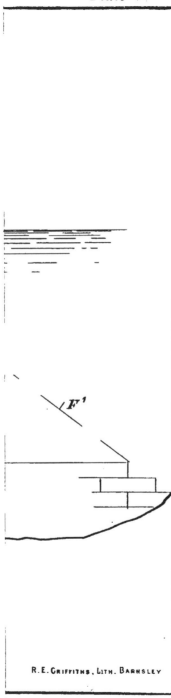

R.E. GRIFFITHS, LITH. BARNSLEY

TO ILLUSTRATE MR. G. B. WALKER'S OBSERVATIONS ON THE PROPOSED NICARAGUAN CANAL, AND A NEW TYPE OF LOCK FOR SHIP CANALS.

MIDLAND INSTITUTE OF MINING, CIVIL, AND
MECHANICAL ENGINEERS.

GENERAL MEETING.

HELD AT THE INSTITUTE ROOMS, BARNSLEY, ON WEDNESDAY, MARCH 20TH, 1889.

C. E. RHODES, Esq., President, in the Chair.

The minutes of the last meeting were read and confirmed.

ADJOURNED DISCUSSION ON MR. G. BLAKE WALKER'S PAPER ON "ELECTRICITY AS A MOTIVE POWER, WITH SPECIAL REFERENCE TO ITS APPLICATION TO HAULAGE IN MINES," ALONG WITH MR. ALBION T. SNELL'S PAPER ON "AN ELECTRIC LOCOMOTIVE FOR MINES."

The PRESIDENT: The business is the adjourned discussion on Mr. George Blake Walker's paper on "Electricity as a Motive Power." You have all, I think, had in your hands the Transactions since our last meeting at Leeds, when a somewhat lengthy, and to my mind, very interesting discussion took place. It was thought then that if the discussion was printed and circulated, it might be the means of starting another, probably equally interesting. The subject is of the greatest interest to everyone connected with colliery operations; the papers read and the observations thereon have added to our knowledge and information, with reference to the application of electricity for haulage and pumping purposes.

Mr. G. B. WALKER: I was unfortunately unable to be at last meeting, but I see a statement on page 365 of the Transactions, where Mr. Snell stated that he was running a motor in a box with an inflammable mixture for the purpose of testing the dangers of sparking, and he adds "I have not finished these experiments, but they are hopeful." Did it transpire what that hopefulness consisted in,—to what extent we are justified in supposing that the sparking was innocuous in an explosive mixture.

Mr. W. HOOLE CHAMBERS: I think the next sentence shows that, when he says that he had the motor running half an hour without exploding the gas.

Mr. G. B. WALKER: Was it gas mixed to a fine explosive point?

The PRESIDENT: He says that a candle would explode it in a few minutes.

Mr. G. B. WALKER: A candle should explode it instantly.

The PRESIDENT: I asked him to let me have a motor to test in the apparatus in which we test safety lamps. If it would have stood a test of that sort it would have been a great point. We should have seen where we were. Hitherto I have not had an opportunity. What his own arrangement is I do not know, but it is impossible in my opinion to get a mixture that is reliable without you have some velocity to mix the gas and air together, because to mix it by diffusion alone is unsatisfactory.

Mr. COBBOLD: I suppose that he did not mean to say that there was any difference between the electric spark of his motor and one of the gas lighter.

The PRESIDENT: Yes; that is his point—one is high tension, the other low tension. He says the sparks given off in the ordinary way from the machine when running are high tension sparks, but the sparks given off when the cable is cut or that is given off from these electric lights, are low tension sparks.

Mr. NASH: You have just reversed them. He (Mr. Snell) says the sparks given off from the motor are low tension, but those from the electric gas lighter are from an induced current, and of high tension.

The PRESIDENT: It was pointed out by Mr. Chambers, of Darfield Main, that when a piece of stone or something fell down the shaft and cut the cable in two, the spark, or heat, or flame given off was of such virulence as practically to melt everything round it.

Mr. COBBOLD: Yes, that would set anything on fire; and melt an ordinary steel table knife.

The PRESIDENT: I do not attach much importance to the risk of the wires being cut, because I think it is possible to have the wires so far apart that they would be innocuous.

Mr. NASH: That does not make any difference. If one of them gets cut, and you have the dynamo running you have the spark as though both were cut.

The PRESIDENT: There is no reason why they should not be covered so that they could not be cut. If they were far apart you would have no risk of them coming together and giving off a spark.

Mr. McMURTRIE: That was what did the damage—they were short circuited. If they were apart that would not have happened.

Mr. T. W. H. MITCHELL: Mr. Snell says the low tension spark would not light the gas, though it might fire the covering.

Mr. NASH: I do not see that Mr. Snell gave us any explanation in case the pumps were not required to do the whole of the work for which the dynamo was arranged with a 66 ampère current, if there was only 40 ampères required, what was he to do with the other 20 ampères.

Mr. T. W. H. MITCHELL: Pass it through the resistance coil.

Mr. NASH: That would be getting 30 per cent. out of his engine instead of 40 per cent. If he has got no mode of gathering that extra current with the changing weight of hauling engines, there would be great loss.

Mr. McMURTRIE: There would be by the man altering the speed of the dynamo.

Mr. NASH: Yes; but you have the dynamo at the top, and you have not sufficient power for your whole work unless the dynamo is running at full speed.

Mr. McMURTRIE: If you only want less work you get it by driving at less speed.

Mr. NASH: Yes; but with an endless rope you sometimes have ten and sometimes twenty corves on; you want greater and less power, but you have always the full power being given out at the dynamo.

Mr. W. HENRY CHAMBERS: That would be regulated by the governors on the engine. Mr. Snell said if the wires were short circuited they would know at once, as the engine would run away.

Mr. NASH: That is when you take the whole of the power off, but when you take part of the power off where is your load going to be. It is all right if you take all the power off, but suppose you are running with 60 ampères, and only require 40, if you have not a break or something on the motor to run it down you will have to go any speed your engineman at the surface likes. Suppose a run got off the road and at once stopped, what are you going to do?

Mr. T. W. H. MITCHELL: The governors would do that.

Mr. W. HENRY CHAMBERS: I had misunderstood Mr. Nash. I thought he referred to the engine at the top, running the motor at such a speed that it would be run to waste, and I did not think that would be the case. Now, I understand Mr. Nash to say it is the engine down below that he wishes to regulate the speed of?

Mr. NASH: Yes. The dynamo keeps it at one regular speed with the pumps, but if it had a varying load you would have to vary the motor down below, and he has no power or command to do that without rapping to the engine man at the top to go at a different speed, whereas if a run was off the road all the damage would be done.

Mr. T. W. H. MITCHELL: He explained that he had that power by means of these coils.

The PRESIDENT: There is no more danger of the engine running away than with an ordinary hauling engine. If you get the load broken with an ordinary engine, you have to rap to the engine man to slacken his engine. I do not say the man with the motor can do it, but the man on the incline can.

Mr. NASH: I understood that he communicated with the man at the bottom, who would have to communicate with the man at the top.

The PRESIDENT: I understood you would communicate with the surface at the same time as the other, but that is a matter of detail which would not take much getting over.

Mr. G. B. WALKER: I am afraid I rather agree with Mr. Nash. I think there will be during part of the day a considerable amount of electricity wasted through the resistance coils. I do not quite see how that is to be avoided, unless the flow of electricity can be checked, and that can only be checked, I fancy, from the originating dynamo.

Mr. NASH: That is what I think.

Mr. G. B. WALKER: Of course if you have accumulators, by means of a switch you can regulate the flow; the resistance coils hinder the flow; but when you have the dynamo absolutely creating a fixed amount of electricity, you must either stop the flow by reducing the speed of the engines, or let it be wasted by passing it through the resistance coils.

Mr. NASH: The question is whether accumulators can be so arranged that in any case when the full power is not being used by motors it could be charging the accumulators, and they be used for some other purpose. Whether if you only needed 40 instead of 60 ampères, that the accumulators could be used at any time, to store the other 20 ampères, so that when it had run into them it could be used at some other point.

Mr. COBBOLD : I agree with Mr. Nash. I suppose it is the same working an engine as working so many lights. In our own installation we have 600 lights, and they are divided into circuits of 200 each. If someone turns off a circuit of 200 lights, the indicator of the engine immediately jumps up to something beyond what it should be. The only way the man at the engine can reduce that and save his wires being fused, is to alter the slide of his resistance coil. Then I understand that he is wasting power. In order to avoid that, for he knows they are not going to turn 200 lights off for fun—he goes to the engine and turns off the valve a bit. But with a hauling engine with a varying load it would be difficult for him to regulate the engine without losing some power.

Mr. NASH : That is where I see the risk of great loss unless you can alter the power with the varying load. If you require at any portion of the journey the full ampère current, you must have the full current on all through such journey, or otherwise you cannot get the power when required.

The PRESIDENT : Take the ordinary hauling engine run by steam or compressed air. You run your engine to bring a certain load, and the steam necessary to do the work is thrown on the engine : if a portion of the load is lost through any cause, the engine runs away and has to be steadied by the engineman. With the endless rope you put on as many corves as the engine is capable of drawing with a given amount of steam, and as the load varies the steam is varied to suit. I assume that, though electricity be the direct motive power, still the engine producing it would be the means by which the power could be controlled.

Mr. MCMURTRIE : If the gradient altered, in that case you would lose some power.

The PRESIDENT : You do not alter the steam engine for every varying incline ; all she does is to go faster at one place than the other.

Mr. HEDLEY: Have they any means of communicating with the engine house?

The PRESIDENT: Yes.

Mr. HEDLEY: Then the man at the top has the power to regulate his engine accordingly.

Mr. NASH: He cannot handle that as he can steam; the man at the bottom has it to handle through the motor.

The PRESIDENT: If anybody could use steam direct they would be foolish to use electricity. There may be some points against electricity, but to my mind, if sparking could be got over, the fact of being able to take electricity anywhere you like, overcomes an enormous amount of objection of the character raised.

Mr. NASH: The only question with me is to get the motor so fitted as to have ready command over it, but here you only appear to have command of the engine on the top. You do not seem to have command of the motor to run it at any speed you like without wasting a lot of electricity, which means waste of money and waste of effective power.

The PRESIDENT: I hope at the next meeting Mr. Snell may be able to give us some explanation of what they may have in their minds for obviating that. I have no doubt they will have thought of it. But assuming for all practical purposes—a haulage system requires a 30 inch steam cylinder, and the engine is constantly running: something goes wrong in the pit, and your corves get off the road, and suppose this current of electricity is directed into these other channels, I do not see myself that that would tell particularly against the adoption of electricity. Although you might be wasting that power for the time being, you must be wasting it when you do not want it. You would get the useful effect of your engine when it was running on the road and working its load, and whilst the power would be wasted when the engine was not doing any work, the same thing would apply to a steam engine.

Mr. G. B. WALKER: Steam which would otherwise be blowing out of the safety valve.

The PRESIDENT: You cannot alter the power every ten or twelve minutes. If the power was lost, it would not be lost so far as the useful effect went.

Mr. G. B. WALKER: There is no doubt if you are off the road that the load man at the bottom of the pit would ring to the man at the top, who would stop until the electricity was wanted again.

The PRESIDENT: That would be my opinion. If some power was wasted it would not matter so much, unless they wanted the waste power for something else. If it could be spared to use for something else it would be a very good thing; but it would have to be a very vague kind of work if waste power, which they never knew when they were going to have for working, could be used.

Mr. COBBOLD: Mr. Snell says here that he is fitting up an improved haulage plant at the same place, so that he will be able to tell us what use he could make of it.

The PRESIDENT: They have been kind enough to give us a general invitation to go there as soon as it is running. They will have it running for the meeting at Leeds in June, and we shall see the hauling engine at work. Another invitation is to see some coal cutting machine run by electricity, which they hope to have ready at Bowers' Collieries before the same meeting.

Mr. G. B. WALKER: On the kindred subject of lighting mines by the electric light, I believe that up to the present time nothing has been done beyond having some incandescent lamps placed in pit porches and about the top. Portable safety lamps have only, I think, been used in isolated cases, perhaps one or two at a particular colliery, but I have been able to arrange with a company which has been formed for supplying electric mining lamps, to send down 50 to Wharncliffe Silkstone Collieries for a prolonged practical trial, practically at the expense of the company, and we shall be able to put the lamp to such a test as I think it has not yet had. When these lamps are in use, as soon as the little preliminary difficulties are got over, if there are any, and members of the Institute who like to come and see them used, and see how they conform to the necessities of mining, will be very welcome to do so.

Mr. NASH: Is there any practical arrangement upon them whereby the presence of gas can be tested?

Mr. G. B. WALKER: No; for that you must depend on the present safety lamp. The problem I am anxious to solve is the mere lighting question, apart from the detection of gas.

Mr. NASH: They are not much practical use in any case, unless you have a testing apparatus for gas with them, for your men might be suffocated by an outburst, and have plenty of light.

Mr. McMURTRIE: Would it be possible to have a drawing showing the gearing by which the locomotive is driven, as on page 335?

Mr. COBBOLD: I beg to propose that the discussion of the paper be adjourned to next meeting, in the hope that Mr. Snell will be able to answer the questions.

Mr. McMURTRIE: I have great pleasure in seconding it.

Mr. G. J. KELL: I have only read the paper very hurriedly, but I see very important points that want discussing. I think it is not a thing which should be hurriedly gone over, and every discussion should be encouraged. I certainly think we ought to adjourn the discussion until we have an opportunity of hearing Mr. Snell's reply to the various questions that have been put to-day.

The discussion was then adjourned.

MIDLAND INSTITUTE OF MINING, CIVIL, AND MECHANICAL ENGINEERS.

GENERAL MEETING.

HELD AT THE VICTORIA HOTEL, SHEFFIELD, ON TUESDAY, APRIL 16TH, 1889.

C. E. RHODES, Esq., President, in the Chair.

The minutes of the last meeting were read and confirmed.

FEDERATION OF THE MINING INSTITUTES.

The PRESIDENT: The first business we have is this scheme of federation of the various Mining Institutes. It is a matter which the members of the North of England Institute, and one or two other Institutes, have given a considerable amount of time to; and there is no doubt a good deal to be said in favour of the scheme if it can be carried out. All that has been done has been placed before the various members of the Institute. There is nothing really new that I can say on the subject, but I should like to hear any expression of opinion from those present with regard to it.

Mr. NEVIN: As I read the proposals they are that each federated Institute pay the sum of 15s. per associated member per annum to the general fund. How will that affect this Institute with a guinea subscription. It will leave only 6s. per member—will the subscription have to be raised?

Mr. G. B. WALKER: I was present when the subject was discussed. It was suggested that as our subscription was so very small, those members who became federated members should pay 5s. a year more than they do now.

The PRESIDENT: A number of us would be federated members on account of belonging to other Institutes where there is not this to pay.

Mr. G. B. WALKER: Then you would cease to be members of those other Institutes, and get all their publications through the Federation.

Mr. NEVIN: I suppose we should have to choose which Institute we should elect to belong to?

Mr. G. B. WALKER: That question was only discussed in a very general way. There will, no doubt, be a good deal of difficulty with it sooner or later.

The PRESIDENT: It is a point which would have to be considered by the members of this Institute. It almost seems to me an anomaly that a federation scheme of this sort should be carried out as affecting the whole Institute, and then individual members have to pay an extra subscription in order to entitle them to belong to it. As the Midland Institute, we ought all to belong to it, or all be out.

Mr. G. B. WALKER: Not necessarily so. You see there is a provision for those who do not wish to join. We shall only contribute 10s. 6d. in respect of such members as do not wish to become members of the federated Institute, and who, therefore, receive nothing but the Midland Institute Transactions.

The PRESIDENT: We should be 10s. 6d. out of pocket by every member who did not belong to it.

Mr. G. B. WALKER: It is hoped that all members would belong to it, as they would get so much more value for their money by getting the Transactions of the federated Institute, but if they prefer to confine themselves to their own local Institute they can do so, and retain all the advantages they possess at present.

The PRESIDENT: It simply means an increased subscription of 5s. Our subscription at present is barely sufficient to make ends meet and tie, but we should reduce expenditure somewhat.

Mr. G. B. WALKER: Our publishing expenses would cease.

The PRESIDENT: If they ceased our good as an Institute would cease. It must cease if we are simply going to fall back on the federated Institute.

Mr. G. B. WALKER: Why? We go on as at present and they publish for us in exchange for this contribution.

The PRESIDENT: We shall have to publish our Transactions.

Mr G. B. WALKER: No; they do it all for the money. The question is whether the 6s. is enough to pay the other expenses apart from publishing. The proposal is to publish in two forms, the important papers and the unimportant papers and discussions. The Institutes will receive a copy of the complete Transactions for every member who joins the federated body; and they will also receive their own proceedings which will be extracted from the larger volume, to send to those members who do not join the federated Institution. They will probably be able to do the printing more cheaply having it all together than can be done at present, when it is done in several different places.

Mr. NEVIN: Then really it will leave us 10s. 6d. per member, as we are to pay 10s. 6d. per member, and our subscription is a guinea, and the associated members will pay 5s. extra.

Mr. G. B. WALKER: That is not exactly correct. Each Institute is to pay 15s. in respect of its federated members and 10s. 6d. in respect of its non-federated members—the first class receiving the publications of five Institutes, the latter class receiving only the publications of the Institute to which they belong.

The PRESIDENT: What does a man who belongs to this Institute and does not wish to join the federated Institute, get for his guinea?

Mr. G. B. WALKER: Just what he does now.

The PRESIDENT: They would print and bind our Transactions separately as we do now; or should we have to do that?

Mr. G. B. WALKER: They will extract our papers and discussions and send us as many copies as we have members who do not join the federated Institute.

Mr. HABERSHON: A non-federated member would have no copy of the Transactions of the other Institutes? He would simply have our own Transactions bound altogether?

Mr. G. B. WALKER: He could no doubt obtain any particular paper by paying so much extra for a special number.

Mr. NEVIN: He might separate them in binding as they do the law reports.

The PRESIDENT: Suppose a man belongs to three Institutes, what will he do?

Mr. G. B. WALKER: He will cease to do so.

The PRESIDENT: Suppose he now pays his three guineas to the North of England Institute?

Mr. G. B. WALKER: They will be the losers. A member of the Midland Institute will have nothing to gain from belonging to the North of England Institute.

The PRESIDENT: If I am to get the North of England Transactions, the Midland Transactions, and three or four others, as a federated member, what reason have I to go on paying to the North of England?

Mr. G. B. WALKER: The North of England Institute expect a loss by reason of that. It will be well if gentlemen who do not now belong to the Midland Institute do not join us in order to get the benefit of our cheaper subscription.

The PRESIDENT: In the event of federation, all subscriptions to each Institute should be made equal, and then allow each so much for working its own Transactions and expenses.

Mr. NEVIN: I think that will have to be done ultimately.

The PRESIDENT: It seems to me we have 65 to 30 the best of it; I suppose all that has been thoroughly discussed at your meeting.

Mr. G. B. WALKER: I think it was pretty well thrashed out, but the conclusions we came to are all more or less tentative; but they seemed to be the best basis it was possible to arrive at for a start. No doubt the thing will be modified as time goes on.

Mr. NEVIN: Have these recommendations of the joint committee to be approved by the separate Institutes?

Mr. G. B. WALKER: A meeting should be called at which the members may discuss what has been done, and, if they approve, endorse it.

Mr. NEVIN: The question of the extra 5s. will have to be taken at the annual meeting.

The PRESIDENT: I think it would be advisable to move this be sent to every one of the members, and that notice be given to them that the question of this federation will be discussed specially at the next meeting at Barnsley.

It was then resolved that in the notice of the next meeting at Barnsley, it should be announced that the proposed scheme of federation, as suggested by the joint committee of the various Institutes, will be specially brought under discussion; and that the Council desire to point out that in the case of members joining the federation there would only be an available margin of 6s. for the ordinary expenses of the Institute; and this being insufficient, it seems that in the case of members joining the federation, the subscription must be increased by 5s., or such other sum as may be decided upon, and members joining the federation will get the benefit of the Transactions of all the other Institutes in the federation, which will give them an amount of information, and opportunities of acquiring it, which are entirely out of their reach at present.

ADJOURNED DISCUSSION "ON MR. G, BLAKE WALKER'S PAPER ON "ELECTRICTY AS A MOTIVE POWER, WITH SPECIAL REFERENCE TO ITS APPLICATION TO HAULAGE IN MINES," ALONG WITH MR. ALBION T. SNELL'S PAPER ON "AN ELECTRIC LOCOMOTIVE FOR MINES."

The PRESIDENT : The next business is the adjourned discussion on Mr. G. Blake Walker's paper on "Electricity as a Motive Power," along with Mr. Albion T. Snell's paper on "An Electric Locomotive for Mines." We have had two meetings now, and the discussions have been pretty prolonged on this question. I should be glad to know if anybody has any new information to give us, or any new ideas to impart, as it is a subject of such great interest.

Mr. NEVIN : Are you using the locomotive now, Mr. Walker ?

Mr. G. B. WALKER : No ; it is away in London. Experiments are being made with various classes of accumulators.

The discussion was again adjourned.

Mr. G. B. WALKER : I am very sorry there is no paper down for reading to-day. The Institute will never flourish unless members will take the trouble to contribute papers. I am sure there is no lack either of subjects or ability, if only more members would take the trouble to do a little in the way of papers. I am really ashamed of being so constantly before the Institute as I have been lately. However, rather than those members who have come here to-day should have their trouble for nothing, I have brought with me a few notes on a subject which I think is of some interest, and if you care to hear a few remarks *viva voce* I shall be glad to give them.

The PRESIDENT : I am sure we shall be very much obliged to Mr. Walker if he will favour the Institute with his communication.

Mr. NEVIN : I am sure we shall listen to it with very great pleasure.

ON THE PROPOSED NICARAGUAN CANAL, AND A NEW TYPE OF LOCK FOR SHIP CANALS.

Mr. G. B. WALKER: During the last few months the question of constructing a ship canal to place Sheffield and the South Yorkshire coal-field in connection with the Humber has been attracting a good deal of attention, and the possibility of making such a canal at a reasonable cost has been a good deal debated and called in question. It may, therefore, be interesting to know what are the developments which have taken place in the course of the last two or three years with a view to solving the great problem of connecting the Atlantic and Pacific Oceans by means of canals or railways across the Isthmus of Central America. A glance at the map shows that there are four points at which Central America becomes exceedingly narrow, and, fortunately, at two of those points the narrowness of the isthmus is also accompanied by a very considerable reduction in the height of the chain of mountains—the Andes. Schemes have been proposed in connection with all these narrow portions. Of course, the first and greatest is the Panama Canal, the construction of which has lately been stopped owing to the immense sum which it has cost and the great difficulties which have been encountered. You will see from the map of Central America that in the north— the lower extremity of the Gulf of Mexico—there is an isthmus about 100 miles wide which is called the Isthmus of Tehuantepec. The elevation of the highest point across the Isthmus of Tehuantepec is only 600 feet above the sea level. It is here that Captain Eads proposed to establish a ship railway; an essential feature of the scheme being that the ships are bodily lifted by hydraulic process on to travelling cars and transported across the isthmus on rails. The heaviest gradient proposed is 1 in 120, the greater part of the line being level.

The Panama Canal, which has been partially cut, and, for the present, at any rate, abandoned, is situated at the shortest distance across the isthmus. It is about 40 miles in length, and traverses in part the lower valley of the river Chagres, along which, for some 16 miles, an easy route can be obtained without locks. The idea originally was to cut through the mountain range without locks at all, but in the later scheme, with locks, there is a distance of 16 miles in which only one lock occurs. Then the rise over the Andes is very rapid indeed, and altogether there are 11 locks, of which several are very close together. The excavation actually done is only a small portion comparatively of that which is required, as the accompanying diagram will show. The amount still necessary to complete the Panama Canal is said to be some fifty millions, and in spite of the large sum already expended it is doubtful if it will ever be made.

The Panama Canal is divided into five sections, commencing from the Atlantic end of the work. The first section extends for the first 14 miles. On January 1st, 1888, there remained to be excavated a total of 8,626,000 cube metres, of which 4,350,000 metres had been removed during the first seven months of the year; if the work proceeded at the same rate, all the excavation on this section would probably have been finished by this time. From Colon to the 11th mile the depth attained below the sea level varied on August 25th last from 14 feet to 28 feet; width on water level 131 feet, and on the bottom 72 feet. From the 11th to the 14th mile the work was down to sea level. On this section the deviation works for the Chagres river were complete for a bottom width of 114 feet to a depth of 13 feet, except for a length of about half a mile. The heaviest part of the harbour works at Colon were completed, and most of the navigation channel had been dredged to a depth of from 26 feet to 29 feet. The dredging plant on this section comprised: Seven dredging machines of 250 HP., with transporters; one sea dredge of 250 HP., one of 180 HP., and two of 60 HP. each. The plant for the dry excavation included 600 wagons, 9 locomotives, 2 steam navvies, and about 30 miles of permanent way.

The second section includes from the 14th to the 27th mile. The first work executed on this section was to lower the whole level to about that of the Chagres river, some 33 feet above the sea. For this work were employed 1,100 wagons, 12 locomotives, 6 excavators, and some 31 miles of permanent way; there were also on the side of the canal 9 excavators of 180 HP. each. The total amount of excavation on this section remaining to be done on January 1st, 1888, was 6,952,000 metres, of which, on August 25th, 1,499,000 had been completed.

The third section is from the 27th to the 33rd mile. Between 27 and 29 miles the earthworks were completed. On January 1st,

1888, there were 4,000,000 cube metres to be removed, and on August 25th 710,800 cube metres had been taken out. The plant on this section was: 2,500 wagons, 45 locomotives, about 70 miles of permanent way, 150 steam cranes, and 120 rock drills.

The fourth division is from the 33rd mile to the 38th mile. On January 1st, 1888, there remained to be done 5,133,000 cube metres of excavation, which had been reduced on August 25th by 810,000 metres. Until that date the work had been almost wholly dry cutting, but since then a great deal of dredging has been done on both sides of the Culebra in forming artificial lakes with the waters of the Obispo and Rio Grande. On this section there are about 85 miles of railway, 2,600 wagons, 70 locomotives, and 35 steam navvies; for the dredging operations there are 8 dredging machines, 18 spoil barges, 8 tugs, 2 transporters for the dredgers, and 4 elevators.

The last section extends from 38 miles to 46·3 miles. On August 25th there had been 2,028,000 cube metres excavated from the 5,547,000 metres remaining on January 1st, 1888. The navigable channel at Panama had been opened to a depth of 36 feet below sea level, a width of 131 feet, and a length of 4 miles, while the end of the canal for a length of $1\frac{1}{4}$ miles was finished to a depth of 29 feet 6 inches below mean sea level. The plant comprised 7 dredges of 180 HP. each, 2 large dredges with long transporters, 2 sea dredges, 14 hand dredgers, 6 steam spoil barges, 10 tugs, and 3 transporting apparatus.

The following table gives particulars of all the excavation that had been done upon the locks up to August 25th, 1888:—

No. of Lock.	Total Excavation.	Excavation to August 25th, 1888.	Remaining.
	Cube Metres.	Cube Metres.	Cube Metres.
1	124,000	50,000	74,000
2	127,000	60,000	67,000
3	126,000	60,000	66,000
4	119,000	50,000	69,000
5	121,000	20,000	101,000
6	128,000	40,000	88,000
7	116,000	50,000	66,000
8	117,000	60,000	57,000
9	137,000	60,000	77,000
10	132,000	30,000	102,000
	1,247,000	480,000	767,000

The total weight of ironwork for the locks is 37,000 tons, of which 26,000 tons have been ordered; the contract also includes 25,000 tons of cement. About 6,000 tons of ironwork have been shipped to Colon, and 24,000 cube metres of broken stone for the béton have been prepared.*

But the scheme to which I wish to draw your attention is that in connection with Lake Nicaragua, which is the one favoured by the American Government, and which will probably be constructed in substitution for the Panama route. At this point, as you will see by the map, the isthmus is nearly severed by a large lake and the river flowing out of it into the Atlantic; the distance remaining between this lake and the Pacific being very short—I am not quite sure, but I think something under 50 miles. The highest point in the bit of land between Lake Nicaragua and the Pacific is only 140 feet; therefore the amount of work to be accomplished in cutting a canal through there would not be very excessive. But the scheme which I will now describe to you is a great improvement on any previous system of canal making, and certainly far in advance of anything suggested hitherto. You will perhaps hardly realise, till it is explained, how very simple this plan is. It consists of a system of making canals without excavation, by utilising the natural boundaries of the river valleys, and by raising the level of the water. Any width and depth of water can be obtained,—it is a question simply of damming up the valleys. In the case of this Nicaraguan scheme the physical features of the country lend themselves in a remarkable way to the achievement of this design, for which locks of a very special and ingenious kind are required.

The river San Juan, which flows from Lake Nicaragua to the Atlantic, is for the most part already a fairly navigable river; but at one point it is impeded by rapids, and at its mouth it forms a delta and exceedingly difficult bar. After leaving Lake Nicaragua, it falls 110 feet in 100 miles, traversing for the most part a valley bounded by hills of moderate height, and forming, where it enters the Carribean Sea, a delta through which it wanders by several different channels. Five miles north-east of Greytown, which is

---

* The above data are taken from an article in No. 1,202 of *Engineering*, based upon the last official report.

situate at the northern extremity of the delta, two low hills of rock base approach within a very short distance of each other, and afford an opportunity of damming back the waters of the San Juan at this point by means of a lock and by throwing a dam across its present course. The excavation of five miles of canal through soft alluvial soil between Greytown and Lock No. 1, which has a lift of 65 feet, constitute the only excavating work required throughout the whole of the waterway. Above Lock No. 1 a low lying district would be submerged, forming a lake extending to a point in the valley of San Juan 35 miles from Lock No. 1, where Lock No. 2, having a lift of 60 feet, is proposed to be placed. Lock No. 2 will not only give a depth of 30 feet of water over that part of the river now obstructed by rapids, but will raise the level of Lake Nicaragua 15 feet. On leaving Lake Nicaragua, through which the line of navigation passes for a distance of 76 miles, Lock No. 3, with a lift of 50 feet, raises the water level to a sufficient height to surmount the ridge which is the highest point between the two oceans. Observe, there is no excavation at all. By utilising the existing valleys the water level is simply raised and sufficient height obtained to pass over the coll of the ridge, after crossing which Lock No. 4, the largest of the series, having a lift of 120 feet, brings the vessel within 55 feet of the Pacific level. The distance from Lock No. 3 to Lock No. 4 is about 11 miles. The gorge of a small river flowing into the Pacific at Brito is here used as the bed of the canal, and a lock of 55 feet (No. 5) gives access to Brito on the Pacific. Within this lock a small lake is formed, which will constitute a magnificent natural dock. The estimate for the construction of this magnificent waterway is put down at 23,000,000 dollars, or £6,000,000. Messrs. Bruce and Abernethy have made a report stating that they consider the estimate amply sufficient to complete the work.

Having seen what is the scheme by which the isthmus is to be crossed, I will now point out the wonderful locks by means of which these great elevations are accomplished. I have here designs showing the construction of the lock invented by Colonel Blackman, an American engineer. The lock illustrated has a length of 750 feet between the gates, a breadth of 100 feet, and a maximum lift of 120 feet. The lock chamber is to be constructed of substantial masonry, and all the details of the most gigantic description. The gates are of the sliding type of pontoons, and are wedge shaped in form, with the broad part at the bottom, being constructed of ship's plates, supported by wrought iron framing. They are actuated by capstans or hydraulics. By the adoption of the design for the gates, sufficient buoyancy is obtained to float the gates and to neutralise very considerably the lateral pressure of the water. The gates are opened and closed by means of racks and pinions, actuated by a turbine, and can be opened or closed in two minutes. The depth of water over the mud sill is proposed to be 40 feet, which with an allowance of 10 feet for shrinking of the water in the upper levels, would leave 30 feet available under all circumstances. The system of emptying and filling the lock is most elaborate. Precautions to prevent injury to the bottoms of vessels being necessary in the presence of such exceptional head of water, to meet this necessity the following arrangement is adopted: Two 18 feet culverts or ducts communicate with the water above the lock, and after descending to a level below the bottom of the lock, extend from end to end on either side. The admission of water to these culverts is controlled by gigantic sluice valves. The lock is crossed throughout its length by 84, or more or less, 3 feet sub-ducts or metal pipes, the upper surfaces of which are pierced with 2-inch holes, say 4 in width and 80 in length, for each pipe, making a total of 320 holes per pipe, or 26,880 for the whole lock. By this distribution the water is admitted into the lock without violence or possible injury to vessels. It is calculated that the lock can be filled in four minutes, although it is not intended to use such a speed. If it is desired to economise the water used, it is proposed to do this by means of a series of tanks arranged in steps or terraces on the slope of the lower lock dam. By making use of the well-known law by which water will always seek its own level, the water is successively transferred to the highest of these tanks, and with the fall of the water successively fills each of the lower tanks. the loss of water being equal to only half the contents of one tank on each side, say 50 per cent. By reversing this process and emptying the tanks the water is transferred from them to the lock, the deficiency being made up from above. Such, very shortly, is a description of the Blackman lock, by means of which it is, in the opinion of the most competent engineers we have—namely, Sir James Bruce and Mr. Abernethy—in this country, as well as some of the best American hydraulic engineers, possible to lift ocean-going steamers of the largest size through heights up to 120 feet vertical. By the adoption of such locks as this, it is believed that the canalisation of the Nicaraguan Peninsula can be successfully accomplished, at a cost of £6,000,000 sterling. If so, an alternative and far cheaper and more satisfactory method of joining the Atlantic and Pacific Oceans is that which I have been briefly describing.

It seems to me, in regard to the Sheffield Ship Canal, there would be great advantage in the application of such locks. For instance, at Swinton, where you have five locks in a tier, vessels, instead of passing slowly through five locks, would be lifted at one operation. You would get one big lock, capable of taking a big vessel, and lift it through the whole height and possibly over the railway clean away, without the necessity of lowering the short masts, which would be left standing.

The account I have given both of the Nicaraguan Canal scheme and of the Blackman lock is of course the merest outline sketch, but such as it is I trust it is not altogether without interest.

Mr. NEVIN: There does not seem to be much difficulty in the high lift. You could not adopt the principle of filling up the river valley here; land is too valuable.

Mr. G. B. WALKER: But still if you can get a type of lock which dispenses practically with with an enormous pressure of the water against the lock gate, and so makes them workable, you have done a great deal towards facilitating the construction of ship canals.

Mr. NEVIN: That pontoon system is in use in some large docks for the lock gates. They are made hollow and take a great proportion of their own weight.

Mr. G. B. WALKER: Yes; I brought this forward thinking there was nothing else to be done at the meeting, or I should not have thought it worth naming.

Mr. NEVIN: I must say we are much obliged to you for bringing it forward.

Mr. JEFFCOCK: Anything of this sort is of interest at this moment, when people are thinking of ship canals so much about the country. I shall be very glad to propose that the best thanks of the Institute be given to Mr. Walker for his communication, and that it be printed in the next Transactions.

Mr. HABERSHON seconded the motion, which was carried unanimously.

The meeting then ended.

ERRATA.

On page 356, Part CI. :—

    Line 2 from bottom, *for* " 33,000 foot-pounds = 1 HP. per minute," *read* " 33,000 foot-pounds per minute = 1 HP."

On page 357 :—

    Line 1, *for* " watts per minute," *read* " watts."
    Line 2, *for* " work " *read* " power."
    Line 3, *for* " 4·4 " *read* " 44·0."
    Line 4, *for* " quantities " *read* " work."
    Line 7, *for* " quantities " *read* " work."

MIDLAND INSTITUTE OF MINING, CIVIL, AND MECHANICAL ENGINEERS.

GENERAL MEETING.

HELD AT THE INSTITUTE ROOMS, BARNSLEY, ON WEDNESDAY, MAY 22ND, 1889.

C. E. RHODES, Esq., President, in the Chair.

The minutes of the last meeting were read and confirmed.

The following gentlemen were elected members of the Institute, having been previously nominated :—

Mr. THOMAS GILL, Colliery Manager, Car House Colliery, Rotherham;

Mr. JOS. WM. GILLOTT, Engineer, May-Day Green, Barnsley.

THE FEDERATION OF MINING INSTITUTES.

The PRESIDENT: My view is we have nothing whatever to lose by this federation. The only persons who are likely to lose, as far as I can see, is an Institute like the North of England, where you have a heavy subscription.

Mr. GERRARD: Although I should have preferred a much larger scheme,—a Mining Institute of Great Britain, having head-quarters with local centres,—I shall support as far as possible this federation, and consider withdrawing from the North of England Institute.

The PRESIDENT: You would have no advantage in remaining, except the credit of belonging to an Institution of this standing. The scheme seems to me to be bound to work in favour of an Institute like this, because, as I understand it, we shall have the whole of our printing expenses taken off our hands, and we shall have the advantage for a very slight increase in our subscription of receiving the Transactions of all the other Institutes.

Mr. GARFORTH: Of the cream of the other Institutes. Is it possible to get a guarantee fund, and then look upon the subscriptions from the Midland Institute members as being able to cover all our expenses when the printing is taken off? What is the income of the Midland Institute?

Mr. T. W. H. Mitchell: Including donations towards fan experiments, over £200; subscriptions amounting to £176 8s. 0d. are included in this amount. The amount paid for printing last year was £65, but it would be more than that, because there is a balance due to the printer of £67.

The President: That means that 50 per cent. of expense is for printing. We pay over to the Federated Institute how much?

Mr. T. W. H. Mitchell: 15s. per member; so that we should pay 15s., and should, as it were, save 10s., and for the extra 5s. contributed by members of the Federated Institute, they would obtain not only these Transactions, but the Transactions of every other Institute.

Mr. Garforth: How many members have we now?

Mr. T. W. H. Mitchell: 155.

The President: It would mean an increased subscription to those who join the Federated Institute of 5s. or 6s.

Mr. Gerrard: Would it not be better for all the members to join? Would it not reduce the amount?

The Peesident: You could not without the consent of individual members. We could not commit anybody to increased liability except with his consent.

Mr. T. W. H. Mitchell: Could the Council do it by proposing an increased subscription at the annual meeting—by proposing that the subscription be increased to 25s.

The President: I do not think you could do it without practically terminating the membership of every individual member.

Mr. Garforth: If it takes £109, we have 145 members which at 15s. each would produce £109. Could we say we will have no expenses except the £109 due to the Federated Institute.

The President: The only way would be to put before the members a statement showing the increased subscription which would have to be paid to become members of the Federated Institute. If you could show that the increased subscription of 5s. or 6s. per annum would entitle them to all the advantages of membership in the North of England, the South Wales, and every other Institute in the federation, it would have to be for individual members to say whether they would belong to the federation, or remain simply members of this Institute.

Mr. GARFORTH: That is one proposal. Another is that no Transactions be printed in connection with this Institute, and that we contribute £109 towards the Transactions of the Federated Institute; so that from this moment, if a member does not choose to belong to the Federated Institute, he gets no more Midland Transactions, but what is far better, he gets the cream of the Transactions of every other Institute.

The PRESIDENT: You cannot do it by putting the members in bulk.

Mr. GARFORTH: We shall assume, unless we get a protest from each member to the contrary, that the Midland Institute, as a body, is willing to incur a liability of £109.

The PRESIDENT: You could not commit the Institute as a body to anything like that. You can only accept an individual member's individual liability.

Mr. GARFORTH: I pay my guinea. I do not expect more Midland Transactions, but prefer that 15s., or as much as required, goes towards the £109. The difference between a guinea and 15s., say a matter of £40, is to pay rent, Secretary's salary, &c.

The PRESIDENT: But suppose the membership fell off?

Mr. GARFORTH: I should ask each member by circular first.

Mr. NASH: It says each Federated Institute shall pay 15s. per associated member to pay the publication expenses of the federation, and the local Institutes shall receive their own Transactions for members who do not receive the whole of the Transactions, and pay 10s. 6d. each for such non-associated members.

Mr. GARFORTH: I was present at the meeting in London when the project was discussed. It seemed to me to raise the standard of this and other Institutes. If the Midland Institute does not join, the scheme may fall through for years to come.

Mr. NASH: We can join. We are not bound to guarantee a certain number, but simply say that the Institute will agree to the scheme, and see how many will become associated members.

The PRESIDENT: Shall I move, "That this meeting, having considered the federation scheme as set out by the North of England

and other Institutes, agrees to join, and recommends individual members to do so upon the basis set out in the circular issued."

Mr. GARFORTH: Could you say that unless you hear by a certain time, you will assume each member agrees to become an associated member?

Mr. GERRARD: I think something further is wanted. You have to contribute to the funds of the Federated Institute 15s. per associated member. How can you ascertain, without consulting the members, how many will become associated members?

The PRESIDENT: Would not such a motion cover that. Suppose only three joined, and those three paid that 15s. over from this fund, those three would get the benefit of the federation scheme.

Mr. GERRARD: How do you arrive at that?

The PRESIDENT: They would be the three who sent in their adhesion. We submit the scheme to all the members.

Mr. GERRARD: Then I am with you, and would second a motion that we join, and that some effort be made to ascertain how many of our members will become associated members.

Mr. TEALE: At an increased subscription. Derbyshire is a guinea and a half.

Mr. NASH: Ours should be a guinea and a half, and we should have half to work the Institute with, and the other half for the federation.

Mr. T. W. H. MITCHELL: You would save two guineas for the North of England, a guinea and a half Derbyshire, and so on.

Mr. GERRARD: The difficulty is to know who will become associated members.

Mr. NASH: We do not guarantee any fixed number, but we should let them know at a future date.

The PRESIDENT: Then I will move, "That this meeting is of opinion that the Midland Institute should join the federation scheme, and will notify to the Secretary of the scheme, as early as possible, the number of members who will do so upon the terms set out in the circular issued.

Mr. GARFORTH: You know the difficulty of getting a printed slip back?

The PRESIDENT: This is not to be sent to members. There would have to be a proper circular sent to members, asking them about joining, and stating that we had agreed to join.

Mr. GERRARD: I will second the proposition.

Mr. NASH: Cannot we make it a guinea and a half subscription?

The PRESIDENT: You cannot do that.

Mr. TEALE: Any alteration will have to take place at the annual meeting.

Mr. T. W. H. MITCHELL: How can we join the federation scheme without an increased subscription? The circular must say there will be an increased subscription.

The PRESIDENT: The circular should embody full particulars,— the benefits to be got, and the increased subscriptions involved.

Mr. NASH: Does not that come in, that we cannot alter the subscription?

The PRESIDENT: We cannot except by consent of individuals. If individuals agree to join by consent, the increase follows logically; but we cannot increase the subscription of any member.

Mr. W. HOOLE CHAMBERS: An increase of subscription would necessitate an alteration of rule, which will have to be made at the annual meeting, notice being given at the previous general meeting.

[The President at this point left the meeting, and Mr. Gerrard took the chair.]

Mr. TEALE: Those who remain as they are, have all they had before. They can be asked which they will join.

Mr. NASH: Whether they will belong to both branches of the Institutes.

Mr. GARFORTH: If we begin by 60 joining the federation, I do not think it will be long before we get the whole number to join the Federated Institute.

Mr. TEALE: You should have this alteration of subscription carried.

Mr. T. W. H. MITCHELL: There is no necessity to alter the guinea. If they are willing to pay the guinea and make a presentation of 5s. a year extra to this Institute to belong to the federation, that settles it.

Mr. NASH: Will there be sufficient to work the Institute?

Mr. GERRARD: The first question is shall we join the Federated Institute?

The motion proposed by the President was put and carried unanimously.

Mr. GERRARD: Now it is open to any member to propose any resolution with regard to obtaining this 5s. required for the Association, to say whether an alteration shall be made in the rule or the subscription at the next annual meeting. If notice is given of any such intention it need not be binding, but it brings us within the letter of the rule, so that it can be discussed.

Mr. BONSER: As I am a member of three Institutes, I shall be a distinct gainer of three guineas a year. I shall save two subscriptions, and get all the advantages I have for £3 or £4.

Mr. GARFORTH: That is so; besides getting better papers.

Mr. GERRARD: A notice of motion for the annual meeting would be the best way of raising the question.

Mr. W. HOOLE CHAMBERS: I think most decidedly it would be a great advantage to members to join, and I really think the best way would be to alter the amount of the annual subscription to meet it. A notice of motion to do so would raise the subject, and any person who felt an interest or who did not feel disposed to advance the subscriptions, would attend the annual meeting and have the matter explained and see what the advantage would be.

Mr. GARFORTH: I shall be glad to give notice of motion, " That at the annual meeting I shall propose that the annual subscription of this Institute be increased from one guinea to 30s. for the ensuing twelve months; such increased subscription being necessary to enable this Institute to join in the federation scheme proposed by the North of England Institute of Engineers, and which, if carried out, will it is felt greatly increase the usefulness of the Midland Institute to the profession and business in which so many of our members are so deeply interested, for the following amongst other reasons —

A greater number of subjects will necessarily be brought forward by the Amalgamated Institutes than can possibly be produced by any one Society.

A much wider and more extended interchange of ideas must take place.

Information from other districts not now available will be brought before this Institute, and consequently meetings which sometimes fail for want of papers, will be made more interesting and useful."

Mr. GERRARD: Let that be put on the minutes as a notice.

Mr. GERRARD : The next business is a paper on "Artificial Foundations and Method of Sinking through Quicksand," by Mr. W. E. Garforth. Mr. Garforth has only had a few days in which to prepare this paper. It was undertaken at the urgent request of the President, and in compliance with his wish, that Mr. Garforth would prepare something to lay before the members at this meeting. It is a most important matter this of sinking through quicksand, inasmuch as the extension of the South Yorkshire coal-field is a prominent topic just now. We are all grateful to Mr. Garforth for the trouble he has taken and the interest he shows in the Institute.

Mr. GARFORTH then read his paper as follows:

ARTIFICIAL FOUNDATIONS AND METHOD OF SINKING THROUGH QUICKSAND.

BY W. E. GARFORTH.

IN selecting a site to sink a shaft, or construct foundations for surface works, troublesome ground is naturally avoided, for reasons too obvious to mention. At the same time instances can be given where, in order to gain certain advantages of position, in connection with existing works, or to obtain reduced rates for the carriage of goods, it has been necessary to sink a shaft and erect buildings on quicksand. The increased outlay on such works has been fully repaid by a permanent saving in the cost of production.

The writer has recently had an experience of this kind, and at the President's request he now ventures to lay before the Institute a description of such works. It certainly seems as desirable to have on record a description of difficult work, and how difficulties have been overcome, as it does to have an explanation of the latest scientific invention, especially as Engineering is not an exact, but rather an empirical science. Such being the case, it is unnecessary to advance reasons to shew the desirability of getting the best information on works which members of this Institute may be called upon to execute. The foregoing remark specially applies to foundation work, for experience has shewn that the most perfect engine or machine has failed to produce the maximum of useful effect when the foundations have been imperfect. It has likewise been proved that the permanence of engineering structures depends almost entirely on the manner in which the foundations have been laid.

In the following remarks it is intended to describe:

A.—A method of sinking through quicksand by means of cast-iron tubbing and lowering screws.

B.—An arrangement for suspending self-contained pumping engines in a sinking shaft, to deal with one thousand gallons of water per minute.

C.—The construction of certain foundation works, and the method adopted to prevent the same being affected by heat.

The object of the works in question was to sink a shaft 320 yards deep to the Silkstone or Middleton Main seam of coal, and to erect certain colliery plant, consisting of winding engines, boilers, chimney, and workshops.

Before finally deciding upon a site for permanent works, it is always advisable in unproved ground, to take the precaution of sinking a number of trial holes. To bore a hole 3 in. diameter, to a depth of 20 to 30 feet, is a trifling matter compared with the advantages usually gained by additional information; by substituting certainty for uncertainty; and if in consequence of better information the site is changed, then by a large saving in material, workmanship, time, and anxiety.

Drawing No. 1. shews the surface levels, the position of the bore holes, together with the thickness and nature of the subsoil. As the difference between an eligible site and one that was not suitable for surface plant consisted in only a few extra feet of quicksand, the position of the pit, as shewn in the section, was decided upon.

Drawing No. 2 shews the sand was met with four inches below the level of the ground, which likewise proved to be the height of the standing water. After the centre of the pit had been determined, it was deemed advisable to erect the pillars for supporting the longitudinal timbers, and pillars for the headgear, as far as possible from the sides of the pit, in order to prevent them being undermined by the sand slipping, and to avoid any additional lateral pressure and friction on the tubbing. The main timbers were each 53 ft. long by 17 in. square, trussed with wrought-iron tie-rods 2 in. diameter. [Old winding ropes of the necessary strength would have answered the purpose of tie-rods equally well.] The corner timbers which supported the main timbers were composed of railway sleepers 9 ft. × 10 in. × 5 in., built in the form of a chock. The chock, which presented a surface of 81 superficial feet, was tied together by iron rods one inch diameter passing through each corner, and fastened by nuts and washers. In this form the pillars or chocks were stronger and better able to resist the unequal and severe strains to which they were subjected than pillars of brick or stone, besides being cheaper. The relative first cost was:—Timber, £34, as compared with brick, £80 (strengthened with hoop-iron bond); stone, £104. By using timber there was the additional advantage of getting the pillars erected in a few hours, whilst in the case of brick or stone the mortar would have required time to set. The wooden pillars practically cost a very small sum, as the sleepers were afterwards used in the colliery sidings.

The transverse beams (each 14 in. square) rested upon the longitudinal baulks. On the former were placed the gearing and screw arrangement for working the four lowering screws, each 14 ft. long, 3 in. diameter, the thread and nut being cut square. Owing to the swampy nature of the ground, the segments had to be temporarily erected on timber. When the circle was complete and bolted together, the screws were attached, and numbered respectively 1, 2, 3, 4. The object in numbering the screws was a matter of convenience to assist in maintaining the perpendicular position of the tubbing. In case the sand was watery on one side, or a boulder stone caused an obstruction, it was only necessary to call out to the screw-men (on the surface stageing) the number of the screw or screws which required checking. The water in the pit maintained its level, and so served as a guide when compared with the internal flange of the tubbing. The lowering operation was continued until the friction of the sand prevented the casing sinking; an excavation was then made at the sides, and the sand thrown towards the centre of the pit to prevent the quicksand boiling up and forming a partial cavity, which was usually followed by a sudden and unequal strain being thrown upon the screws. If the tubbing had been allowed to deviate from the perpendicular line, consequent on the yielding nature of the quicksand, an immense weight would have been necessary to force it down, with the imminent risk of breaking some of the segments.

At a sinking near Wrexham, the writer saw in one sinking shaft, three different sets of tubbing cracked and rendered almost useless. The serious loss in time and money, and the permanent inconvenience of the reduced diameter of shaft, might have been saved had lowering screws been used, and the vertical ribs of the segments been made stronger.

After the cutting ring and tubbing had been sunk to a certain depth, other rings were added, until the cutting edge reached the stratified measures, shewn on Drawing No. 2. On reaching the stratum of blue bind, the pumps were run at an increased speed, to keep the pit bottom free from water, when the sinkers were able to cut away the ground from under the cutting edge, after which the tubbing was gradually sunk to the required depth. Although there was not much fear that the casing would sink after the lowering screws had been removed, or a settlement take place at some future time owing to the treacherous nature of the quicksand, still as an extra precaution, and to expose a greater surface, footplates were inserted under the cutting edge.

The two principal objects in using the tubbing herein described were:—Firstly, to complete the sinking as quickly as possible; and secondly, to prevent the surface water leaving the quicksand, as otherwise the various buildings,—engine-house, workshops, chimney, &c.,—would have been endangered. Both objects were attained. The quicksand was sunk through in four days and a half. The surface water is still maintained at the same height in the quicksand.

The detailed particulars of the tubbing are given in Drawing No. 4. The segments were 4 ft. 3. in. long, 2 ft. 6 in. deep, $1\frac{3}{4}$ in. thick in the web, strengthened by horizontal and vertical ribs. The latter were made extra strong to resist the tensile strain to which this description of tubbing is specially subject when the sand boils up or " blows" and allows the casing to sink suddenly and unequally.

As an instance of this, it may be mentioned that at a sinking in North Wales, of which the writer had charge, during the most difficult part of the work the quicksand would sometimes rise 15 or 16 feet in a few seconds. This sudden and unequal strain immediately subjected two of the lowering screws to a very severe tensile strain, whilst the remaining screws were unaffected; a few hours afterwards the latter screws were in tension and the former loose. In this sinking, where the quicksand was 66 ft. thick, the tubbing, when completed, was out of the perpendicular only to the extent of the thickness of the horizontal flange, that is to say, whilst the water in maintaining its level touched the lower part of the flange on one side of the pit, it covered it on the other,—a difference of only two inches.

At another sinking in Lancashire, with which the writer was acquainted, (the subsoil being 176 feet thick) the sand and water would sometimes suddenly rise to a height of 34 feet.

The cause of these sudden risings or "blows" of sand is primarily due to an excess of external pressure. The effect has been to break the strongest tubbing, necessitate months of delay, and seriously increase the cost of the sinking. The remedy has been found, in two instances within the writer's knowledge, to be in leaving sufficient water in the shaft (to form a kind of hydrostatic balance) and excavating the sand under water. This latter operation was effected in one instance by perforating holes in the sand by means of a shell, and afterwards reducing the whole depth of the pit by a skeleton cutter. In the other case a drum or tube with a clack was forced into the sand. Both arrangements were carried on by the sinkers working the shell and drum from a staging placed immediately above the water. A description of the details of such work would greatly exceed the limits of this paper.

The tubbing in question being subject to a tensile strain, whilst the tubbing frequently used to dam back water is in compression, it will be understood why the thickness of the ribs and section of the segments generally are increased. According to Professor Rankine, cast-iron in tension is equal to 16,500 lbs. (avoir.) per square inch; in compression 112,000 lbs., or nearly seven times stronger.

Another reason for increasing the strength of the segments was in case of additional weights being required to sink the casing. In the Welsh sinking before referred to, it was necessary during the latter portions of the sinking to apply a weight of 300 tons, and in the Lancashire sinking 450 tons. It is impossible to distribute pig-iron and chains as evenly as to increase the thickness of the segments.

The writer believes it is preferable to be above rather than below the proper thickness, as the extra weight gives greater tensile strength; compensates for air-holes and other imperfections to which castings are liable; gives a greater substance to withstand the effects of corrosion; and generally leads to a reduction in the price of the castings owing to the increased weight.

Before proceeding to cast the segments it is advisable to test the iron-founder's mixture of metal,—a bar 1 in. square and 4 ft. 8 in. long, or 4 ft. 6 in. between supports, should carry a weight of 500 lbs. suspended on a knife edge in the centre. The first ring with the lining pieces should be bolted together, and the exact diameter ascertained, before proceeding with the remainder. The castings should not be painted until inspected, and other points should be carefully watched, on the principle that the strength of a chain is only equal to the weakest link.

During the sinking, feeders of water exceeding one thousand gallons per minute were met with. To deal with this quantity, it would have been necessary to erect two pumps with working barrels each 17 or 18 in. diameter and stroke 4 ft. 6 in. long. For important reasons it was out of the question to delay sinking until suitable engines, L legs, and the attendant pumping appliances could be erected. Assuming the machinery could have been purchased second-hand, too much time would have been lost in preparing foundations on such difficult ground. To meet the emergency, self-contained direct-acting engines, made by the Pulsometer Engineering Company, were fitted on cradles, and, by means of chairs and adjusting screws, were suspended from the main transverse timbers. The largest pump delivered 35,000 gallons per hour, and measured 9 ft. 9 in. long, and 1 ft. 9 in. wide. The internal diameter of the shaft was 14 ft. 5 in., which allowed sufficient space for working the hoppit (holding 50 cwt). A light staging was fixed between the pumps, so that one attendant could work three pumps.

All the piping for steam, exhaust, and water, was of wrought-iron. The pipes were kept in position by being cramped to a wire rope, suspended from the surface timbers, and lowered, by means of screws, as the sinking progressed. The lower end of the steam pipes was connected to the engine by flexible tubing, which latter proved of great service. The water-bearing strata continued to a depth of **54 yards**.

The water was tubbed back by cement lining 3 in. thick, supported on each side by brickwork, with an arrangement for allowing the water to escape during the time the cement was setting. This kind of tubbing was used in preference to cast-iron, as the latter corrodes after a lapse of years. There are on record several instances of underground workings being flooded in consequence of the cast-iron tubing having burst, owing to corrosion. Where good cement is used, the tendency is for the lining to harden. The first cost of cement lining is much cheaper than cast-iron, but in such an important matter first cost should not be considered too seriously. It is important to adopt that which will be most durable during the full life of the colliery. In using cement there is, however, more risk than in using cast-iron. Inferior cement and careless workmanship during one shift, may make the difference between a dry and wet shaft, which cannot afterwards be rectified for fear of disturbing other portions of the work. By using riddled soil as back filling there is a chance small fissures may be closed. If the cement lining is not of sufficient thickness and strength, there is the further risk that it may be cracked during the ordinary settlement of the brickwork.

The direct-acting pumping engines before referred to answered every requirement. The pit bottom was kept drier than by ordinary shaft pumps, and when working on air the severe shocks which are so objectionable in shaft pumps, especially those with a long stroke, were not experienced. The saving in time between having to erect engines, L legs, &c., and using direct-acting pumps was much in favour of the latter; there was also a considerable saving in the cost. An additional advantage was obtained by being able to use the small pumps for other purposes when the sinking was completed. Large clack pieces, working barrels, &c., have often remained unsold for years after the colliery sinking has been completed.

After the means before described had been accomplished, to maintain the water at its proper level (in which case a quicksand becomes a fairly good foundation) the work connected with the boilers, engine house, &c., was proceeded with.

For the boiler foundation the sand was excavated to the depth requisite to place the boilers and steam pipes in proper position with the winding engines. Owing to the watery nature of the quicksand, and to prevent any concrete being wasted, rough timbers were placed on the sand. On these the first layer of concrete, 15 in. thick, was laid over the entire surface. By the time this operation was completed, the concrete first laid had set sufficiently for the second being laid thereon, and so on until a thickness of about seven feet was attained.

As it was feared (indeed actually proved) that the heat from the boiler flues might, after a number of years, destroy the nature of the cement, it was deemed advisable, to prevent any wasting action to keep the cement as far as possible below the flues. Fire bricks might have been used to counteract the heat, but they were not considered sufficiently strong, even with hoop-iron bond, to resist the unequal strain due to the quicksand. Blocks of stone 6 ft. 6 in. by 4 ft. 6in. were used in preference. To further strengthen the concrete and stone work, a number of iron rails (70 lbs. section) were carried in a longitudinal and transverse direction (4 ft. 3 in. apart) over the entire surface, thereby forming a kind of lattice work. The rails were bent at each end to embrace alternately the concrete and stone work. Before the stones were placed in position, the spaces between the rails were filled in solid with concrete and reduced to a smooth surface, on which the stones were laid. Afterwards a number of rails were placed horizontally, and carried outside the foundation, the intervening spaces being wedged tight and joints grouted. The application of these rails has given to the foundation an uniform support, which, in works of this kind, is as valuable as an unyielding one, with the advantage of being much cheaper. On the upper surface of the stones several cast-iron ventilating tubes were placed, 4 ft. apart, as shewn on Drawing No. 3, extending the full width of the boilers, and so arranged that they can be lengthened when additional boilers are erected. The object of these tubes was to prevent the heat affecting the stone work; and by connecting them with the chimney, a greater quantity of cold air can at any time be passed through them. From Drawing No. 3 it will be noticed that the end of the flue nearest to the chimney, also the main flue, are so bridged that the space beneath can be ventilated, and so prevent the heat affecting the foundation. The precautions which have been described as being necessary for the boilers were not required for the foundations of the other superstructures. The engine house and workshops were placed on beds of concrete 6 ft. thick, composed of layers about 15 in. thick. No crack or unequal settlement has hitherto been perceived.

There are several matters of detail connected with the timber and concrete which might with advantage be mentioned. Comparisons might also be made between the pneumatic and other systems of sinking through quicksand. But to carry out the intention of the present paper it will probably suffice to state the results of certain experiments made by the writer on the strength of mining timber, and to give a brief description of a new process of constructing foundations and excavating by water jets.

The following tests have been made by means of a hydraulic press capable of exerting a pressure of five hundred tons:—

Diameter of Prop.	Length of Prop.	Area of Prop.	Deflection.	Breaking Strain in tons.	Pressure per square inch on Ram.
Inches.	Feet.	Inches.	Inches.	Tons.	Pounds.
6·0	5·00	28·274	1·00	61·50	900
6·0	5·00	28·274	1·25	34·15	500
6·0	5·00	28·274	1·12	43·17	640
6·0	5·33	28·274	1·12	19·00	278
6·0	5·11	28·274	1·25	27·27	398
6·5	5·33	33·183	1·50	45·00	660

A chock formed of 24 pieces of elm timber, each piece 2 feet long by 6 inches square, cut perfectly true, gave the following result:—

Pressure per square inch on Ram.	Deflection.	Crushing Strain in Tons.
Pounds.	Inches.	
580	7·00	—
600	8·00	—
620	9·00	—
640	10·37	43
877	—	60

In comparing the strength of a chock with a prop, it may be mentioned that a prop $7\frac{1}{4}$ inches diameter, 6 feet long, broke at 137 tons.

From the foregoing it will be perceived that in actual practice props of precisely the same dimensions vary as much as 80 per cent., and that a chock formed of pieces of elm 6 in. square and 2 ft. long collapsed at a pressure of 60 tons. These results shew that the formula given by Tredgold and other authorities is not reliable as regards ordinary mining timber. It is difficult, after minutely examining the broken props, nature of wood, weight, concentric rings, and obtaining information about the locality and growth of the timber, to explain the great difference shewn by the tests.

With reference to the chock, the cause of failure was apparent. When the pressure was applied the pieces of timber which gave way first were shewn to contain shakes and knots, and these proving defective the stability of the whole chock was affected. The good pieces seemed almost untouched, although the pressure was sufficient to cause water to ooze out of the pores of the wood. Before the test the various pieces when built up seemed a fair sample of the chocks generally used in the pit.

As regards the new method of constructing foundations and excavating by water jets, it is important that a description, however brief, should be given, not only because the process is ingenious, but principally that any member who contemplates the erection of new works on difficult ground should be informed that the cost of laying a foundation on quicksand has been reduced approximately from forty shillings to less than four shillings per cubic yard.

The system is applicable to sinking through sand, marl, and other subsoil; constructing foundations for a large engine-house or other colliery superstructure; or for the retaining walls at the sides of rivers and canals where coals are loaded. The new process has been carried out in connection with the harbour works at Calais. As the writer has had occasion to visit Calais several times a year, he has had the advantage of seeing the system in actual operation, and of having the details fully explained by the resident engineer.

The works at Calais consist principally of large retaining and dock walls, sunk to a depth of about 30 feet below the level of low water. Although the walls when finished present an appearance (for some hundreds of yards of a straight and unbroken line, yet they really are formed of a number of detached pillars, erected separately in the following manner: Each pillar is 26 feet square, with an octagonal hole in the centre about 8 feet diameter; the bottom portion of the pillar is tapered from the inside to form a kind of cutting edge, and for other reasons hereinafter mentioned. The work is formed of rubble masonry faced on the outside, and set in hydraulic cement. The pillar is built to a height of 12 or 14 feet, or to such further height as may be found necessary to increase the weight for sinking purposes. Sufficient time is then allowed for the cement or mortar to harden, during which interval other pillars are erected. Spirit levels are placed at the corners of each pillar, and a graduated scale or staff fixed at the side. A short distance from the pillar three small direct-acting pumps are placed, similar in design to the sinking pumps described. The pumps are fixed on timber, and supplied with steam from two vertical boilers. [Assuming the attendance to be the same, there is an advantage in having small pumps and boilers, as they serve as duplicates in case of accident, and the main work of lowering the pillars is not interfered with.] These pumps work at a pressure of 30 lbs. per square inch, and supply four or eight jets of water as required. The main delivery or supply pipe, with the necessary branch pipes, is carried close to the pillar. Flexible pipes are then connected, and by an arrangement of valves the several jets of water can be discharged or stopped at pleasure.

A centrifugal pump, worked by a small portable engine (which can also be used for the mortar mill), is fixed at a convenient distance from the pillar. The flexible suction pipe is carried down the centre of the pillar and immersed in the sand and water. By hanging loosely from the top the workmen use it as an agitator to keep the sand in suspension. The flexible pipes with iron nozzles attached and connected to the force pumps are carried down the corners of the octagonal hole to the bottom of the pillar, which is tapered or bevelled from the inside—the object being to reduce the resistance; to prevent any angular corners acting as an obstruction; and to allow the jets of water to act directly upon the sand under the entire surface of the pillar. As the water issues from the pipe in the form of jets it cuts away the sand and holds the same in suspension. Before the sand has time to settle, the centrifugal pump lifts the sandy water to the top of the pillar. By watching the discharge pipe, the colour of the water is seen to vary according to the quantity and nature of the matter (sand or marl) held in suspension. This discharge also serves as an important guide for other purposes. The centrifugal pump not only lifts the sand, but by keeping the water (which has been discharged by the jets inside the pillar) at a low level, the sand is kept in a loose state. [With a large head of water sand may be consolidated.] By allowing jets of water to impinge and cut the sand immediately under the pillar, and afterwards pumping the water with the sand held in suspension, in the manner described, the pillar gradually sinks. The graduated staff indicates the depth of such sinking. If the sand under one side is harder than the other, which often proves to be the case, the difference is soon perceived on the spirit levels. The positions of the flexible pipes and jets are thereupon altered, and in a few minutes' time the perpendicular position of the pillar is regained. The operation is carried on in the simplest manner conceivable: the same machinery and tackle are applicable to every pillar; only a few unskilled men are required; and the cost, as already stated, does not exceed four shillings per cubic yard. As the process may shortly be used in Yorkshire, when the work will be open for the inspection of members of this Institute, it seems unnecessary to describe further details.

Mr. T. W. H. MITCHELL: I beg to move a vote of thanks to Mr. Garforth for his paper, from which a great deal may be learnt; and also move that it be printed in the Transactions.

Mr. W. HOOLE CHAMBERS: I am glad to second it. It has been a most interesting paper. There is a deal of information in it which will be of very great use to anybody who has any work of a similar character to undertake in future.

The resolution was carried.

Mr. GARFORTH: I am much obliged to you.

Mr. GERRARD: The rest of the business must be postponed.

Mr. TEALE: I move that it be postponed.

The motion was carried.

MIDLAND INSTITUTE OF MINING, CIVIL, AND
MECHANICAL ENGINEERS.

GENERAL MEETING.

HELD AT THE QUEEN'S HOTEL, LEEDS, ON TUESDAY, JUNE 18TH, 1889.

C. E. RHODES, Esq., President, in the Chair.

VISIT TO ALLERTON MAIN COLLIERIES.

Previous to the meeting a number of members, by kind permission of Messrs. T. and R. W. Bower, had the privilege of inspecting the Electrical Coal-cutting Machines in operation at Allerton Main Collieries. Mr. Blackburn, Manager at the Collieries, accompanied the members, and explained in detail the working of the machinery.

The minutes of the last meeting were read and confirmed.

THE FEDERATION OF MINING INSTITUTES.

The PRESIDENT: In reading the minutes of last meeting the Secretary has read a resolution which Mr. Garforth is going to propose as to an increase of the subscription from a guinea to 30s. Some members probably are not aware of all that has passed with reference to this proposed federation scheme; and, perhaps, they are also unaware how our present subscriptions are disposed of. The Secretary has been kind enough to get out the cost of printing for this Institute, and he finds it represents 15s., or rather more, per member per annum. All we get for that 15s. is the rather meagre Transactions which embody the result of, I am sorry to say, a great number of our meetings through our simply not having the papers necessary to make our meetings and discussions as interesting and as useful as we could wish. I fear we shall never get that in an Institute such as this, consisting of a comparatively small number of members, in most instances confined entirely to one district. If we join the federation scheme, all members of the Institute will get their own Transactions printed; and in addition will get the most important part of the Transactions of every other Institute that joins in the the federation scheme,—that is, the valuable papers. Every thing that passes, discussions and that sort of thing, will not be printed, because it would mean such an enormous volume that it would not pay for publishing. You would thus have all the valuable papers month by month—for a very slight increase in the subscription—of the North of England, the South Wales, the Staffordshire, and our own Institute, and our subscription is the lowest of any of them. If members present would bring those facts before members of the Institute who may not understand the point, and show the several advantages, it might smooth the way a little. We have no power to pass a resolution of this sort; we do not want to stuff it down the throats of members; but we desire to bring it before them in a fair and open way. We, as a Committee, think it would be desirable to carry out the scheme; but it is for the members to say whether it should be so or not. I think the reasons Mr. Garforth gives for joining the federation scheme are fair and logical; and I shall be glad if those who are here will look at the question in a broad spirit, and discuss it with any members outside who may raise the question with them.

Mr. GARFORTH: I am much obliged to you Mr. President, for bringing it forward. It would be a good thing if we could have this ventilated before the next meeting, so that if it resolves itself into this, that we cannot bind the whole of the members of the Institute to this scheme, yet out of the 145 members 100 might begin by joining the federation scheme, and I do not think it would be long before the other members, seeing how valuable and important the Transactions of the other Institutes are, would join us.

The PRESIDENT: Unless it is unanimous the passing of that resolution is not binding on any individual member, but only on those who elect to join in the federation scheme.

Mr. GARFORTH: If, after next meeting, we have not a vote in favour of joining we can then put it to members to join individually.

The PRESIDENT: That would be so.

Mr. T. W. EMBLETON: No member is compelled to join the association. It is this way—he continues to pay a guinea here or 30s. elsewhere.

Mr. NASH: It is all paid to this Institute, and we remit 15s.

Mr. BENNETT: How would it be to send a circular to each asking them what they would do.

The PRESIDENT: We could send a slip out with the voting paper, and if they are not going to be present ask them to sign their name, for or against, and send the slip back.

Mr. BENNETT: I move that the Council be instructed to send out the resolution and particulars to members.

The motion was seconded and carried.

DISCUSSION ON MR. G. BLAKE WALKER'S PAPER ON "THE PROPOSED NICARAGUAN CANAL, AND A NEW TYPE OF LOCK FOR SHIP CANALS."

The PRESIDENT: The next business is the discussion of Mr. G. Blake Walker's paper on "The proposed Nicaraguan Canal, and a new type of Lock for Ship Canals." I am sorry to say that Mr. Walker is not here, and it is awkward discussing a paper of this kind without the presence of the author. Ship canals are occupying a good deal of attention in South Yorkshire. I suppose in West Yorkshire you are not so much interested, as you have got one that answers your purpose already. In South Yorkshire we take a great deal of interest in it, and look for great benefits when we get it. I am not in a position to discuss this matter, and unless someone has something to say, I should suggest it be adjourned.

Mr. JOSEPH MITCHELL: I look on it as a statement of facts, which gives nothing to discuss.

Mr. NASH: The principle embodied in it is hardly applicable to English canals, as the value of the land absorbed would swamp the other benefits derived from it.

The PRESIDENT: It is a communication of facts, which are very interesting, but I do not think we can discuss it or express an opinion about it.

Mr. EMBLETON: I think the President is not aware what is doing in this district respecting the Aire and Calder Canal We expect before long the Aire and Calder Canal Company will be able to bring up to Leeds vessels containing 300 or 600 tons. They have already a Bill introduced into Parliament to get authority to improve their waterway; and it is the intention, I believe, ultimately to do what I say, so that Leeds instead of Hull will be the port for Leeds.

The PRESIDENT: We do not hope to get beyond small compartment boats for a long time in South Yorkshire. Is it your pleasure that the discussion be adjourned or closed?

Mr. EMBLETON: I do not think it is any use discussing Mr. Walker's paper unless some member has personal knowledge of the canal so as to show that the account Mr. Walker has been so good as to give us is of a thing that is likely to be carried out.

Mr. NASH: I believe Mr. Walker's idea was to show the principle, more than the actual work done, as it might be advantageously used.

The PRESIDENT: Then I move we go to the next business, and that is the adjourned discussion on Mr. G. B. Walker's paper on "Electricity as a Motive Power."

ADJOURNED DISCUSSION ON MR. G. BLAKE WALKER'S PAPER ON
"ELECTRICITY AS A MOTIVE POWER."

THE PRESIDENT: We have had one or two meetings on this subject of electricity, and our last meeting at Leeds was a very interesting one. The subject was then adjourned to enable members to take advantage of a kind invitation from Mr. Bower's manager to visit their collieries and see a coal-cutting machine in operation by electricity. They had already had the opportunity of seeing a pumping machine driven by this power at Messrs. Locke & Co's. colliery, and it was there promised we should have the opportunity of seeing a hauling engine at work. It is not yet in operation, but the invitation holds good to the Institute whenever it is. Mr. Brown is present, and I have no doubt we can trespass on his good nature to repeat his invitation to us to go as soon as he has got the engines at work. I have not had the opportunity to-day of seeing this coal-cutting machine in operation, but I may say that so far as I am concerned there is no doubt in my mind as to the utility of electricity to almost any purpose. The main question, however, raised at our last meeting was whether the sparks coming from the electric motor were dangerous or not. I should be glad to hear the opinions of members who have seen the coal cutting-machinery.

Mr. NEVIN: I think about twelve of us went over to West Allerton to day. We found that they are driving the dynamo working the coal-cutting machine from a pair of 18-inch cylinder engines, which also are doing the haulage of the West Allerton Pit. These are run 60 revolutions a minute. The dynamo is run from 600 to 700 revolutions per minute, and gives a current of 240 volts and 32 ampères. The machine itself is exactly the same as that which they have had for some time driven by a rope. I think it has already been described in the Transactions—a rotary cutter bar furnished with teeth, which bar can either be put in a line with the road on which the machine travels or can be turned at right angles into the coal, cutting to a depth of 3 feet 6 inches. We had only a short time in the pit, and during that time the machine was timed to cut a yard in something under 4 minutes, which would be 15 yards an hour. The dynamo of the coal-cutting machine is directly on the "cutter" shaft, so that the "cutter" makes a revolution to each revolution of the dynamo. The weight of the machine is about a ton; height 2 feet above the rails, and width about 2 feet. I do not know the length. The whole thing seemed to work very smoothly, with very little jar or noise.

Mr. NASH: It requires two men to look after it, one to attend to the machine, and the other to work the crab moving it along the face.

The PRESIDENT: How long has the electric machinery been at work?

Mr. BLACKBURN: About 18 months.

The PRESIDENT: How do you find that compares with the old-fashioned way you had of driving it with ropes?

Mr. BLACKBURN: We are a long way in bye, and ropes are no use, the friction is too great.

The PRESIDENT: What do you estimate the cost per ton of holing the coal?

Mr. BLACKBURN: Holing the coal costs about 1½d. per ton.

The PRESIDENT: That includes all machinery?

Mr. BLACKBURN: All machinery, save, of course, the cost of the plant,—all labour and machinery.

The PRESIDENT: Does that include fuel for driving the engine at the surface?

Mr. BLACKBURN: No.

Mr. NASH: Nor your interest on the electric plant?

The President: So that to the 1½d. per ton would be added the proportion of fuel and wear and tear?

Mr. Blackburn: We have not discovered much difference in the cost of fuel, because we are bound to run the engines for hauling, and the amount of steam is not much over and above that we use ordinarily with the haulage.

The President: It must take some driving?

Mr. Blackburn: Yes.

Mr. Nash: The engine is working about 12 horse-power; it is usually worked about 10; that is the extra cost it is using in work.

Mr. Blackburn: The coal that we fire with would cost about a shilling a day if we had to buy it elsewhere.

The President: What would be the size of the engine necessary to drive the dynamo to work this one machine?

Mr. Blackburn: A pair of 10 inch cylinders.

The President: Would that drive more than one?

Mr. Blackburn: Not more than one.

The President: Then we may take it that a 10 inch cylinder is required for every one of these machines; how many sets of men would the machine do for?

Mr. Blackburn: I do not say that, but you would not think of putting less than a 10 inch cylinder down.

The President: Then 15 yards an hour is what it holes?

Mr. Blackburn: 20 yards an hour.

The President: In ten hours it would hole a face of 200 yards, which would turn out how much,—200 tons?

Mr. Blackburn: In our seam, over 200 tons.

The President: What is the thickness of the seam?

Mr. Blackburn: 5 ft. 6 in.; it would yield 270 tons.

Mr. Nash: At a low estimate, it would hole 120 yards a day 3 ft. 6 in. deep?

Mr. Blackburn: It averages 3 ft. 3 in.

Mr. Nash: And it takes 5 inches out at the face and 3 inches at the back of the holing?

The President: Given the utility and adaptability of electricity for driving a coal-cutting machine, we revert to the old question we had under discussion before, namely, sparking, and Mr. Blackburn said he would think over the question.

Mr. BLACKBURN: I do not think there is much difficulty about sparking. I think all had an opportunity of seeing what sparking there was to be seen, and I do not think much was seen.

Mr. NASH: There was not a great amount of sparking, but there was some. The question is if the slight sparks we saw would fire gas in an outburst?

Mr. BLACKBURN: The machine can be easily covered over; it was running open to-day.

Mr. NASH: Quite so; but in the Barnsley bed, I do not think it would be safe to run it as you was running to-day.

The PRESIDENT: Mr. Snell certainly did not give as hopeful an account of his experience as I had expected. I think if they had quite obviated the danger of sparking, we should have heard from them, but we have not heard since he was here. Six months ago he promised me a dynamo to test in an explosive mixture, but have not yet got it. It seems to me that is the important question. In some seams it possibly does not matter, that is a matter of opinion for those immediately interested; but in seams such as we are working, where you have done away with powder and everything in the way of open lights, and gone to a large amount of expense to introduce lamps and that sort of thing, it seems an anomaly to run a machine which throws off sparks, if those are dangerous. If they are not, let us get to know.

Mr. NASH: Certainly, a comparison of the motors—the one at Locke's, and the one working the coal cutters—shows much improvement. I should say that with the one we saw to-day in bye there was not one-thousandth part of the sparking that there was with the one we saw at the pit bottom, which must be a step in the right direction.

Mr. BLACKBURN: Mr. Brown can express an opinion on the question.

Mr. BROWN: I cannot say whether the sparks will ignite the gas or not. As to the machine we saw to-day, you might say there was no sparking, but there is a spark when there is a tremor of the machine, which Mr. Blackburn attributed to the dither of the rails. If there is a spark which comes and goes it would not be safe. Our machines are larger and give off more heat. There may be some improvement in their machine, but there will have to be something more done before the machines can be used in the returns. They will have to be made safer. With regard to running one practically sparkless, that is not good enough; it must be absolutely sparkless. I doubt very much if the sparks given off will not ignite gas. If their machine at Allerton was put into an explosive mixture I should like to be away. I have tried the dynamo at the top, and have tried to burn tissue paper and it will not char it even; but that does not say whether it would ignite gas or not.

The PRESIDENT: Every test is of value, but I think experiments of a most crucial character should be made, and if we find the sparks are dangerous, find some means of applying the power so as to obviate it. We cannot go on in the dark simply on somebody's assertion that there is no risk, or it is of such rare occurrence that it is not worth while taking it into consideration. If electricity is to be applied, it will be applied in all sorts of places underground; and, therefore, it should be demonstrated clearly and distinctly whether there is danger in its adoption or not, and if there is danger provision taken to meet the danger.

Mr. BLACKBURN: I have run my hand on the commutator, close to the brushes, and the effect of that spark never told upon my naked flesh at all; but if you were to put your hand in the flame of a candle or anything which would instantaneously light gas, it would be immediately felt.

The PRESIDENT: You take our fan belt and you can run your fingers along it and feel nothing but a tickling sensation, yet these sparks will light the gas. I have seen our man repeatedly do it; and have seen the same thing done at Mr. Shaw's, at Kirby.

Mr. GARFORTH: I have done the same thing.

Mr. BROWN: If you take an electric gas-lighter, the spark will go into your finger and you cannot feel it; but it will light the gas for all that.

The PRESIDENT: Their contention was that it was a low tension spark, and that a high tension spark was required to light the gas.

Mr. GERRARD: We shall have to prove, as we have done with safety lamps and the new explosives, whether the sparks are harmless. A mere expression of opinion from visual observation is not conclusive. One man says that the sparks are harmless; another says they are not; another man says he thinks there are sparks; another says there are not. If you put that to the same conclusive test as you have safety lamps and some of these new explosives, that is, submit the electrical motor to some highly explosive atmosphere, and fully work the electric motor, not trust it to mere ordinary conditions of working where you have these low tension sparks, but submit it to severe tests, to irregular work such as we know takes place, where the brushes are burnt up, where you have it setting fire to the woodwork, and so on, showing that under certain conditions you have a larger spark given off, which in all probability is highly dangerous; then, having experimented with your motor in an explosive atmosphere, with the two classes of sparks, you will have to my mind, conclusive results. You would then be able to speak decisively to these inventors and say "You must meet this point or we shall not put this appliance into our mines." When we have removed fires and open lights, when we have taken out all explosives of a flame producing kind, we will not put in these electrical appliances until we know they are absolutely safe. It is no use discussing this backwards and forwards; we have arrived at a point when someone ought to do something to prove whether it is dangerous or not. I think it is a proper subject for the Institute to take up in the same manner as we have safety lamps and high explosives, to see whether there is any danger attaching to these electrical motors or not.

The PRESIDENT: I agree with that. There is no doubt a great deal to be said on the economic side of the question, but you cannot get over this. By introducing safety lamps, doing away with naked lights, and with explosives, the cost of wages alone in getting coal has been increased in South Yorkshire—I think I am well within the mark,—3d. or 4d. a ton. I think Mr. Embleton's range of experience will confirm these figures. After an increased expenditure of that sort, after all sorts of legislative enactments to compel us to take precautions of every kind even if we do not want to do so of our own will, after that to act on the *ipse dixit* of somebody, and stick in a machine of which you know nothing as to its safety, is a step backwards we cannot and ought not to take. I think we should have an opportunity of testing this question, and it can be done at a small cost. If Messrs. Immisch had let us have a motor, we could have had it tested, and I should have been able to lay the result before the Institute, as I had hoped, before to-day, and have had this question thrashed out.

Mr. GERRARD: I should be sorry to say a word which would hinder improvement or advancement. I am of opinion that electricity can be applied for haulage and pumping operations near the shaft bottom where you have little or no danger of an explosive atmosphere. But we are proposing to introduce this power to the working faces, to pump water, or to cut coal, and I think before we do that we ought to know with authority whether it is safe or not. It will not do upon what a man thinks; but we ought to have facts before us. I have seen experiments made with low tension sparks, with regard to fine fibres and so on, which could not be singed nor set in flame. I am prepared to believe that low tension sparks are harmless, but I should like to see it proved more conclusively in an explosive atmosphere.

Mr. EMBLETON: I think it is clear from the discussion that it is not in the power of the Institute to say that these machines are safe in gas. In order to satisfy the Institute, the trials you have spoken of should be made; and made so satisfactorily as to show whether any spark will light the gas or not. The new power appears to be very economical, but it should be tested so that we could assure ourselves that in using that power in any case there is safety.

The PRESIDENT: I should not like Mr. Blackburn or Mr. Brown to think that we are not fully alive to their ingenuity and enterprise in adopting electricity for the purposes they have done. They have been the pioneers in proving that it is specially adapted for underground work, both for economical application and because it can be conveyed rapidly to every part of the mine. All I want to know is to be assured that it is safe to go in for electricity, not only at the pit bottom, but at other places. I may want to apply electricity to places where we have had a feeder of gas and water running over four years. I have no compressed air, no endless haulage, no means of pumping the water except by lading, or by putting two or three ponies to a pump. But I wish to know before I put electricity there whether it is safe or not. That is why I am anxious to thrash the thing out, —I wish to prove it.

Mr. BLACKBURN: The question of sparking is one we have worked at $1\frac{1}{2}$ years, and we have reduced it 70 per cent. in less than three months, and I will stake my existence you cannot get an explosion.

The PRESIDENT: Then you will have done a great deal of good.

Mr. BLACKBURN: We have a machine that we are experimenting with on purpose to destroy the existence of sparks.

Mr. NEVIN: I should like to suggest to Mr. Blackburn to get the people who are making this dynamo to send one to Mr. Rhodes for him to test it.

Mr. BLACKBURN: We make our own; Mr. Rhodes can bring his apparatus and do what he likes to test it. We have had experiments and lit the gas with a naked light when the sparks would not touch it. We are now experimenting ourselves, and I have no doubt we shall be able to bring out means by which we shall obviate sparking altogether.

Mr. BROWN: That is the one thing needful.

Mr. EMBLETON: The President could give you instructions how to make a testing apparatus, and the safety of the machine could be ascertained.

Mr. BLACKBURN: We will do that.

The PRESIDENT: If it could be ready before July it would be a good thing.

Mr. BLACKBURN: We are experimenting and cannot tell how long it might take.

Mr. GARFORTH: You would do well to take the machine to Aldwarke and get the benefit of Mr. Rhodes's experience, as I do not think anything you could do in twelve months would result in so good a test or gain the same confidence, as the apparatus nearly all the members have seen.

Mr. BLACKBURN: One thing I would suggest to the President, that a small motor or dynamo would not test it. You cannot get a small machine without dangerous sparks. You must have a machine of certain size to make the sparks innocuous.

The PRESIDENT: You think the sparks from a small machine would light the gas?

Mr. BLACKBURN: Yes; you must have it a certain size and firm to render the sparks harmless. The mode we employ is different to any applied hitherto, and is based upon our experiments. I think within three months we can let you have what experiments you like upon a machine and you will find it harmless.

Mr. BROWN: I do not see why a small machine should be more likely to cause sparks than a larger one.

Mr. EMBLETON: Then a small machine will give off sparks; a larger one will do so in less degree?

Mr. BLACKBURN: Yes. There are certain calculations to be made as to the receiving and giving off power at a certain balance, when you have sparks on one side or the other. If you get a small machine with a big current you must have dangerous sparks.

The PRESIDENT: If you had a large machine working several coal-cutting and other machines, and half of the power was thrown off suddenly, would you not have danger in that case, just as you would in a smaller motor with a smaller quantity of power?

Mr. BLACKBURN: In a compound machine it is like you pulling an elastic band, you have no effect until you get the pressure. They only give off the amount of power you require. If you do not require power, they do not give any; if you require a small amount of power you get it; and when you require a great deal you get it.

The PRESIDENT: That would get over the question raised by Mr. Nash. There was a long discussion, and great disadvantage was suggested as likely to arise from the impossibility of regulating the electric power. Mr. Snell said that he had not seen his way to quite get over that. It seems now we have got a step further; and you can now practically regulate the power just as the "governors" do on the steam engine?

Mr. BLACKBURN: Yes.

Mr. T. MITCHELL: I understood Mr. Snell, there was no more electricity generated than the work required to be done.

Mr. BROWN: That is so.

The PRESIDENT: He said it would be genérated at the surface, but he was going to try some governors to regulate it.

Mr. BROWN: You must have misunderstood him.

The PRESIDENT: It was fully raised as you will see at page 384 of the Transactions.

Mr. BLACKBURN: The current regulates itself. If you do not want the electricity, it is not generated; if you do want it, it is generated according to demand. There is no more danger of it running away with the load than of an ordinary hauling engine doing so. It varies with the demand made upon it. If you are running the dynamo, you can charge accumulators, which can be used for anything you like; at the same time, if you do not use the accumulators, there is no current to be wasted. As to communication, the man at the engine and the man at the other end know exactly the same thing. If you are two miles away it makes no difference, the same communication is made at each end.

The PRESIDENT: Then you do not think if one got a small motor, and ran it in a testing box two feet square, that that would be a test you would care about accepting.

Mr. BLACKBURN: Provided all the calculations were made to agree as well as these experimented calculations agree, I do not say one would not be as effective as the other; but seeing the calculations may not be got at accurately if you get a small machine, you may get a dangerous spark. We have learnt from experience you must have a certain amount of power at one end, and a certain amount of power at the other, and by a calculated arrangement you can save your spark, or even, so far as these gentlemen have seen to day, do without it altogether. But that is not what we are going to try at; you shall make your spark, try to fuse the brush if you like, and we will destroy it.

The PRESIDENT: Something of that sort will have to be done; because I am certain it is impossible to box in any machine so as to prevent danger. I do not believe it possible by any system of casing or boxing to avoid communication from any mixture that is inside your box, where the dynamo is run, to that outside. No safety lamp gauze can stand an explosion in a box three feet by two feet. It would be blown to rags; and I am not certain that the box would not too, judging by our experience with our testing boxes before we got loose flaps. I know when we first started at Aldwarke we had iron boxes, and it shattered them.

Mr. BLACKBURN: We shall not leave any area for any gas when we have finished; we shall not box it.

The PRESIDENT: The reason I raised the question of boxing was because it was raised by a gentleman named Atkinson.

The PRESIDENT: I think Mr. Blackburn is on the right track if he can stick to it.

Mr. BLACKBURN: I think we shall manage it. We have improved so far, and that is very satisfactory. Our commutators wore out at first very quickly, and they had to be turned up frequently. We can run them now 12 months without turning them up.

The PRESIDENT: I am sure we are much obliged to Mr. Blackburn for coming here, and for showing his machine at work. I hope that he will add to our indebtedness by giving us any further knowledge that he may get. We shall appreciate it not only as a great kindness, but as a matter of interest and benefit to everybody connected with mining. I hope that he will join us at Barnsley, and dine with us on the 31st of July.

Mr. BLACKBURN: Thank you, Mr. President.

Mr. NEVIN: I have pleasure in proposing that the best thanks of the Institute be given to Messrs. Bowers and Mr. Blackburn for allowing us to inspect their machinery and dynamo.

Mr. EMBLETON: I have great pleasure in seconding that.

The motion was carried unanimously.

MIDLAND INSTITUTE OF MINING AND MECHANICAL ENGINEERS.

TO ILLUSTRATE MR. W. E. GARFORTH'S PAPER ON ARTIFICIAL FOUNDATIONS AND METHOD OF SINKING THRO' QUICKSAND.

LONGITUDINAL SECTION BETWEEN NOS. 2 AND 3 BORE-HOLES.

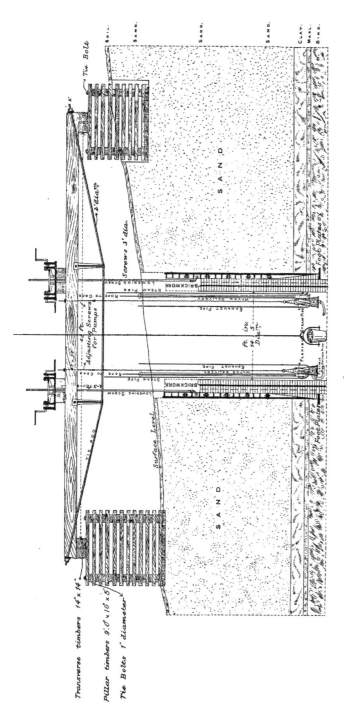

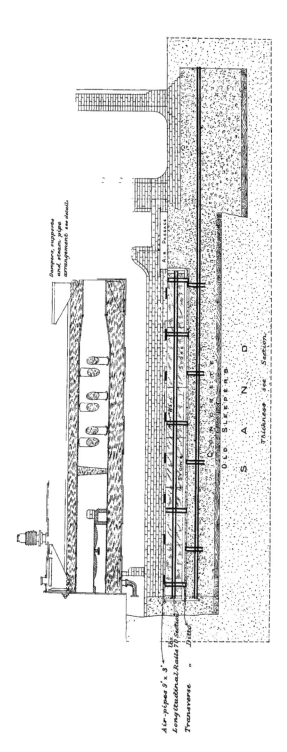

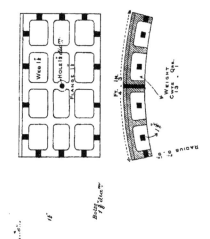

MIDLAND INSTITUTE OF MINING AND MECHANICAL ENGINEERS.

TO ILLUSTRATE MR. W. E. GARFORTH'S PAPER ON ARTIFICIAL FOUNDATIONS AND METHOD OF SINKING THRO' QUICKSAND.

DETAILS OF CAST-IRON CUTTER & TUBBING.

SCALE OF FEET.

INDEX TO VOL. XI.

	PAGE
Accounts, Statement of	22, 217
Angus, Jas., elected member	341
Annual Dinner	30, 285
Annual Report of Council	23, 212
Annual Meeting	11, 212
Artificial Foundations, and method of Sinking through Quicksand	407
Ashworth's Patent Hepplewhite Gray Safety Lamp, exhibited	147
Bancroft, R. E., elected member	317
Barnby, R. C., elected member	11
Belk, W. J., elected member	341
Bennett, John, elected member	31
Blackburn, W., elected member	172
Brierley, W., elected member	31
Brown, E., elected member	317
Brown, John, Memoir of	315
Chambers, Hy., elected member	253
Chambers, J. E., elected member	75
Coal Mines Regulation Act, 1887, Considerations arising out of Sections 51, 52, 53, and 54	75
Discussion	97, 117, 135, 146
Cocking, A. T., elected member	11
Cooper, H., elected member	11
Council's Annual Report	23, 212
Cradock, G., elected member	31
Dates of Meetings	272
Defective Detonators	336
Discussion on Clarke's Arrangement for Arresting the Fall of Colliery Cages in cases of Breakage of the Rope	12
Discussion on the Committee's Observations on the Final Report of the Royal Commission on Accidents in Mines	12, 73
Discussion on the Easterly Extension of the Leeds and Nottingham Coal-field	11
Durnford, H. S., elected member	317
Easterly Extension of the Leeds and Nottingham Coal-field (*Supplementary Paper*)	63
Discussion	68, 97

INDEX.

	PAGE
Election of Officers	20, 219
Election of President	237
Electric Locomotive for Mines	333
Electricity as a Motive Power, with special reference to its application to Haulage in Mines	317
Discussion	341, 359, 381, 393, 422
Experiments with Roburite	124
Fan Experiments, discussion	137, 171, 173, 301
Federation of Mining Institutes, discussion	99, 119, 173, 273, 389, 401, 419
Foreign Mining Rents and Royalties	149
Discussion	163, 231, 270
Garforth on Artificial Foundations and method of Sinking through Quicksand	407
Gascoyne on the Easterly Extension of the Leeds and Nottingham Coal-field	63
Discussion	68, 97
Gill, T., elected member	401
Gillott, J. W., elected member	401
Hargreaves on Tonite as an Explosive when used with a Flame-destroying Compound	239
Discussion	242, 263
Hedley, S. H., elected member	99
Honorary Members	2, 202
Hydro-carbon Explosives and their value for Mining Purposes	101, 138
Discussion	122, 144, 157, 170, 198, 225, 263
Inaugural Address of the President	31, 253
Institution of Civil Engineers, Vote of thanks to	199
Life Members	2, 202
Liversedge, W. G., elected member	212
McAdoo, John, elected member	359
McMurtrie, G. E. J., elected member	119
Mellors, Jas., elected member	172
Members, List of Life Members	2, 202
,, ,, Honorary Members	2, 202
,, ,, Ordinary Members	3, 203
Memoir of John Brown	315
Method of Sinking through Quicksand	407

INDEX.

	PAGE
Mining in the Middle Ages	175
Discussion	221
Mining Institutes, Federation of, discussion	99, 119, 173, 273, 389, 401, 419
Model of new Safety Cage, exhibited	135
Musgrave, Hy., elected member	119
Nash on Foreign Mining Rents and Royalties	149
Discussion	163, 231, 270
Notes on Matters of Current Interest	247
Discussion	267
Officers, Election of	20, 219
Ordinary Members	3, 203
Peasegood, W. G., elected member	31
Pollard, John, elected member	99
President elected	237
President's Inaugrual Address	31, 253
Proposed Nicaraguan Canal and a new type of Lock for Ship Canals	394
Discussion	421
Retiring Officers, Vote of thanks to	221
Retiring President, Vote of thanks to	28
Rhodes on Defective Detonators	336
Roburite Experiments	124
Rules	7, 208
Safety Lamp having a "Shut-Off" Appliance, exhibited	117
Snell, A. T., elected member	172
Snell on an Electric Locomotive for Mines	333
Statement of Accounts	22, 217
Tonite as an Explosive when used with a Flame-destroying Compound	239
Discussion	242, 263
Vote of thanks to the Institution of Civil Engineers	199
Vote of thanks to Retiring Officers	221
Vote of thanks to the retiring President	28
Visit to Allerton Main Colliery, Kippax	419
Visit to St. John's Colliery, Normanton	359
Walker on Considerations arising out of Sections 51, 52, 53 and 54 of the Coal Mines Regulation Act, 1887	75
Discussion	97, 117, 135, 146

INDEX.

	PAGE
Walker on Electricity as a Motive Power, with special reference to its application to Haulage in Mines	317
Discussion ... 341, 359, 381, 393,	422
Walker on Hydro-carbon Explosives, and their value for Mining Purposes ... 101,	138
Discussion ... 122, 144, 157, 170' 198, 225,	263
Walker on Mining in the Middle Ages	175
Discussion	221
Walker, Notes on Matters of Current Interest	247
Discussion	267
Walker on the proposed Nicaraguan Canal, and a new type of Lock for Ship Canals	394
Discussion	421
Wordsworth, R., elected member	317
Wroe, James, elected member	137
Wroe, Jonathan, elected member	137

BINDING

BINDING SECT. JUL 1 6 1971

TN Midland Institute of Mining
1   Engineers
M54   Transactions
v.11

Engineering

PLEASE DO NOT REMOVE
CARDS OR SLIPS FROM THIS POCKET

UNIVERSITY OF TORONTO LIBRARY

ENGIN STORAGE